Law and Society: An Introduction

Steven E. Barkan

University of Maine

Upper Saddle River, New Jersey 07458

Library of Congress Cataloging-in-Publication Data

Barkan, Steven E.,
Law and society : an introduction / Steven E. Barkan. — 1st ed.
p. cm.
Includes bibliographical references and index.
ISBN-13: 978-0-13-194660-6 (alk. paper)
ISBN-10: 0-13-194660-9 (alk. paper)
1. Law—Social aspects—United States. 2. Sociological jurisprudence. I. Title.
K370.B37 2008
340'.115—dc22

2007047116

Editorial Director: Leah Jewell
Marketing Manager: Lindsey Prudhomme
Marketing Assistant: Jessica Muraviiov
Production Manager: Kathy Sleys
Creative Director: Jayne Conte
Cover Design: Bruce Kenselaar
Cover Illustration/Photo: Getty Images, Inc.
Full-Service Project Management/Composition: Chitra Ganesan/GGS Book Services
Printer/Binder: R.R. Donnelley & Sons, Inc.

Credits and acknowledgments borrowed from other sources and reproduced, with permission, in this textbook appear on appropriate page within text.

Pearson Education LTD. London
Pearson Education Singapore, Pte. Ltd
Pearson Education, Canada, Ltd
Pearson Education–Japan
Pearson Education Australia PTY, Limited
Pearson Education North Asia Ltd
Pearson Educación de Mexico, S.A. de C.V.
Pearson Education Malaysia, Pte. Ltd
Pearson Education, Upper Saddle River, New Jersey

10 9 8 7 6 5 4 3 2 1
ISBN-(10): 0-13-194660-9
ISBN-(13): 978-0-13-194660-9

CONTENTS

PREFACE

Welcome to the field of law and society! Most people who hear someone is going to study law naturally assume the person is talking about going to law school. However, there is another way to study law, and that is from the perspective of the social sciences. This perspective views law as an essential social institution that shapes society and is also shaped by it, and it emphasizes that the manner in which law and the legal system actually work often differs greatly from how they are supposed to work.

This perspective lies at the heart of this new book. As its subtitle implies, the book is intended as an introduction to law and society, not as an encyclopedic account. The study of law and society is multidisciplinary, and so is the book. In its pages you will find the work of anthropologists, historians, law professors, political scientists, psychologists, and sociologists. You will find material on the many functions that law potentially serves, but you will also find material on how and why law sometimes fails to achieve these functions. Because law affects so much of our lives and the world around us, the nine chapters of the book will help readers understand one of the many social forces that shape our behavior.

The book was written primarily for advanced undergraduates, but it will also be useful for graduate students and faculty as well. It is intended as the main text for the law and society courses and introduction to law courses commonly taught in political science and sociology departments and in criminology, criminal justice, justice studies, and legal studies programs and departments.

Writing this book has been a labor of love, as the old cliché goes. I first became interested in law and society as an undergraduate sociology major after I was arrested in May 1972 for being part of a group of Vietnam war protesters who chained ourselves to the federal building in Hartford, CT. Our three-day trial later that summer, in which we acted as our own attorneys, ended in a hung jury, as at least one or two jurors evidently thought we should not be convicted even though we clearly had broken the law. The trial made me decide not to go to law school (much to my parents' chagrin!), but it also made me want to learn how law helps or hurts social movements and more generally, how law and society affect each other. I wrote a senior-year paper about my trial that eventually became my master's thesis and then my first journal article and the basis for my first book.

Much has changed since that fateful day for me in 1972, but my interest in law has only grown. I have taught a law and society course for about 25 years, and this book reflects much of what I have learned during this time thanks to the work of law and society scholars in the United States and elsewhere. I hope that students who read this book and take the course in which it is used will end up being as fascinated by and about law as I have been since my undergraduate days. I envy the new journey into the world of law and society they are about to take.

This book would not have been possible without the help, influence, and/or support of several people. Norman Miller introduced me to sociology during my first year in college and never stopped forcing me to think like a sociologist. I owe him an unpayable debt for any success I have had in my career. Forrest Dill taught me law and society in graduate school and was my mentor and friend. His untimely death shortly after I began my career left all of us who knew him very empty.

On a happier note, my colleagues at the University of Maine have provided a wonderful environment in which to work for the almost three decades I have been in our great state. I thank the editorial and production staff at Prentice Hall for their faith in this book and for all their efforts to help it get into print. Several reviewers read earlier chapters and made some wonderful suggestions that improved the book. Any mistakes that remain are, of course, my responsibility. These reviewers are Jeffrey Ulmer, Pennsylvania State University; Lloyd Klein, Bemidiji State University; Edna Erez, Kent State University; and Sarah Gatson, Texas A & M University.

My deepest thanks, as always, go to my wife, Barbara Tennent, and our sons Dave and Joe for their love and for everything they do to enrich my life. Like my previous books, this text is dedicated to my late parents, Morry and Sylvia. Even though I didn't go to law school or become a doctor, at least I became a Doctor of Philosophy who studies and writes about law. My parents ended up being very happy with the choice I made, and I think they would have enjoyed reading my latest effort.

CHAPTER 1

The Social Nature and Significance of Law

Chapter Outline

The title of this chapter is also the major theme of this book's introduction to law and society. You will see that law is a social phenomenon and thus highly significant for many reasons. To begin doing so, consider some issues and events from the last few years that demonstrate the importance of law for so many aspects of our lives.

Many legal matters made headlines in 2007. In July of that year, a group of veterans of the wars in Iraq and Afghanistan filed a federal lawsuit that accused the U.S. Department of Veterans Affairs of "shameful failures" in its treatment of veterans from the two wars and in particular criticized the department for failing to treat post-traumatic stress disorder (Weinstein 2007). Earlier in 2007, a judge ruled that New Orleans survivors of the Hurricane Katrina flooding could proceed with a lawsuit against the U.S. Army Corps of Engineers for mistakes it allegedly made that worsened the flooding and the damage the hurricane caused (Schwartz 2007). Any number of U.S. Supreme Court rulings in 2007 could be cited, but in two cases with far-reaching implications, the Court made it easier for prosecutors to exclude jurors with qualms about the death penalty from capital cases (Liptak 2007), and it limited students' freedom of speech when it upheld the suspension of an Alaskan high school student for displaying a banner that said "Bong Hits 4 Jesus" (Lane 2007). In just two of countless other legal developments in 2007, Florida passed legislation that would restore voting rights to most felons who had served their sentence (Goodnough 2007), and the Eli Lilly pharmaceutical company agreed to pay up to $500 million to settle thousands of lawsuits for people who said they became ill after taking one of its drugs for mental illness (Berenson 2007).

Two years before all these legal developments, the heartbreaking case of Terri Schiavo dominated the headlines (Campo-Flores 2005). As a young woman in Florida, Schiavo suffered a severe eating disorder that eventually led to a heart attack in 1990, when Schiavo was only 26. She survived the heart attack but went into a persistent vegetative state. For years her husband, Michael, had wanted to end her life, but her parents just as stubbornly wanted to keep her alive, and they battled her life out in the courts. In March 2005, the Congress and White House enacted legislation to allow the case to be heard in the federal courts. Federal judges refused to order Schiavo's death, and she finally died days later (Goodnough 2005). In the wake of her death and the national debate it created, a newspaper columnist marveled that "amid all the anger and anguish, one crucial fact was never in doubt. Terri Schiavo's fate would be determined lawfully. The courts would determine whether her feeding tube should be reconnected, and their decision, whatever it finally was, would be obeyed" (Jacoby 2005:D11).

Another controversial issue during this decade has been the use of torture, abuse, and harsh legal treatment by U.S. military and government personnel of suspected terrorists and detainees from the wars in Iraq and Afghanistan. One critic said these practices all reflected "a culture of low regard for the law, of respecting the law only when it is convenient," and quoted former Supreme Court Justice Louis D. Brandeis's famous words about the importance of the rule of law: "Our government is the potent, the omnipresent teacher. For good or ill, it teaches the whole people by its example. Crime is contagious. If the government becomes a lawbreaker, it breeds contempt for the law; it invites every man to become a law unto himself" (Lewis 2004:A25).

While questions of law and terrorism rocked the nation this decade, other issues with legal underpinnings also won national attention. Gay marriages, which became a hot issue during the 2004 presidential election after the Supreme Judicial Court in Massachusetts declared them legal in that state, continued to arouse controversy, as several states passed legislation or constitutional amendments that ban such marriages (Elliott 2005). In a related issue, a California court ruled in August 2005 that lesbian couples who conceive and raise children must be treated as parents under the law even if they had not registered as domestic partners; this ruling gave lesbian couples the same legal rights and responsibilities as unmarried heterosexual partners (Christie 2005).

Racial issues were also legal matters during the past few years. In 2004, police in Charlottesville, VA, the home of the University of Virginia, asked

197 African American men to provide DNA samples for a case involving a serial rapist. Critics said the request was coercive, even if it was technically voluntary, and added that it undermined civil liberties and smacked of racial profiling. The city's police chief replied that the rapist's victims had identified him as African American and that the police were simply trying to bring him to justice. Bowing to community pressure, the police eventually suspended the DNA searching (Glod 2004).

Various laws have also been used in novel ways in the past decade. For example, after the terrorist attacks of September 11, 2001, three dozen states enacted laws that provided the death penalty or other serious punishment for terrorist-related activities. Although these laws obviously had 9/11-style terrorism in mind, the laws' wording lent itself to other situations. A New York City prosecutor invoked his state's new law against terrorism to charge members of a juvenile gang with engaging in shootings "committed with the intent to intimidate or coerce a civilian population." Civil libertarians said that the terrorism law was being misused and that the definition of terrorism was becoming overly broad (Garcia 2005:A4). In another case, police in the New Hampshire town of New Ipswich charged an illegal immigrant from Mexico with criminal trespassing after federal officials declined to arrest the man on immigration charges. Civil libertarians criticized the arrest for exceeding the boundaries of the trespassing statute. A judge eventually dismissed the charges on the grounds that immigration was a federal issue, and the state's attorney general decided not to appeal the judge's ruling (Belluck 2005).

Other legal developments during the past few years abounded. In Denver, voters approved a new law legalizing possession of small amounts of marijuana; the vote made Denver the first major city to do so. As jokes about Denver's nickname of "Mile High City" became commonplace, the city's mayor reminded Denver residents that marijuana use was still illegal under state law and that arrests would therefore continue (O'Driscoll 2005). In New York, New Jersey, and Washington, D.C., new laws banned the use of hand-held cell phones while driving; by mid-decade, more than 400,000 drivers had received tickets and warnings under these laws since New York passed the first one in 2001 (Radsch 2005).

THE SIGNIFICANCE OF LAW FOR SOCIETY AND FOR OURSELVES

Ripped from the headlines, these issues and events are but a few of the many with fundamental legal underpinnings and implications that could have been discussed. Together, they deal with matters of life and death, of the proper powers of the U.S. government after 9/11, and of new family structures and roles in the contemporary era. Some examples also concerned issues of race, ethnicity, and gender, and some also involved attempts to use new laws, or to adapt existing laws, to deal with changing social and political conditions.

In these and so many other ways, law is an intrinsic part of our society, and much of how our society works cannot be understood without appreciating the role that law plays.

Law is important not only for our society but also for us as individuals. In our daily lives, law affects how we behave or do not behave and possibly some of what we think, and it also governs many aspects of our lives. Driving a car is an obvious example: the side of the road on which you drive, the speed at which you travel, stopping at a traffic light or stop sign, and other aspects of driving too numerous to mention are all behaviors governed by law. If you have ever signed a lease for an apartment, paid taxes, gotten married or divorced (or at least ever contemplated marriage), then you again are following legally prescribed procedures. Complex legal regulations also govern the design, production, and sale of the many products that you purchase regularly: food, clothing, electronics, and textbooks like this one. Depending on how law is defined (more on that later), the rules governing your campus are also legal in nature. Most or perhaps all campuses have codes of conduct governing plagiarism and other forms of cheating, sexual harassment, and other behaviors, and they have established procedures for addressing violations of these codes. These codes parallel the rules and procedures found in the larger society. Although most of the examples in this paragraph have involved law regulating the relationship between the government (or campus) and the individual, the apartment lease example reminds us that law also helps to manage the interaction between two (or more) individuals or organizations. Ideally law eases interaction rather than aggravating it, but sometimes law may worsen a problem more than helping to resolve it.

In all these respects and more, law plays a fundamental role in our society and in our lives, and it does the same in other societies across the world. As prominent law and society scholar Lawrence Friedman (2004a:3) observes, "In the world we live in—the country we live in—almost nothing has more impact on our lives, nothing is more entangled with our everyday existence, than that something we call the *law*." The significance of law in this regard provides perhaps the major reason for why social scientists study it, just as they study the family, religion, education, medicine, and other social instiutions. Thus, while the term "law" often refers to rules and regulations, the term also refers to the **social institution** of which rules and regulations are an important part. "Law" will be used in both these ways throughout this book.

The Social Functions of Law

As a social institution, law ideally serves important functions for society. One of these functions is to help maintain social order. This is called the **social control** function. Without law, society would be much more chaotic, even it is true that law sometimes aggravates situations rather than improving them, and without the criminal law and criminal justice, in particular, there would probably be much more crime. If you had been asked before taking this course, "What is the most important function that law serves for society?", this is probably the function that would have first come to your mind. In addition to keeping the crime rate lower than it would otherwise be, law also governs many aspects of our lives, as noted earlier, and helps to preserve social order in this manner.

Social order is also threatened when individuals have disputes over any number of issues that they cannot resolve by themselves or may resolve only by resorting to threats, violence, and other behaviors normally considered inappropriate. In both traditional and modern societies, law also provides a vehicle for the possible settlement of disputes among individuals, groups, and organizations. This second function of law has various names, including **dispute settlement** and *conflict resolution*, and it is one that anthropologists have especially emphasized in their studies of law in traditional societies.

Although social control and dispute settlement, and the more general preservation of social order, may be the most familiar functions of law, other important functions also exist. One of these is **social change**. In many societies, but perhaps especially in the United States, law has often been considered a key tool for bringing about changes in individual behavior, for altering the nature and dynamics of certain social institutions, and for effecting broad changes in the larger society. In the stories of events from mid-decade, we saw examples of law being used to produce social change in such areas as gay marriages and the use of cell phones. The controversy today over gay marriages is at least as legal in scope as it is religious and moral, and efforts by courts and legislatures to ban or allow gay marriages have had, and will continue to have, important consequences for individual gays but also for the family as a social institution.

Another important function of law, at least in a democracy, is the *preservation of civil rights and civil liberties* and more generally of *individual freedom*. In the United States, this function is a key element of our Constitution and Bill of Rights and the subject of innumerable rulings by the U.S. Supreme Court. The term **rule of law** captures this idea and more formally means that no one in a democracy, no matter how wealthy or powerful the individual may be, is above the law. A major reason that President Richard Nixon was forced from office in 1974 was that he had put himself and his office above the law with various illegal actions that came to be called the Watergate controversy. In a memorable statement, President Nixon once declared, "When the President does it, that means that it is not illegal," but it was precisely this belief that the U.S. House of Representatives rejected when it was about to impeach President Nixon just before he resigned. The importance of the rule of law is a major reason for why certain Bush administration actions after 9/11 raised so much controversy.

A final function of law is the *expression of a society's moral values* on important social and political issues. Ideally, law should reflect a social consensus on appropriate and inappropriate behavior and on the structure and functioning of a society's basic social institutions. Law in this sense not only reflects a society's moral values but also helps to transmit these values from one generation to the next.

The Dysfunctions of Law

Law is ideally functional for society in all the ways just discussed, but law also has its dysfunctions as well. Perhaps the most serious dysfunction is that it may *create and perpetuate inequality*. Many scholars believe that law, even in a democracy like the United States, benefits individuals and groups with wealth and power over those without wealth and power, in particular the poor, people of color, and women. In this way, law is said to help perpetuate the inequality that already exists in American society. As Richard L. Abel (1989:vii), a former president of the Law and Society Association, put it, "All law is inescapably two-faced. It reproduces and justifies existing inequalities and injustices; yet it also embodies ideals and offers mechanisms through which they can be pursued." Although various reforms may have created a more equal legal playing field in recent decades, evidence supporting the inequality dysfunction of law is sufficiently abundant to warrant its discussion in Chapter 7.

In a related dysfunction, law may *reflect the moral values of influential social groups* rather than a consensus in society over these values as assumed just above. To the degree this happens, law does not express the moral values of groups without power and influence, and members of these groups may be punished if their behaviors are legally banned. An historical example of this dysfunction is the passage of the Eighteenth Amendment in 1919 that banned the manufacture, sale, and possession of alcohol. As Chapter 6 will recount in greater detail, the forces behind Prohibition were individuals and groups from white, rural, and Protestant backgrounds who thought that alcohol use was a sin and who also disliked the kinds of people who tended to drink: poor Catholic immigrants living in urban areas (Gusfield 1963). Prohibition thus reflected the moral values of a very influential social group at the time, rural white Protestants.

In another if less serious dysfunction, law may also *complicate matters more than it eases them*. For example, although law ideally helps individuals and other parties to settle various kinds of disputes, there is ample evidence that law may aggravate the situation and thus do more harm than good. For this reason, disputing parties may avoid using the law for dispute settlement, and studies by anthropologists provide a rich understanding of how this process works in the kinds of societies they examine. In a related example, law may also involve over-regulation and rigidity and in this sense lock people and organizations into inefficient procedures that may impede them from accomplishing various goals.

One of this book's themes is that law as it actually functions is often very different from how it should ideally function and from what you might see on

TV and in the movies. The best TV shows and films show law in all its complexity and flaws, but too many portray an overly simple, idealized version of law that exists in theory but not in practice. Dysfunctions of law remind us that law has its downside as well as its upside and that a complete understanding of law requires an appreciation of its positive contributions to society but also of the ways in which it may fall short of our idealized conception of it.

THE STUDY OF LAW AND SOCIETY

Because law is a social institution that both reflects and influences the society in which it is found, social scientists have studied law since the nineteenth century. This section outlines the major assumptions of the social science study of law. The discussion just above of the social significance of law points to some of the major emphases in the study of law and society.

Traditional and Social Science Views of Law

One way to understand the social science view of law is to contrast it to the *traditional* or *orthodox* view of law. This view, which is taught either explicitly or implicitly in law schools everywhere, may be gleaned by examining the typical questions included in the Law School Admission Test (LSAT) that all accredited U.S. law schools use to assess applicants to their programs. One typical section of the test includes questions that begin with a set of conditions and then ask what outcomes are possible based on these conditions. For example, a question might say that Ann, Bill, Carol, David, and Elaine are to be assigned seats on an airplane and then specify certain conditions, e.g., Ann can sit next to Carol but not next to David or Elaine; Carol can sit behind David but not in front of Ann; and Elaine can sit behind Bill but not next to Carol. The question would then specify various seating arrangements, and the LSAT-taker is asked to indicate which seating arrangement is possible based on the question's conditions.

Another type of question starts with a short description of an event or scenario and then may ask several types of questions, including (a) what specific conclusion may be drawn from the description? (b) which one of several facts, if true, would weaken an argument presented by a character in the scenario? and (c) on which one of several assumptions does this argument depend? Thus, and to use an easy example, a scenario might discuss an employee named Joan Smith who will take an umbrella to work if it is very cloudy in the morning or, in the event it is not very cloudy, if the weather forecast is for rain later in the day. A question might then ask,

Joan's decision regarding taking an umbrella assumes which of the following?

a. Umbrellas will not blow away in windy weather.
b. A weather forecast is always correct.

c. Rain is more likely if it is very cloudy in the morning than if it is not very cloudy

d. If it is not very cloudy in the morning, it will not rain later in the day.

As should be evident, these and other types of LSAT questions measure the applicant's reasoning ability, which is to say the applicant's ability to draw logical conclusions from a set of facts and conditions. You might remember a similar, if simple, logical argument from high school mathematics: if A is greater than B, and B is greater than C, then what has to be true? The answer, of course, is that A must be greater than C. This argument may be expressed as follows: *if A>B and B>C, then A>C*. There is only one correct conclusion to be drawn from the premises, A>B and B>C, and this conclusion, A>C, follows logically from these premises.

The LSAT questions similarly are tests of logical thinking. These questions, and the law school curriculum more generally, reflect the **traditional view of law**. According to this view, law is, and should be, a self-contained system of logic. In this view, legal reasoning consists of nothing more and nothing less than logical thinking based on the facts of a case and the law governing the circumstances of a case. Cases are complex (and certainly much more so than *if A>B and B>C, then A>C*), and reasonable people can and do disagree on the appropriate conclusions to draw from a combination of facts and law (unlike the LSAT questions where only one correct answer exists), but whatever conclusion is drawn should follow logically from a considerate interpretation of the interplay of facts and law in a particular case.

Notice that in the logical argument *if A>B and B>C, then A>C*, nothing can affect the correct conclusion to be drawn from the premises because this conclusion is logical. Thus, if you were given the premises *if A>B and B>C*, your gender will not and should not affect the conclusion you reach. If you are a woman, A is greater than C, and if you are a man, A is still greater than C. By the same token, your race or ethnicity, your social class, your age, the region of the country in which you live, and the part of the world in which you live will also not, and should not, affect the conclusion you draw. Logic is just that, and your many social backgrounds are quite independent of any conclusion that follows logically from a set of premises. By the same token, this conclusion obviously has no effect on any aspect of your life nor on any in the larger society. Like logic, then, law in the traditional view is considered independent (or autonomous) from society. Just as no aspect of society affects logic, no aspect of society affects law; similarly, just as logic does not affect society, law also does not affect society. Social and moral considerations are not appropriate to legal reasoning; what is appropriate is logical deduction.

The **social science view of law** differs from the traditional view in all these respects. As indicated earlier, law—considered either as rules or as a set of established procedures for creating, implementing, and enforcing these rules—is, in the social science view, considered a social institution. Like any other social institution such as the family, education, or religion, law therefore is firmly embedded in society. As such, it has a reciprocal relationship with society: law affects society, and society affects law. Social backgrounds often matter for law, and law often matters for important aspects of our lives. For better or worse, social and moral considerations affect law at almost every turn, and law has social and moral impact.

This general view lies at the heart of the study of law and society and distinguishes this field from the legal scholarship practiced in law schools (Sarat 2004). As law and society scholar Friedman (2005) observes, whereas legal scholarship asks whether a particular legal finding is right or wrong, law and society scholarship is not interested in this question, just as sociologists of religion do not try to determine whether God exists or would want us to behave in a certain way. Instead, law and society scholarship focuses on the social sources, dynamics, and impact of law. In doing so, it uses the theories and methods of several other social sciences, especially anthropology, economics, political science, psychology, and sociology. As this list should suggest, law and society is not a discipline in and of itself but rather "the application of other disciplines to a specific social system" (Friedman 2005:2).

A famous passage by one of the most respected U.S. Supreme Court Justices, Oliver Wendell Holmes, Jr., who served on the Court from 1902 to 1932, aptly summarizes the social science view of law and how it differs from the traditional view:

> The life of the law has not been logic; it has been experience. The felt necessities of the time, the prevalent moral and political theories, institutions of public policy, avowed or unconscious, even the prejudices which judges share with their fellow men, have had a good deal more to do than the syllogism [logical argument] in determining the rules by which men should be governed. The law embodies the story of a nation's development through many centuries, and it cannot be dealt with as if it contained only the axioms and corollaries of a book of mathematics. (Holmes 1938 [1881]:1)

Assumptions in the Study of Law and Society

Although law and society scholars come from different disciplines and understand law with different frameworks, the field as a whole shares certain assumptions that define the field and also help us understand the nature of law from a social science perspective. Friedman (2005) lists several such assumptions.

First and most important, *the legal system is not autonomous.* Friedman (2005:6) says an autonomous system is one "that operates under its own rules, that grows changes, and develops according to its own inner program." This is hardly true of law, he notes, since it would mean that social phenomena such as culture, politics, and social norms would not affect how the law functions. Most law and society scholars would agree that law is at least somewhat autonomous from society, but not nearly as autonomous as traditional legal scholarship assumes. Instead, they believe that the legal system "always bears the deep imprint of the society in which it is imbedded" (Friedman 2005:6).

The remaining assumptions all derive from this first, basic assumption that law is imbedded in society. The second is that *laws and legal decisions may have a potential impact on one or more aspects of society*. In contrast, legal scholarship disregards this impact, as it is concerned primarily with the logic and hence appropriateness of rules and decision-making.

A third assumption is that *major changes in society often bring about changes in the law*. As noted earlier, just as law can have impact on society, society, or, to be more precise, changes in society, can have impact on law.

The fourth and final assumption is that *law, whether considered as rules or as a legal system, reflects the type of society in which it is found*. Anthropological studies, to be discussed in later chapters, provide much support for this assumption, as the type and operation of law found in traditional societies differ in many ways from the type and operation of law in modern societies. And while law has similar characteristics in all modern societies, it still differs in important ways from one society to the next.

What Does Law and Society Study?

The scope of the field of law and society follows naturally from these four assumptions and also reflects the field's conception of law (Friedman 2005). Law in this conception consists of four dimensions, the first two of which were noted earlier in passing. The first dimension is the **substance of law**, or the various kinds of rules that help guide behavior. The second dimension involves **legal procedures**: how and why laws get enacted, disputes get settled, and cases get handled. The third dimension is the **organization of the legal system**, involving the structure and functioning of its many various components. The final dimension is called **legal culture**, which refers to the beliefs and values that people hold about law and the legal system.

With these assumptions and dimensions of law in mind, the scope of law and society scholarship should be clear and comprises at least three very general areas of study. A first area of study is *the origins and impacts of the substance of law*, i.e., the various rules that comprise law. We have many fine accounts of the social, political, and economic origins of particular laws and also of their impact, or lack of impact, for people's behavior, for social conditions such as poverty and environmental pollution, and for the dynamics of various social institutions. A second area of study is the *operation of the legal system*: how and why the various components of the legal system—police, courts, regulatory agencies, and so forth—work the way they do. This has been a prime focus of law and society scholarship for several decades and was a key topic when law and society scholarship began in the United States almost a century ago. A final area of study in law and society focuses on *legal culture*: what the beliefs and values comprising a specific legal culture actually are, the origins of these beliefs and values, and their consequences for dispute settlement and other aspects of the law. Here anthropologists have made major contributions, but there is also an increasing number of studies of how legal cultures are similar or different across the nations of the modern world.

Related to these general areas of study, and recalling the functions and dysfunctions of law discussed earlier, the field of law and society tries to answer many key questions, and later chapters of this book will address several of them. A summary of these questions will set the stage for the later discussions and also help clarify the scope of law and society theory and research.

1. ***How effective is law in controlling human behavior and in affecting organizational behavior?*** Our society constantly passes new laws and changes old laws with the hope and even with the expectation that these changes will have an impact on people's behavior and on what organizations do or refrain from doing. Law and society scholarship finds, however, that this impact is often less than lawmakers and the public assume and that its extent depends on various social and legal factors. Sometimes new and changed laws even have unintended consequences that bring with them new problems or aggravate the very problems the laws were intended to address.
2. ***To what degree do new laws, court rulings, or other types of legal changes affect social institutions or other aspects of society, and what factors affect the impact that law may have in these respects?*** As noted earlier, law has also been used to help address social conditions such as poverty or environmental pollution and to shape the structure and functioning of social institutions like the family. Interest in using law for these purposes again stems from the belief that law will have an impact. Yet law and society scholarship again finds that this impact is variable and may differ from what policymakers and public interest groups expect.
3. ***How effective are legal procedures compared to nonlegal procedures in the settlement of disputes?*** In the United States, lawsuits and other legal actions and procedures seem to be as American as apple pie, but the use of law to try to settle disputes can involve several kinds of complications, including the possibility that a dispute may worsen because law becomes involved. More informal methods of dispute settlement such as mediation are found in the societies anthropologists study, and scholars have considered whether these methods would work if used more often in a large, industrial society such as the United States.
4. ***To what degree do changes in society affect the various dimensions of law?*** This is perhaps not a question that an average citizen or policymaker ponders, but it is certainly one that law and society scholars consider. Evidence that changes in society do affect various legal

dimensions reinforces the social science view of law's embedment in the society in which it is found.

5. ***To what degree does inequality characterize the operation of the legal system?*** This question speaks to a possible dysfunction of law noted earlier and is a growing concern in law and society scholarship. Chapter 7 discusses evidence of possible inequality in the legal system; any evidence of such inequality again reinforces the general understanding offered by the social science view of law.
6. ***How, to what extent, and in what ways does law reinforce or contribute to social inequality or, instead, reduce social inequality?*** This is a corollary to the previous question and again reminds us that law does not operate in a social vacuum. Any evidence that law reinforces or reduces inequality in the larger society once again exemplifies the basic perspective of the social science view of law. More practically, such evidence reminds us that law can have very real impact on the lives and welfare of average Americans.

These are all very general questions, and within each line of inquiry many more specific areas of investigation exist. For example, within the study of law and dispute settlement, Question 3 in the list just above, an important topic is the amount and impact of lawsuits in the United States. Is there a litigation explosion? Are juries out of control in awarding multimillion-dollar settlements? These are just two of the many topics that should be considered in assessing the effectiveness of law in dispute settlement. Note that the questions listed above are not mutually exclusive, and that any one topic in the study of law and society may involve trying to uncover answers to two or more of the questions. To use the law and dispute settlement example again (Question 3), while some scholars ask how effective the law is in settling disputes, other scholars ask whether the use of law to settle disputes benefits the "haves" in society more than the "have-nots" (Galanter 1974), an issue that deals with possible inequality in the law (Questions 5 and 6 above).

LAW AND JUSTICE

Notice that the word *justice* in its meaning of moral rightness or fairness has not yet appeared in this chapter. That is because, as noted earlier, law and society scholarship does not try to determine whether a particular legal ruling or practice is right or wrong. Although traditional legal scholarship does try to determine this, the criteria it uses are purely or primarily logical; whether a legal ruling is just or unjust or fair or unfair for any party in a case is irrelevant in legal scholarship. If a ruling is logically defensible, then it is a "right" ruling, regardless of its possible unfairness or unjustness.

The concept of justice, then, does not fall in the province of law and society scholarship nor in that of traditional legal scholarship. Instead, it is a key topic in legal philosophy, which is beyond the scope of this book. That does not mean, however, that the concept of justice is irrelevant to the everyday world of law and society and to participants in the legal system; far from it. As Edwin Schur (1968:18) observed four decades ago,

> [I]t is impossible to avoid the socio-ethical dilemmas that continuously arise in the operation of legal systems. Individuals and groups within a society, including legal functionaries themselves, find it necessary to make ethical choices regarding legal matters. . . . These choices—which, in turn, influence the actual shaping of the legal system and hence help to determine that which we are concerned with studying empirically—are not easy to make. There are always conflicting interpretations and alternative conclusions that can be reached.

Thus, although the field of law and society does not normally consider questions of justice, the topic of law and society cannot be properly understood without appreciating two related aspects: (1) the ethical choices that legal actors make regarding questions of justice, and (2) the ways in which these choices sometimes conflict with the logical thinking that is thought to guide legal decision-making.

An actual court case will illustrate this conflict. A Massachusetts resident named Joseph Dasha was diagnosed with a fatal brain tumor in June 1988 at a Portland, ME hospital. He had surgery to remove the tumor and then thirty radiation treatments that damaged his brain and left him in a mostly vegetative state. In March 1991, a doctor examined a tissue sample from Dasha's tumor and determined that it was benign, meaning that the radiation treatment that left him in a vegetative state had been unnecessary. After Dasha's family filed a malpractice suit in May 1992, the case was heard by the Maine Supreme Judicial Court. Maine state law at the time provided a three-year statute of limitations for bringing a medical malpractice claim. Because Dasha was diagnosed in June 1988, his three-year window for filing a claim expired in June 1991, only three months after his tumor was discovered to be benign. Because his suit was not filed until May 1992, it exceeded the statute of limitations and therefore, under a strict interpretation of this statute, could not be considered. In contrast to Maine's law, most other states say the window for filing malpractice suits starts from the discovery of a medical mistake, not from the time of the initial diagnosis that led to the alleged malpractice. If these states' timetable had applied to Maine, Dasha's family would have had until March 1994, three years after the discovery that his tumor was benign, to file their suit. But under Maine's law, they were eleven months too late.

In their suit, Dasha's family argued that the statute of limitations should not have started until after the discovery of the diagnostic mistake. In doing so, they invoked a legal doctrine, *equitable estoppel*, which is used in fraud cases and takes into account the actions of the defendant in determining when any statute of limitations should begin. Rejecting this argument, the Maine Supreme Judicial Court ruled 5-1 that the lawsuit was invalid because the statute of limitations had expired, with the majority opinion noting that the medical misdiagnosis was "not the equivalent of fraud sufficient to support the assertion of equitable estoppel" (Associated Press 1995:G5). The lone dissenter argued that the Court should ordinarily carry out the intent of the state legislature in deciding a case but at the same time needed to avert any injustice that the legislature did not anticipate. This justice argued that the medical misdiagnosis was the equivalent of fraudulent conduct and that the statute of limitations should not have begun until after the conduct was discovered.

As should be clear, the majority opinion in this case followed the traditional view of law's assumption of logical decision-making. The statute of limitations had expired, and the medical misdiagnosis was not fraudulent as that term is usually conceived, however mistaken it otherwise was. Thus, the case could not be brought to court. The minority opinion took into account a consideration, injustice, that logic normally finds irrelevant. The judge who wrote this opinion was concerned about the injustice resulting from a refusal to let the case come to court and thus was more inclined to believe that equitable estoppel should have applied. A strict application of logic in this case let to an outcome that, however logical, was, of course, very unfortunate for the plaintiff and his family. Although we cannot know for sure, it is very possible that the five justices who joined in the majority opinion regretted the outcome but concluded that the law left them with no other choice.

The Case of the Speluncean Explorers

The tension between the traditional view of law and questions of justice is perhaps most compellingly explored in a hypothetical story, "The Case of the Speluncean Explorers," published in 1949 by Lon L. Fuller (1949), a Harvard Law School professor. Fuller was known for writing many parables about the law that were designed to teach lessons about legal reasoning and other legal issues. His Speluncean Explorers case is perhaps his most famous parable and is still studied today because the issues it raises remain relevant. It is well worth reading on your own, but a summary will have to suffice here.

The case involved four defendants who were members of the Speluncean Society in the hypothetical nation of Newgarth. As its name implies, the Speluncean (from *spelunker*) Society was an organization of people interested in cave exploring. Five members of this club were exploring a cave when a landslide occurred and blocked the entrance. Dozens of rescuers tried to save them, but repeated landslides, one of which killed ten rescuers, impeded their efforts. They finally broke through into the entrance 32 days after the original landslide. On the twentieth day of their captivity, the trapped explorers communicated with the outside (through what Fuller in 1949 called a "portable wireless machine") and told the rescuers that their meager food supply had already been exhausted and they were now starving. The explorers were informed that their rescue was at least ten days away and that, further, they would probably starve to death before they could be rescued. The explorers then asked if they could survive if they ate the remains of one member of their group and were informed that this action would indeed help them survive. When they heard this, the explorers then asked whether they should cast lots to determine which explorer should be eaten, but no one from the outside was willing to provide a response.

The rescuers heard nothing more from the trapped explorers until they found them on the thirty-second day. Four of the explorers were still alive, and one was dead, having been eaten by his companions. The surviving explorers were charged with murder and later testified that their dead companion had proposed casting lots, using a pair of dice he had with him, for one of them to be put to death and eaten. He lost the roll of the dice and was killed and eaten, enabling the other four to survive. After the four survivors were convicted of murder and sentenced to be hanged, they appealed their case to the Supreme Court of Newgarth, whose five justices each rendered a different opinion.

Chief Justice Truepenny concluded that the four defendants had indeed committed murder and that Newgarth's murder statute, which said in part, "Whoever shall willfully take the life of another shall be punished by death," did not allow for any exceptions. While the chief justice was troubled by the circumstances of the case, he wrote that freeing the defendants would undermine respect for the law. At the same time, he recommended that the Supreme Court ask the Chief Executive of Newgarth for clemency for the defendants.

The second justice, Foster, wrote that he was shocked that the defendants had been convicted in the first place. He had two reasons for feeling this way. First, the defendants were in a "state of nature" after they were trapped in the cave and thus outside of the legal order as normally conceived. Second, their act of cannibalism was in effect self-defense against starvation. For these reasons, Justice Foster voted to overturn their convictions.

Justice Tatting wrote that he was upset that the defendants had been prosecuted but that he also agreed with Chief Justice Truepenny that they had indeed committed murder, and did not feel that their action was the equivalent of self-defense. Justice Tatting then added that to apply the law of murder in this case would simply not be just. The case left him so confused that he could not decide whether to uphold or overturn the defendants' convictions, and so he abstained.

Justice Keen had little doubt. His opinion stated that the role of the courts in Newgarth is to follow the legislature's intent. In this case, the legislative intent of the murder statute was very clear, said Justice Keen, and he voted to uphold the convictions.

The fifth and final justice, Handy, urged the Court to use common sense. The defendants' action did not constitute murder as the legislature conceived of this crime when it wrote the murder statute. Moreover, the defendants have suffered enough, and 90 percent of the public in an opinion poll said they wanted the defendants pardoned or to receive only token punishment. Justice Handy added that the chief executive was very unlikely to pardon the defendants (having heard this from his wife's niece, who is a close friend of the chief executive's secretary). Justice Handy voted to overturn their convictions.

Combining all five opinions, the Newgarth Supreme Court's vote in the Case of the Speluncean Explorers was 2 to uphold the convictions, 2 to overturn them, and 1 abstention. The tie meant that the convictions were upheld, and the defendants were ordered to be hanged.

In this case we see a vivid, if unsettling, example of the conflict between the traditional view of law and questions of justice. The two justices, Truepenny and Keen, who both voted to uphold the convictions followed the traditional view of law in applying logic, however coldhearted and unemotional it may seem in this case, to the circumstances that led to the defendants' act of killing. The law prohibited willfully taking someone's life and provided no exceptions, and these judges decided that the defendants had taken someone's life willfully as the law defines that term. Following this reasoning, their convictions for murder were appropriate and should not be overturned. In contrast, the two justices, Foster and Handy, who voted to overturn the convictions hewed more closely to the social science view of law in considering the justness of the outcome if strict logic was allowed to govern the Supreme Court's decision-making. This does not mean that these two justices' vote to overturn was illogical, only that their vote to overturn the convictions was guided by certain nonlegal considerations that justices Truepenny and Keen rejected or ignored. Finally, Justice Tatting's abstention reminds us of the proverbial animal that is equally distant from two equally attractive items of food. Not being able to decide which way to turn and go eat, the animal stays frozen in place and starves to death. Justice Tatting was unable to resolve the tension between the traditional and social science views of law and so abstained, but with his abstention the Court's tie vote meant that the convictions and death sentence were upheld.

Although Fuller's Speluncean case was hypothetical, he reportedly based it on two actual cases from the nineteenth century that both involved people in lifeboats (Katz 1987). The earlier of the two was *U.S. v. Holmes* (26 F. 360 [ED Pa. 1842]) and involved a ship sailing from Liverpool, England, to Philadelphia, PA. The ship hit an iceberg near Newfoundland and promptly sank. Only two lifeboats were available for eighty passengers and crew. Forty-one people, both passengers and crew, crammed into the larger of the two lifeboats, and nine corroded into the much smaller remaining lifeboat. There was no room for the thirty remaining people, most of them children and none of them crew, who went down with the ship. The smaller lifeboat made it safely to the Newfoundland coast, but the larger boat was too heavy and barely able to move. When a storm began to fill the boat with water, the mate ordered his sailors to make the boat lighter to help it float. In response, at least two sailors, including one named Holmes, threw at least eight people overboard and did the same to two more people the next morning. These actions prevented the boat from sinking, and it was finally rescued a few weeks later.

Holmes was eventually prosecuted in Philadelphia for voluntary manslaughter. His defense argued that the killings were necessary to keep the boat from sinking and thus legally justified. The judge told the jury that the sailors should have placed themselves in a lottery with the boat's passengers to determine who should have been thrown overboard. The jury convicted Holmes, who was sentenced to six months in prison.

The other lifeboat case was *Regina v. Dudley and Stephens* (14 QBD 273 [1884]), which involved an Australian yacht that sank and left four survivors in a small lifeboat. Their total food consisted of two cans of turnips. They ate the turnips and a turtle they captured but soon became dehydrated and began starving. Nineteen days after they entered the lifeboat, the yacht's captain, Dudley, suggested that they hold a lottery and kill and eat the individual who lost, but two of the men refused to go along and so the lottery was not held. Then Dudley took Stephens aside and said that one of the four, a man named Parker, was so sickly he was about to die and that they should kill him now. Dudley then did so, and the three survivors ate Parker's remains for four days until they were finally rescued by a passing ship. The three men were then prosecuted for murder in Australia. Although they argued that their act of killing was necessary for them to survive, the judge convicted them of murder and sentenced them to be hanged. However, a pardon by Queen Victoria saved their lives.

In both these cases we see many of the same issues that Fuller presented in his fictitious case of the cave explorers. In all three cases, a narrow, logical interpretation of the law would have meant, and did mean in the two actual cases, that the defendants would be convicted of homicide in what obviously were very unusual circumstances. Acquittals in the cases would have meant bending logic just a bit and allowing nonlegal considerations to affect the verdict. There was no right or wrong verdict in any of these cases, just a difference of opinion regarding what factors should or should not matter.

Law and Logic in Tumultuous Times

Cases involving starvation and homicide are not the only types in which the logic of the law is in tension with just results. In American history, this tension is perhaps most vividly seen in trials with very important political origins and potential consequences. One of the earliest such trials occurred in 1735 and involved a New York newspaper publisher named John Peter Zenger who printed several articles critical of the royal governor of the New York Colony. He was eventually arrested for seditious libel, a law that prohibited publishing any material that, even if true, might arouse public sentiment against the government. His attorney, Andrew Hamilton, argued that the articles should not be considered libelous because they were true, and he urged the jury to, in effect, disregard the law by acquitting Zenger for that reason. The judge instructed the jury that it should not consider the possible truth of Zenger's publications and that their only role was to decide whether he had indeed published them. Facing a terrible dilemma, the jury chose to listen instead to Hamilton and found Zenger not guilty; it evidently disliked the law under which Zenger was charged and decided that the truth of his articles was in fact a valid defense to prosecution under the law. His acquittal is regarded as a landmark in the history of the freedom of the press in the United States (Finkelman 1981).

More than a century later, juries faced a similar dilemma of logically applying a clear law with a possible unjust result. These juries participated in trials of abolitionists charged with violating the Fugitive Slave Act of 1850 by helping fugitive slaves escape on the Underground Railroad or by breaking captured slaves out of Northern jails. Although in almost every case the defendants had clearly violated the law, many Northern juries in effect rejected the law and declared the defendants not guilty. Historians regard these acquittals as important victories for the abolitionist movement, as they limited government efforts to repress abolitionism and also indicated widespread public antipathy in the North to slavery (Mabee 1969).

In the Zenger and abolitionist cases, juries had to make some very difficult decisions. The defendants were all accused of violating laws that the jurors evidently considered repugnant to American ideals of freedom and democracy. They could either follow the letter and logic of the law and convict the defendants, yielding what they would consider an unjust result, or they could disregard the law and acquit them. The acquittals they did return are honored in history and reinforced certain American ideals but still represent the tension, we have seen before, between the logic of the law and just outcomes. According to Schur (1968:22–23), cases such as these and the brain surgery malpractice case discussed earlier all illustrate "recurrent questions confronting legal systems, questions that must be taken into account by anyone attempting to understand how such systems operate." He adds that they further illustrate that legal issues are not always "simple and clear-cut" (p. 23) and that human feelings sometimes, for better or worse, influence the legal decisions that people make.

THE PLAN OF THE BOOK

This chapter has introduced several topics and issues in the study of law and society that remainder of the book will address. Chapter 2 discusses the always vexing question of how to define law and provides an historical overview of social science perspectives on law and society. These perspectives will help provide a theoretical underpinning for the rest of the book. Chapter 3 examines types of law within the United States and "families" of law across the world. Although many students think about the criminal law and criminal courts when they think about law, many other types of law exist that are also very important for our society. At the same time, the different families of law across the world reinforce the basic social science view that law reflects the society in which it is found. Chapter 4 discusses the related topics of legal culture and dispute processing. As noted earlier, dispute settlement is one of the major functions of law; at the same time, the choice of methods to try to settle disputes is influenced by the society in which we live, our own views about the law, and other factors. Accordingly, this chapter will examine the types of dispute processing and the advantages and disadvantages of using law to resolve disputes. In doing so, the chapter will incorporate findings from anthropological studies of law and from studies of law in contemporary societies.

Chapter 5 looks at the fascinating topic of law and social control. This chapter explores the degree to which law can effectively control behavior of various types, and it also critically explores the use of law to enforce morality through the banning of behaviors that are commonly called victimless or consensual crimes. Chapter 6 examines the impact of law on society and the impact of changes in society on law. The discussion includes work from the founders of sociology as well as from contemporary studies of the mutual influence of law and society. Chapter 7 turns to the issue of law and inequality. As indicated in the law and society questions outlined earlier, this chapter examines the degree to which inequality based on social class, race and ethnicity, and gender characterizes the American legal system and the degree to which law may promote or reduce inequality in the larger society. Chapter 8 covers the legal profession. It summarizes what is known about the origin and history of the legal profession and of its current dynamics and controversial issues. Law schools and their ways of teaching about the law are also discussed. Finally, Chapter 9 discusses courts and juries. Here the dynamics of courtroom workgroups and of related issues such as plea bargaining are emphasized. Though rarely used, the jury remains one of the most important, if controversial, legal aspects of a democratic society. Accordingly, this chapter discusses the history of the jury, the dynamics of jury decision-making, and the issue of jury nullification.

Summary

1. Several issues and events from the middle part of this decade illustrate the importance of law in the United States and other societies across the world. Law is also important for the daily lives of average citizens. Although the next chapter has more to say about how to understand law, the term "law" generally comprises a set of rules and regular procedures for creating, implementing, and enforcing those rules.
2. The social science view of law differs in important ways from the traditional view of law. The latter views law as a self-contained system of logic, while the former views law as deeply embedded in society.

As such, law both potentially reflects and influences the society in which it is found.

3. The field of law and society is based generally on several assumptions. These include: (a) the legal system is not autonomous; (b) laws and legal decisions may have a potential impact on one or more aspects of society; (c) changes in society may effect changes in law; and (d) law, whether considered as rules or as a legal system, reflects the type of society in which it is found.
4. Law as conceived in the field of law and society consists of four dimensions: (a) the substance of law, or the various kinds of rules that help guide behavior; (b) legal procedures; (c) the organization of the legal system; and (d) legal culture, the beliefs and values that people hold about law and the legal system.
5. The scope of law and society scholarship comprises three very general areas of study: (a) the origins and impacts of the substance of law; (b) the operation of the many dimensions of the legal system; and (c) legal culture.
6. The field of law and society tries to answer many key questions. Some of the most important of these questions include: (a) how effective is law in controlling human behavior and in affecting organizational behavior? (b) to what degree do new laws, court rulings, or other types of legal changes affect social institutions or other aspects of society, and what factors affect the impact that law may have in these respects? (c) how effective are legal procedures compared to nonlegal procedures in the settlement of disputes? (d) to what degree do changes in society affect the various dimensions of law? (e) to what degree does inequality characterize the operation of the legal system? and (f) how, to what extent, and in what ways does law reinforce or contribute to social inequality or, instead, reduce social inequality?
7. Questions of justice are ordinarily not part of the study of law of society nor of traditional legal scholarship. At the same time, the actual functioning of law cannot be fully understood without appreciating the role that questions of justice do, in fact, sometimes play in legal decision-making.

Key Terms

Dispute settlement
Legal culture
Legal procedures
Organization of the legal system
Rule of law
Social change
Social control
Social institutions
Social science view of law
Substance of law
Traditional view of law

CHAPTER 2

Understanding Law and Society

Although there is a general social science perspective on law as spelled out in the previous chapter, there are, in fact, many ways of understanding law from a social science perspective. This chapter reviews how social scientists have tried to understand law since the study of law and society began about 150 years ago.

WHAT IS LAW?

Before discussing the various social science perspectives on law and society, it would naturally be helpful to have a precise definition of law. Although it is only a three-letter word, "law" is surprisingly difficult to define. As the previous chapter noted, law usually refers to a set of rules and regulations *and* to the social institution that creates, implements, and enforces these rules and regulations, but the definition and conception of law are, for better or worse, far more complicated than that. As Friedman (2005:3) notes, "There is . . . no general agreement about a definition, nothing that commands a general consensus. Nor can there be. Law is not a thing in the real world that can be described with any precision. There is no such thing as a purely objective definition of law. What we call law depends on why want to call something law."

Perhaps the key problem lies in distinguishing law from custom. For many scholars, this does not pose a problem. To them, laws are official rules passed, implemented, and enforced by government, and customs are unwritten norms that are enforced only informally by family and friends. Thus, Donald J. Black (1976:2), a prominent sociologist of law, has famously defined law as "governmental social control." To Black and other scholars who adopt this view, other types of rules may be rules, but they are not law.

Many scholars think this view of law is too restrictive. Their concern stems from their observation that nongovernmental groups and organizations have their own systems of rules. Colleges and universities have various codes of behavior regarding such things as academic dishonesty and sexual harassment. Voluntary organizations such as the Boy and Girl Scouts, your local Parent Teacher Association, and the Rotary Club and other civic groups all have their own rules and regulations, with appropriate sanctions if these rules and regulations are violated. Limiting the concept of law to governmental social control, say these scholars, overlooks and obscures the fact that all these nongovernmental systems of rules are the equivalent of law: they look like law, they sound like law, and they probably even smell like law. They all "operate what look like miniature legal systems. They make rules, they enforce them, and they have regular procedures" (Friedman 2005:4). Any decision not to define these nongovernmental systems of rules as law means that they do not fall within the purview of law and society research. To these scholars, this result would be unfortunate, as the study of these systems of law may increase understanding of governmental social control and also of the theoretical and research questions outlined in the previous chapter. For these reasons, many law and society scholars favor a broader definition of law that encompasses the rules, regulations, and associated procedures of governments as well as those of nongovernmental groups and organizations. To these scholars, law is both **official** (governmental) and **unofficial** (nongovernmental).

If law can be both official and unofficial, it can also be both **formal** and **informal** (Friedman 2005). Law is formal when it involves rules and regulations that are written and enforced according to written guidelines. Much governmental law and nongovernmental law is of this character. At the same time, the enforcement of rules and regulations often involves much discretion, as decision makers (e.g., the police in governmental law or a campus code enforcement officer as an example of nongovernmental law) have the ability to decline to invoke certain rules and sanctions even if they believe that an infraction has occurred. By letting an offender off with a warning, they are acting more informally and exercising the informal powers of law.

So far we have indicated that the definition of law that is adopted affects what is studied under the law and society rubric. But the adopted definition of law also has important implications for the very fundamental question of whether law exists in a particular society. This is a question that anthropologists began to raise almost a century ago in their studies of small traditional societies. At the time of these studies, traditional societies were popularly viewed as savage and chaotic. Against this view, the early anthropologists found processes of social control and ways of processing disputes that paralleled those in modern societies. The traditional societies' processes of social control and dispute processing may have looked very different from those in modern societies, but they nonetheless had the same result in achieving these essential functions for any society. The early anthropologists concluded that law did in fact exist in these traditional societies even if its substance and operation differed in many respects from law in modern societies (Malinowski 1926).

Recall part of Freidman's (2005:3) observation cited just above: "What we call law depends on why want to call something law." In this regard, it is likely that the early anthropologists wanted to find law in the traditional societies they studied, and so they defined law in a way that would allow them to conclude that law did in fact exist in these societies. If these scholars had concluded that law did not exist in the societies, then they in effect would have been concluding that these societies were lawless. Such a conclusion would have reinforced the then popular view that these societies were savage. In their efforts to counter this view, anthropologists probably wanted to find law, and they did so. By acknowledging this possibility, we do not at all mean to suggest that their conclusion was inappropriate or that they were unduly biased. Rather, we simply mean to reinforce what was said above: that the definition of law cannot be entirely objective and that any attempt to define law is far from simple. As a noted anthropologist once wrote, "To seek a definition of the legal is like the quest for the Holy Grail" (Hoebel 1954:18). Another scholar shared this pessimism: "Those of us who have learned humility have given over the attempt to define law" (quoted in Vago 2006:8).

Selected Definitions of Law

With this backdrop in mind, and to further illustrate the difficulty of defining law, we now turn to and critically examine several well-known definitions advanced by scholars from various fields.

An early definition comes from John Austin, a British legal philosopher of the nineteenth century. Austin, whom we shall discuss again later in this chapter, thought that law was nothing more and nothing less than the *command of a sovereign* (quoted in Schur 1968:26). This definition has been influential, but upon closer inspection its limitations become clear (Schur 1968). First, it limits law to what "sovereigns," or, to use the American context, the executive branch of government, declare, and thus ignores the vast amount of law that emanates from, among other sources, bills passed by legislatures, rulings by judges, and the rules of administrative or regulatory agencies. Second, this definition fails to recognize that sovereigns can never be subject to the law itself. Third, and relatedly, it fails to recognize that their commands can be unjust or harmful (Hart 1958).

A definition related to Austin's is Donald J. Black's (1976) view, cited above, that law is *governmental social control*. This definition has the advantage of incorporating branches of government beyond the sovereign/executive, but, like Austin's, it may be interpreted as implying that government cannot be subject to law and as failing to recognize that governmental social control can be unjust or harmful.

A third definition comes from A. R. (Alfred Reginald) Radcliffe-Brown, an early British anthropologist: *Law is the maintenance or establishment of social order . . . by the exercise of coercive authority through the use, or possibility of use, of physical force* (quoted in Schur 1968:74–75). This definition obviously highlights the social order function of law while neglecting its other functions (see Chapter 1) and suggests that law is effective because of the threat of physical force it entails. We return to this assumption shortly. From an anthropological standpoint, an advantage of the definition is that it allows for law in many traditional societies because it does not mention the need for police or some other agency found only in a modern society; a disadvantage is that it does not allow for law in traditional societies in which physical force is not used.

A fourth definition is that of American anthropologist E. Adamson Hoebel, who wrote that *a social norm is legal if its neglect or infraction is regularly met, in threat or in fact, by the application of physical force by an individual or group possessing the socially recognized privilege of so acting* (Hoebel 1954:28). This definition again highlights the threat or use of physical force and adds to Radcliffe-Brown's definition by specifying that such physical force is to be applied by an individual or group specifically charged by their society with this function. Hoebel's definition again allows for law in traditional societies, since it does not characterize the party with the privilege of applying force in ways that limit the definition to modern societies.

Another definition, and perhaps the most famous in the law and society literature, comes from German sociologist Max Weber, who wrote about a century ago: *An order will be called law if it is externally guaranteed by the probability that coercion (physical or psychological) to bring about conformity or avenge violation will be applied by a staff of people holding themselves specially ready for that purpose* (Rheinstein 1954:5). Weber's definition again highlights physical force and adds the condition that such force will be applied by "a staff of people holding themselves specially ready for that purpose." This phrase, and especially the term "staff," has been interpreted in different ways. Some observers interpret it as equivalent to the police or another modern agency; if it were interpreted in this manner, Weber's definition would then restrict law to modern societies and imply that law does not exist in traditional societies. However, other observers say it is significant that Weber used the term "staff" and not the term "state" or something similar. With this interpretation, they add that his definition does indeed allow for law in traditional societies, as the staff of people could be a group of people within such a society who are entrusted to hold themselves "specially ready" to use physical force if necessary.

Taken together, the last three definitions and especially Weber's imply that several characteristics help distinguish whether a rule achieves the status of a law or instead remains a custom (Vago 2006). First, the pressure to comply with the rule is external to the individual. Second, this pressure involves coercion and force and especially physical force. Third, the party that applies such force is socially authorized to do so. In contrast, customs are more commonly thought to be obeyed because the individual wants to obey them or because the individual fears informal social control, such as disapproval or ridicule, from family members and friends.

Physical Force and Obedience to Law. The emphasis on physical force in all these definitions has been criticized for implying that physical force is the primary reason for people to obey law. Although it is true that many people indeed obey the law because they are afraid of physical force (in a modern society, possible arrest and imprisonment) if they violate it, there are in fact many other reasons why they obey the law generally and certain laws more specifically (Tyler 2006). For example, when was the last time you murdered someone? Undoubtedly, very few, if any, readers of this book have ever murdered anyone. Why not? If we are to believe Weber and the other theorists, the major reason we have not murdered anyone is that we *fear arrest and legal punishment* if we do so. In contrast, the major reason why you have not committed murder is probably that you consider it *morally wrong*.

Another reason many people obey the law is that they *feel obliged to obey* simply because it is, in fact, the law. This is especially or primarily true in a democracy, where laws are generally seen as legitimate and deserving of respect and obedience even if we do not like them.

A third reason we sometimes obey the law is our *concern for peer pressure or other informal sanctions*. Although the field of criminology emphasizes how our peers may induce us to break the law (Warr 2002), we may also have peers and especially family members who help keep us from breaking the law because we do not want to earn their disapproval and all the problems that would entail.

Another reason for obeying the law is that *certain illegal behaviors might harm us* even if we do not get arrested. Here illegal drug use and driving the wrong way on the highway come to mind, as does robbery: if the robbery does not go as planned, the robber may find himself (or rarely, herself) the victim of his own weapon.

A final reason we sometimes obey the law is *habit*: we are used to behaving in a legal manner and do not really ever think about behaving illegally. If you always wear your seat belt when you are driving a car, you might do so primarily because you live in a state that requires seat belts to be worn or because you realize that it is safer to do so. But for many people, seat-belt wearing has simply become such a habit that when they get in their car they buckle up automatically without giving it a second thought.

Which motive for obeying the law is the most salient obviously depends on the behavior in question. Although it is almost certainly true that you have not murdered anyone because you think murder is morally wrong, cheating on an exam might be a different manner. Many students do not cheat because they consider cheating morally wrong, even if it certainly is not as wrong as murder, but others do not cheat only because they are afraid they will be caught and punished. When you have resisted the temptation to park in an illegal spot on campus or at a shopping mall, why did you not park illegally? Some people might be concerned about getting a ticket or being towed away, but others might simply think it wrong, either because of moral reason or because they feel parking regulations deserve obedience.

To cite one other example, have you driven late at night and come to a deserted intersection where there is a traffic light? If the light is red, do you stop for it and wait for it to turn green, do you stop and wait a moment and then proceed, or do you barely slow down? If you always wait for the light to turn green, why do you wait? Are you afraid that a police car will suddenly emerge from behind a billboard or from underground like the machines in *War of the Worlds*? Are you afraid that if you run the light, an huge truck will suddenly emerge from nowhere and split your car in half? Do you just wait out of habit? Do you think it is morally wrong to go through a red light? Or do you think that this type of law is generally deserving of your respect even if you could go through the light safely?

The issue of coercion and obedience aside, one other attribute that all the definitions of law outlined above have in common is that they are *positivist* definitions. That means they describe what the law is but not what the law should be. Some scholars put much more stress on what the law "should be." To sociologist Philip Selznick, the essence of law is **legality**, or the rule of law. Legality, he says, means that no one, not even people in the highest echelons of government, is above the law: "Official actions, even at the highest levels of authority, is enmeshed in and restrained by a web of accepted general rules. Where this idea exists no power is immune from criticism and completely free to follow its own bent" (Selznick 1961:95). Conceived this way, legality is, of course, one of the cornerstones of democracy. Some of the controversies discussed at the beginning of Chapter 1 concerned certain U.S. government actions in the aftermath of 9/11 that, in the opinion of many observers, involved governmental officials and agencies putting themselves above the law.

Lon Fuller (1964), whose parable on Speluncean explorers was discussed in the previous chapter, similarly observed that law is more than a mere rule or command of a sovereign or anyone else in a position of power. For rules or commands to constitute law in his view, Fuller argued that they must adhere to the "inner morality" of law. The inner morality of law in turn consists of several criteria, and a rule or command is not law unless it satisfies these criteria. The most important of these criteria include the following:

1. ***Rules must be general.*** Rules should apply to everyone, and no one should be permitted to violate rules with impunity. This, of course, is what Selznick meant by legality.

2. ***Rules must be generally known.*** It would obviously be unfair if rules were passed in secret and people were then punished for violating rules that only a very few people knew about.

3. ***Rules must not be retroactive.*** Rules that ban or govern behavior should not be used to punish anyone who engaged in the behavior before the rule was enacted. This is one of the cardinal principles of law in a democracy.

4. ***Rules must be reasonably clear.*** There are many jokes about legal writing and the difficulty of understanding tax law and other bodies of law, but this criterion simply means that the meaning of a rule should be generally understandable.

5. ***Rules must not be contradictory.*** It would obviously be unfair for one rule to ban a behavior that another rule says must be followed.

6. ***Rules must not require the impossible or the unreasonable.*** To take a silly example, it would be unfair to ban blinking your eyes or scratching an itch in public.

7. ***Rules should be relatively stable through time.*** Rules obviously need to adjust to changing conditions, but they should not change too frequently. If a rule said something for one week, and then was changed the next week and then again the week after that, this process would violate the stability criterion.

Definitions of Law and the Existence of Law

As noted earlier, the definition of law that we choose to adopt has important implications for the significant question of whether and to what extent law exists in a particular society. For example, Fuller (1964:21) says that each of the "inner morality" criteria just listed must be fulfilled for rules to be called laws and for a system of rules to be called a legal system: "A total failure in any one of these [criteria] does not simply result in a bad system of law; it results in something that is not properly called a legal system at all." As should be clear, Fuller, Selznick, and other scholars with similar views hold a conception of law generally corresponding to the type of law found in democratic societies. On the other hand, the type of law found in authoritarian regimes violates Selznick's concept of legality and many and perhaps all of Fuller's criteria. To these scholars, then, law does not exist in these regimes and certainly did not exist in past regimes such as Nazi Germany and Mussolini's Italy. In contrast, "positive law" definitions like those of Black and Weber, discussed earlier, allow for law to exist in authoritarian regimes. What do you think? Does law exist in authoritarian regimes?

Recall that several early anthropologists were probably interested in demonstrating that traditional societies were not lawless. Because social order did exist in the societies they studied, their definitions of law took account of this fact and implied that law did indeed exist in these societies. Keeping their perspective in mind, consider whether law is found in four very different types of traditional societies that anthropologists have studied.

Our first society is that of the Pueblo Indians in the southwestern United States. Two anthropologists examined the aftermath of several dozen accusations of various types of rule violations between 1870 and 1925 (Smith and Roberts 1954). These cases included the following behaviors:

Behavior	Number of Cases
Breach of peace	1
Drunkenness	12
Fighting	5
Homicide	2
Property settlements after death/divorce	22
Rape	4
Slander	6
Theft	16
Witchcraft	18

A tribal council heard all the cases except the witchcraft accusations, which were heard by the priesthood. The Pueblo procedures for hearing the cases involved such steps as a pretrial investigation and a trial with testimony. During a trial, a single question was asked of a witness four times so that the witness had time to reflect on the answer. Lawyers did not exist among the Pueblos and thus were not involved in the process. Most penalties for those who were found guilty of committing an offense involved compensation in the form of jewelry, clothing, livestock, and money.

Is this law? Did law exist among the Pueblos when they were studied? The Pueblo way of dealing with *trouble cases*, to use an anthropological term, certainly differs in key respects from that found in the United States as a whole. At the same time, it involves procedures that sound very familiar to us, and these procedures are regularly used for dealing with Pueblo trouble cases. Most readers would probably conclude that law did indeed exist among the Pueblos.

Now let us consider a second society, the Tswana tribe in southcentral Africa (Schapera 1955). When this society was studied, it had several local divisions. Each division had its own court, as we would call it, with the head of each division also acting as the judge of this local court. The tribe as a whole had a higher court, with the chief of the tribe also acting as the judge of this higher court. The local court heard cases involving property disputes, slander, and the like, whereas the higher court heard cases involving fighting, theft, and other criminal offenses. In either venue, the trial would take place in or near the judge's living quarters, with judge and his advisors sitting together and facing the parties to the case. The judge would introduce the case and ask for statements by the opposing parties. Anyone else who was attending the trial could also ask questions of witnesses. At the end of the testimony, the advisors would discuss their views about the case with the judge, and he would render his verdict.

Is this law? Did law exist among the Tswana tribe when it was studied? Like the Pueblos, the Tswana had regular procedures for dealing with trouble cases that sound familiar to us in some respects and less familiar in other respects. They did not quite have the legal system with which most of us are familiar, but most readers would probably again conclude that law did indeed exist among the Tswana.

The remaining two societies we examine are more difficult to assess. Among the Inuit (Eskimo) peoples of North America, *song duels* have been a regular method of settling disputes (Kleivan 1971). The two parties to a dispute each compose a song that outlines the person's version of the events and also ridicules the other person. They then sing back and forth in front of other members of their tribe. Besides allowing the disputants to vent their anger, the songs also help settle the dispute, as the onlookers would decide which disputant had the better song. The person with the better song as decided by public opinion would win the case, often regardless of what the evidence actually showed.

Is this law? Did law exist among the Inuit when they were studied? Like the Pueblo and Tswana, the Inuit had a routine process for settling disputes that apparently worked for them. Unlike the Pueblo and Tswana, the outcomes of this process apparently did not always relate to the actual evidence in the case but instead depended on who had the best song. Compared to their views on the Pueblo and Tswana, many readers will be reluctant to conclude that law existed among the Inuit, but other readers can certainly make a reasonable argument that law did indeed exist among them. As our earlier discussion indicated, the belief that law does or does not exist in a particular society depends heavily on the definition and conception of law that we have.

The final traditional society is the Ifugao of the northern Philippine mountains (Barton 1919/1969). Here a common method of settling disputes when they were studied was the *ordeal*. When two individuals had a quarrel or dispute, either one could challenge the other to an ordeal. Typical ordeals included removing a pebble slowly from boiling water (with the person who holds onto the pebble longer winning the case), holding a very hot knife (with the person holding the knife longer again winning), wrestling, or fighting with spears. The winner in any of these ordeals was also deemed the winner of the dispute that generated the ordeal, and the winner would often collect a fine for making a false accusation.

Is this law? Did law exist among the Ifugao when they were studied? The Ifugao again had a regular method for settling disputes, but in this case "might made right," as the winner of the case would be the party who was stronger or who had tougher skin, not necessarily the party who would have won the case had evidence been actually considered. Many and perhaps most readers will probably conclude that because the outcome has nothing to do with any evidence, the Ifugao did not have law even if they had a way of resolving disputes.

This is a sound conclusion, although it would also be reasonable to conclude that the Ifugao had law if your definition equates law with a routine method of dispute settlement. But let us consider the first conclusion—that the Ifugao did not have law because the outcome of the case depended on who was stronger or tougher—a bit further. If this was your conclusion, should you then conclude that the United States also does not have law? Do legal outcomes in the United States sometimes depend on who has more money and influence than on whose case the evidence supports? If you were in a legal skirmish with someone named Rockefeller or Gates, who would win the case? Yes, you might win, but isn't it very likely that you might lose, not because the evidence would not support you but because Rockefeller or Gates could afford to spend millions of dollars, if necessary, for the best attorneys and investigators and bring other influence to bear? If they won the case primarily because they had more money and influence, is that very different from the Ifugao? If not, does that mean the United States does not have law? If you believe the United States still has law, should you not also conclude, to be consistent, that the Ifugao also must have law?

These are difficult questions, and perhaps classroom discussion and debate will help you think further about the issues they raise. But they at least remind us of a central point of this discussion: the definition and conception of law we adopt affects decisions concerning the existence or nonexistence of law in any particular society.

EXPLANATIONS OF LAW

Over the decades, law and society scholars have had many different ways of understanding law and its relationship with society. The previous section on the difficulties of defining law illustrated some of these different understandings and emphasized that there is no one right or wrong way of thinking about law from a social science perspective. At the same time, there is much to learn from the different perspectives that law and society scholars have. Accordingly, this section reviews the most important historical and contemporary explanations of law. Many of the assumptions and beliefs of these explanations will be reflected in subsequent chapters.

Early Approaches

Law has been studied from a social science perspective since the nineteenth century, and many perspectives or schools of thought exist. We review early approaches in this section and then turn to contemporary views in the next section.

Legal Formalism. **Legal formalism**, also called *legal positivism* or *analytical jurisprudence*, was quite popular in the nineteenth century and is still very influential today. Its central assumption, and perhaps its key contribution to the study of law, was the belief that law is a self-contained system of logic that is independent of social and moral considerations. This, of course, is the traditional view of law, featured in the previous chapter, that dominates the teaching and understanding of law in law schools today. In developing this view, legal formalists were unconcerned with the fairness and justness of law; rather, their chief concern was whether legal rules were logical, not whether they were fair or just.

Two chief proponents of legal formalism as it first and later developed were John Austin and Hans Kelsen, respectively. Austin (1790–1859) is the British legal philosopher cited earlier in this chapter for equating law with the command of a sovereign. His lectures at the University of London were published in 1832 under the title *The Province of Jurisprudence Determined*, and the views expressed in his lectures influenced the study of law for many decades to come, although his work did not become widely acclaimed until after his death in 1859. As noted earlier, Austin's primary concern was to outline the fundamentals of law and a legal system that were above all else logical; the actual content of law did not matter. Thus, he wrote, "The existence of law is one thing; its merit or demerit is another. . . . A law, which actually exists, is a law, though we happen to dislike it, or though it vary from the text, by which we regulate our approbation and disapprobation" (Austin 1995 [1832]:157). Austin thought that a law was indeed a law regardless of whether it was moral and should be obeyed even if it was immoral. He did not discount the value of religious and moral views in people's lives and for other aspects of society, but he thought these views should not influence the creation or operation of law.

Hans Kelsen (1881–1973) was an Austrian scholar who taught for many years at European universities before moving in 1940 to the United States and taking positions at Harvard Law School and in Department of Political Science at the University of California at Berkeley. In the study of law and society, Kelsen is perhaps best known for his *pure theory of law*. He thought that the conception and study of law had been contaminated by ideological considerations. Thus, he called his theory a "pure theory" because it was "a theory of law purified of all political ideology and all natural-scientific elements" (Kelsen 1967:xviii). Like Austin, he argued that moral considerations should play no role in determining the worth and validity of law. What makes a law valid instead, he said, is simply whether it is generally obeyed: "A legal order is regarded as valid, if its norms are by and large effective (that is, actually applied and obeyed)" (Kelsen 1967:212).

In arguing that law should be divorced from moral considerations, Austin, Kelsen, and other legal formalists take issue with the fundamental assumptions of the **natural law** perspective that began with ancient Greek and Roman philosophers such as Aristotle and Cicero and was further developed in later centuries by notable religious figures such as St. Thomas Aquinas, the great Catholic theologian who lived during the thirteenth century. As noted above, natural law refers to legal principles derived from nature and thought to be binding on human society, as opposed to human or positive law, which is law as written by the human hand. Aristotle, who is credited with coining the term "natural law," wrote that a key distinction between natural and positive law is that the former embodies universal truth while the latter is more temporary and subject to the whims of the particular people who create law. In contrast to Kelsen, who said that the validity of law depends on whether it is obeyed, Aristotle and Cicero wrote that natural law is valid everywhere. Aquinas, perhaps the key thinker in the early development of natural law philosophy, added that law must serve the common good and that laws that do not do so are unjust and, in fact, not laws at all (to return to an issue discussed earlier in this chapter) because they violate this fundamental purpose of law.

The debate between positive law and natural law has endured for centuries and will, no doubt, continue for the rest of our lifetimes and beyond, and certainly has been the subject of countless essays, journal articles, and books (e.g., Delaney 2003; George 1999) that address issues far beyond the scope of this book. Whether you favor a natural law or positive law approach is a matter of your own philosophy and personal taste, even if it is true that most social scientists probably lean more toward the perspective of positive law rather than that of natural law.

Utilitarianism. **Utilitarianism** is a perspective that was developed during the eighteenth and nineteenth centuries, primarily by Italian economist Cesare Beccaria (1738–1794) and English philosopher Jeremy Bentham (1748–1833). They both came of age and authored their important works during a time when the European criminal justice system was, to put it simply, horrid. People would be routinely arrested on trumped-up charges and weak evidence and incarcerated without trial in jails that were hellholes. The use of torture to force confessions was also quite common. English defendants often suffered a form of torture called "pressing" in which heavy weights would be placed on their bodies to force a confession; some died instantly, while others lasted a few more days until they either confessed or died (Jones 1986). English law at the time provided the death penalty for more than 200 crimes, including theft, the writing of libelous letters, and treason. Treason was defined quite broadly and characterized crimes involving servants plotting to kill their masters and women plotting to kill their husbands. Such offenders were often hanged but not before some of them were disemboweled or dismembered (Gatrell 1996).

Beccaria and Bentham were harsh critics of all these practices. Reflecting the philosophy of the Enlightenment, the name given to the popular European philosophy that arose during the eighteenth century, they both thought that people act rationally and with free will and are chiefly concerned with maximizing their pleasure and reducing their pain. In this view, people calculate the potential rewards and costs of their actions and proceed accordingly. Thus, Beccara and Bentham reasoned, the criminal justice system certainly needs to be punitive, but only to the degree that it deters people from committing crime. Given this assumption, they said, the criminal justice system as it then existed was much harsher than it needed to be to convince rational individuals that they should not risk arrest and imprisonment by engaging in crime. Beccaria's and Bentham's views captured much attention and are credited with helping to reform the criminal justice system in Europe and with influencing the thinking of notable figures in the American colonies such as John Adams and Thomas Jefferson (Jones 1986; Vold, Bernard, and Snipes 2002).

Given the utilitarian view that people wish to maximize their pleasure and reduce their pain, it was quite natural that utilitarian thinkers focused on law as a mechanism to influence the way people act. A nineteenth-century legal scholar, Rudolph von Jhering (1818–1892; also known as Rudolph von Ihering) of Germany, made important contributions in this regard, as the translated title of his most important book, *Law as a Means to an End* (Jhering 1968 [1877]), indicates. He said that people often have competing interests and sometimes resort to crime and other antisocial behaviors to achieve their goals. For an effective society, he said, the human will must be constrained: "There is no greater miracle in the world than the disciplining and training of the human will, whose actual realization in its widest scope we embrace in the word society" (Jhering 1968 [1877]:71). Law, he thought, was an important mechanism for accomplishing this goal and, more generally, for changing society in positive directions.

The Historical School. The **historical school** is the name given to the work of several nineteenth- and early twentieth-century scholars who discussed how law and legal systems changed over the centuries as societies developed from ancient times to the time the scholars were writing. Charles Darwin's 1859 publication of his magnificent book, *The Origin of Species*, was a key inspiration for many of these scholars. Although his theory of evolution focused on how and why organisms change over time, it also motivated scholars in other fields to wonder how and why societies had changed over time. To the members of the historical school, law, broadly defined, changed over time as societies changed, and the ways in which law changed reflected the ways in which the larger society changed. In developing and documenting this principle, the historical school established a basic theme of the study of law and society—that law reflects the society in which it is found—that still guides much theory and research today (see Chapter 1).

Friedrich Karl von Savigny. Perhaps the earliest proponent of the historical school was Friedrich Karl von Savigny (1779–1861), a German scholar who taught Roman law at the University of Berlin. Savigny's interest in the history of Roman law and its enduring impact in Italy and elsewhere long after the Roman Empire collapsed led him to appreciate the importance of customs—informal norms—for the development of law. His key belief as one of the founders of the historical school was that law both reflects and expresses the culture, spirit, and values of a people and in this sense was little different from art or music. Taking issue with John Austin's view that law is the command of a sovereign (see earlier discussion), Savigny wrote that because law develops from a people's beliefs and values, it does not develop just from a ruler: "[Law] is first developed by custom and popular faith . . . not by the arbitrary will of a law-giver" (quoted in Schur 1968:31). This view in turn led Savigny to reach two related conclusions for the understanding of law: (1) law imposed by a ruler may violate the beliefs and values of a people, and (2) attempts to impose law on a people with whose beliefs and values it conflicts may prove relatively futile. We shall see evidence of this latter conclusion in Chapter 5's discussion of law and social control.

Although Savigny's views were important and remain influential, critics say he ignored or underemphasized some important factors (Schur 1968). One problem lies in his view that law reflects the beliefs and values of a people. In fact, critics say, some laws are passed in spite of many people's beliefs and values. Moreover, laws may differ between, say, two locations (such as two states within the United States) even if people's beliefs are very similar in both locations, and the residents of both locations may not even be aware of many of their laws. Schur (1968) provides an interesting example of this phenomenon. In the early 1900s, Massachusetts and New York differed slightly in a particular aspect of contract law: whether a contract is considered completed when it is mailed after a party signs it or only when it is received after being mailed. A contemporary discussion of that example (Gray 1909/1963) cited by Schur (1968) observed that it is very doubtful that residents of Massachusetts and New York held different views on this specific issue or even knew that the two states differed in this regard.

Another problem in Savigny's view that law reflects a people's beliefs and values is that it neglects the possibility that law may also influence and even change people's beliefs and values. To take just one example, it is probably true that many people disagreed with seat belt laws when they were first enacted in several states and continue to disagree with them, but that these laws have convinced at least some people that it is important for safety reasons to wear seat belts. Whether and to what extent laws do change people's beliefs, values, and practice is, of course, an empirical question (see Chapter 6), but any evidence of such an effect underscores Savigny's neglect of this issue.

These criticisms aside, the significance of Savigny's views for the development of the study of law and society cannot be denied. As Schur (1968:31–32) observed,

> Savigny's attempt to place law in historical perspective must be viewed as an important step in the development of broadly social

> conceptions of the legal system. . . . [Scholars who attempt] to study the interrelationship between law and the value system of a society, or to examine public attitudes toward particular law or aspects of legal procedure, or to analyze patterns of sociological change, [are] pursuing lines of inquiry suggested by Savigny's work.

Sir Henry Maine. Another key figure in the historical school was Sir Henry James Sumner Maine (1822–1888), better known as just Sir Henry Maine, a professor at the University of Cambridge, England. His lectures were published in 1861 in his celebrated book, *Ancient Law*. As its title implies, this book sought to describe the nature and structure of law in the ancient world and to outline how it had evolved to the nineteenth century. A key change in law over this expanse of time, said Maine, was the change *from status to contract*. By this he meant the following. In ancient times, relationships were governed by power: someone would have more power entering a relationship, and thus in the relationship itself, because of the social standing deriving from the wealth and prestige (or lack of these) of the family into which the parties to the relationship were born. Many relationships, in fact, were not voluntary at all, as the existence of slavery in ancient Rome reminds us. As society evolved, however, relationships based on power (i.e., status) gave way to ones that were more voluntary and governed eventually by verbal and then written contracts. Thus, law in essence changed from status to contract.

The development of law in this way is one of the many changes accompanying the development of modern society, which is larger and more impersonal than its ancient counterpart. Maine believed that the development of modern society would permit people's intelligence and talents to influence their standing in life, as opposed to the situation in ancient societies, where a family's social standing was the key determinant of an individual's success in life, whether or not the individual was intelligent and talented.

Founders of Sociology: Durkheim, Weber, and Marx. The three key founders of the discipline of sociology were French scholar Emile Durkheim (1858–1917), German scholar Max Weber (1864–1920), and the great German theorist Karl Marx (1818–1883). All three, but especially Durkheim and Weber, considered law a key variable for understanding how societies had changed over time as they became more modern.

Durkheim's insights focused on how the type of law characteristic of a society depends heavily on the society's homogeneity (similar norms, beliefs, and values) or heterogeneity (dissimilar norms, beliefs, and values). As traditional societies grew and became more modern and heterogenous, he said, their type of law also changed and, in particular, became less punitive. Weber's insights focused on how legal decision-making became more rational as societies became more modern, with this dynamic reflecting a focus in much of his work on rationality as the key feature of modern society. Although he paid less attention to law than these two founders, Marx still considered law to be one of the many social institutions through which the ruling capitalist class exerts its ideological and practical control over the working class. Specifically, Marx (along with his colleague Friedrich Engels) thought that law helped the rich maintain their dominant position by preserving private property and by convincing the working class that their society was fair and just.

This brief summary cannot begin to do justice to these three majestic figures in the history of social thought, and Chapter 6 examines their work further because of its relevance for that chapter's discussion of law and social change.

Legal Realism. **Legal realism** was a popular school of thought in the United States in the early to mid-twentieth century that greatly influenced the study of law and society then and now. As its title might imply, its main effort lay in understanding how law really works as opposed to how it is supposed to work in theory. In making this effort, legal realism directly challenged the assumptions of legal formalism (see above), which neglected the possibility that legal rules may not work in practice the way they are supposed to work in theory.

Legal realists devoted much attention to the decision making of appellate judges. Recall that legal formalism and traditional legal theory assume that judges' decisions are nothing more and nothing less than the equivalent of logical conclusions. In this sense, judges are said to *find* law in the evidence and circumstances of case. In contrast, legal realists argued that judges *make* law instead of just finding it. By this, they meant that judges sometimes make their decisions according to certain beliefs accompanying their own socioeconomic backgrounds, including their conception of justice, and not just according to legal doctrine. They then reach a decision based on what they consider a fair outcome and, after doing so, write a decision in which they interpret the law in such a way as to justify this outcome.

Juries were another aspect of the legal system examined by legal realists. Here again they found that how juries actually operate sometimes departs from the way they are supposed to operate. Ideally, juries should base their verdicts on an impartial assessment of the evidence and law in a case, and their verdicts should represent the jurors' best attempt to logically apply the law in view of their assessment of the evidence. Instead, legal realists argued, juries sometimes base their verdicts on their biases for or against a defendant rather than just on the evidence.

Although legal realism reached its zenith in the 1920s and 1930s, its origin is often credited to the work during the 1880s of Oliver Wendell Holmes, Jr., who later served on the U.S. Supreme Court from 1902 to 1932. As Chapter 1 noted, Holmes took issue with the traditional view of law espoused by legal formalism. His famous passage included in Chapter 1 is worth repeating here in shorter form: "The life of the law has not been logic; it has been experience. The felt necessities of the time, the prevalent moral and political theories, institutions of public policy, avowed or unconscious, even the prejudices which judges share with their fellow men, have had a good deal more to do than the syllogism [logical argument] in determining the rules by which men should be governed" (Holmes 1938 [1881]:1). In this passage, we see legal realism's emphasis on the importance of nonlegal factors for legal decision-making.

The two most notable legal realists after Holmes were Karl Llewellyn and Jerome Frank. Llewellyn (1893–1962) was a Seattle native who became a law professor first at Yale University and then at Columbia University, where he published his most prominent legal scholarship (Llewellyn 1930), and finally at the University of Chicago. With anthropologist E. Adamson Hoebel, he authored a classic study on law among the Cheyenne Native American tribe (Llewellyn and Hoebel 1941). Frank (1889–1957) was a New York City native who served on the Securities and Exchange Commission before becoming a federal appellate court judge. His two most influential books were *Law and the Modern Mind* (1930) and *Courts on Trial: Myth and Reality in American Justice* (1949). In both books, he argued that scholars need to pay more attention to criminal and civil trials than to the workings of appellate courts in which they had traditionally been interested. In the former book, he became known and even controversial for attributing judges' decisions to aspects of their personality, while in the latter he was quite critical, among other things, of juries for rendering verdicts based on their own prejudices rather than on the evidence and law governing a case.

Legal realism's theme that law in action differs from law in the books reflects this approach's more general view that nonlegal factors affect legal decision-making. This emphasis is certainly part of the canon of law and society scholarship today. It underlies work by political scientists since the mid-twentieth century on the social backgrounds and political ideologies of judges that help understand the bases for their legal decisions (Cross 1997), and it also underlies work by law and society scholars from several disciplines on the nonlegal factors that may affect the outcomes of criminal and civil cases. At the same time, legal realism has also been faulted for exaggerating the subjectivity of legal decision-making, which many scholars consider more objective and informed than the legal realists did.

Sociological Jurisprudence. Developed at about the same time as legal realism, the school of **sociological jurisprudence** drew on the new discipline of sociology to argue that law has social underpinnings and impact. The two leading figures of sociological jurisprudence, Eugen Ehrlich and Roscoe Pound, pursued this theme in slightly different ways. Ehrlich (1862–1922), an Austrian legal scholar, is often credited with being the founder of the sociology of law subfield and published his very influential book, *Fundamental Principles of the Sociology of Law*, in 1913 (Ehrlich 1936 [1913]). He distinguished between *positive law* and *living law*. Positive law, as we saw earlier, simply means the body of law enacted by legislators and shaped by judges. Living law refers to a society's customs. Erhlich emphasized that positive law can be effective only to the extent that it corresponds to living law; if specific laws violate people's customs, these laws are not likely to be very effective. Taking this view one step further, he added that law by itself cannot produce social order unless people were already predisposed to want to obey legal norms because they regarded these norms as legitimate and binding on their conduct.

Roscoe Pound (1870–1964) was a Nebraskan native who served from 1916 to 1936 as Dean of Harvard Law School. Although he greatly influenced legal education at Harvard, he also wrote many articles, reports, and books (e.g., Pound 1930) that greatly shaped the bourgeoning field of law and society. He distinguished between *law in action* and *law in books* to stress his idea that there is a large gap between what formal legal rules say and how they actually work out, or, to put it a bit more simply, between what the law says and what it does. To illustrate this distinction, Pound pointed out that although justice should be blind to differences based on race, social class, and other such factors, the legal system was in fact discriminatory and a setting in which the poor received little justice. Accordingly, he called for a series of reforms in the criminal justice system and other areas of the law.

As another illustration of the distinction between law in action and law in books, Pound observed that legislators may have a specific purpose in passing a statute, but that the actual effects of the statute may be quite different from what the legislators intended. In a recent example of this phenomenon, a 2006 newspaper report noted that a new Iowa law requiring convicted sex offenders to live more than 2,000 feet from a school or day care center had had "unintended and disturbing consequences" (Davey 2006:A1). As intended, the law prevented sex offenders from living in many locations, especially in cities, which obviously have many schools and day care centers within a relatively small area. However, this meant that the sex offenders had to move to rural areas, whose residents also did not want them, or to become homeless. Many sex offenders simply disappeared, with the number of sex offenders who were missing tripling during the six months after the law went into effect, certainly not a consequence that Iowa's legislators had expected or desired.

As this example suggests, sociological jurisprudence's emphases continue to help us understand law from a social science perspective. As Schur (1968:43) observed, "[S]ociological jurisprudence left an indelible mark on American legal thought and also provided suggestive guidelines for social research on the law." As we shall now see, however, the social sciences as a whole were slow to incorporate insights from sociological jurisprudence and legal realism, and several decades passed until law became a prime subject of theory and research in the social sciences.

The Early Law and Society Movement. By the end of the 1930s, the stage was set for the social sciences to embrace law as a key topic for theory and research. Not only was law a social institution in its own right, but, as the historical school showed, the study of law could also help understand society itself. Utilitarianism, the historical school, legal realism, and sociological jurisprudence all had yielded important insights and emphases that the social sciences could have incorporated and expanded to study law. Despite this foundation, however, the social sciences generally ignored law for another two decades until after World War II, when the study of law and society, often called the *law and society movement*, began in earnest (Friedman 1986; Garth and Sterling 1998).

This movement encompassed several academic disciplines and was spearheaded by anthropology. As noted earlier, anthropologists in the early to mid-twentieth century found that the societies they studied had what were then considered to be surprising amounts of social order and social control. Like modern, complex societies, these small, traditional societies had regular methods for settling disputes that often did not involve the use of physical violence. To many anthropologists, these societies were far from lawless, and much anthropological writing emphasized the nature and extent of law that did in fact characterize traditional societies.

One of the earliest and still most influential anthropological legal studies, *Crime and Custom in Savage Society*, was penned by Bronislaw Malinowski in 1926 (Malinowski 1926). Malinowski (1884–1942) was a Poland native who moved to England and became one of the most renowned anthropologists of the twentieth century. His book examined norms, norm violation, and social control among the peoples of the Trobriand Islands near New Guinea in the South Pacific. It emphasized that law existed among the Trobriand people even if they did not have the police, courts, and judges that typify modern societies. To make this case, he defined law as the reciprocal obligations that people have on each other and which, if not fulfilled, would lead to disappointment and other negative consequences. Some scholars later criticized his definition for not adequately distinguishing law from custom and for not recognizing that laws are norms enforceable by physical coercion from an authoritative body (Schur 1968). On this point, Malinowski's definition was especially disputed by another noted anthropologist, A. R. Radcliffe-Brown, whose definition of law, presented at the beginning of this chapter, emphasized physical coercion by a centralized authority. For this reason, Radcliffe-Brown thought law did not exist in the many traditional societies that lack centralized authority and physical coercion (Radcliffe-Brown 1952).

A noted legal anthropologist of the post–World War II period was E. Adamson Hoebel (1906–1993), who taught for many years at New York University, the University of Utah, and the University of Minnesota. Hoebel, co-author with Karl Llewellyn of a book on Cheyenne law noted just above, also wrote another very important book, *The Law of Primitive Man: A Study in Comparative Legal Dynamics* (Hoebel 1954). This book explained Hoebel's views of law based in part on his observation of norms, law, and dispute settlement among the Plains Indians, including the Cheyenne and Comanche. He emphasized both that law was essential for society and that society, or rather social bonds and consensus, were essential for law: "Without the sense of community there can be no law. Without law there cannot be for long a community" (Hoebel 1954:332).

Two other legal anthropologists of this period were Paul Bohannan (1920–2007) and Max Gluckman (1911–1975). Bohannan ended his long academic career at the University of Southern California and won acclaim for his study of African homicide and suicide. Gluckman spent his career at the University of Manchester in England and was active against colonialism. During the 1950s and 1960s, both men were part of a debate among anthropologists on the applicability of modern Western concepts of law to the society and culture of traditional societies (Nader 1969).

Political science also made notable contributions after World War II. Beginning in the 1940s and especially the 1950s, political scientists pioneered sophisticated empirical studies of the decisions of justices on the U.S. Supreme Court and on the federal appellate courts. This body of research began where legal realism left off and examined how factors such as the socioeconomic backgrounds of justices, their religious upbringing, and their membership in either the Democratic Party or Republican Party seemingly influenced their decisions on various types of cases (Grossman 1966; Schubert 1963). For example, in cases involving disputes between labor and management, Democratic judges tended to side with labor and Republican judges tended to side with management. Political scientists during this period also wrote several pathbreaking studies of the Supreme Court's involvement with the political world outside its chambers (Pritchett 1948) and of the social psychology of its decision making (Murphy 1966; Snyder 1958). The work of political scientists in the post–World War II period was significant in its own right and also provided the basis for much law and society scholarship today. Perhaps most importantly, it made clear that judges are not merely "logical machines" but rather very human and thus subject, as most people are, to beliefs and biases stemming from their personal backgrounds and involvement with individuals and groups outside their chambers. To the extent this is all true, law is a very different phenomenon from how it was pictured by the legal formalists of the nineteenth century.

In the 1970s, several political scientists turned their attention to the criminal courts (Eisenstein and Jacob 1977; Jacob 1978). They discovered something that judges, prosecutors, and defense attorneys already knew; this "something" is that the adversary or combat model of criminal justice is largely a myth. In most criminal cases, prosecutors and defense attorneys do not vigorously contest the evidence, as this model implies and as *Perry Mason* and many other TV shows and films about lawyers and courtrooms dramatically illustrate. Instead, political scientists found, consensual decision making characterizes most criminal court proceedings, as judges, prosecutors, and defense attorneys cooperate on plea bargaining and in other ways to ensure the efficient processing of huge numbers of criminal cases. For this reason, these political scientists considered plea bargaining to be a necessary and even useful phenomenonious needs they have. Chapter 9 examines this issue further.

During the 1960s, sociologists published several accounts that criticized the lack of justice in the criminal courts for the poor and people of color (Blumberg 1967a; Clarke and Koch 1976; Mather 1973; Sudnow 1965). They found that defense attorneys were often overburdened and/or underqualified and spent little time on the legal defense of criminal suspects. They also found that the caseloads in the nation's municipal courts were so large that little time could be spent on any one case, with the result that many defendants may be induced to plead guilty even if they had not committed any crime. One scholar charged that law was a "confidence game" in which defense attorneys' chief concern was to collect their fee while doing as little work as possible on behalf of their clients (Blumberg 1967a). As this body of work has been summarized, it

> charged that poor but innocent defendants were being railroaded into pleading guilty by layers who cared more for courts' administrative needs and their own professional needs than for their clients' well-being. Urban courts were depicted as "assembly lines" in which the typical defendant . . . spends at most a few minutes with a public defender or assigned counsel before pleading guilty. [These attorneys] were depicted as undertrained and overworked, and urban courtrooms as dismal, dirty, and crowded settings. (Barkan 2006b:510)

Beginning in the 1970s, sociologists and criminologists began to conduct quantitative analyses of the factors affecting whether criminal defendants were found guilty or not guilty and the harshness of the punishment they received if they were found guilty. Early studies found, contrary to what many scholars might have assumed, that race did not affect sentence lengths once factors such as the seriousness of the charges and of a defendant's prior record were taken into account (Hagan 1974; Kleck 1981). Although this view was popular for a while, sentencing research has continued into the present, and later studies began to reveal the subtle ways in which race and other extralegal factors sometimes do indeed affect sentencing. For example, some studies found that race affects sentencing more often in minor cases than in serious cases (Spohn and Cederblom 1991) and more often when the race of the victim is considered than when the race of the defendant is considered (Baldus, Woodworth, and Pulaski 1990). Chapter 7 discusses this body of evidence further.

Contemporary Perspectives

The law and society movement has certainly continued since the 1970s and flourishes today. The current movement both draws on and expands the emphases of the early law and society movement and of the disciplines and schools of thought from which the early movement grew. Although contemporary law and society work comprises many perspectives, certain perspectives stand out because they have either shaped scholarly thinking or challenged key assumptions of traditional law and society scholarship. We now turn to these perspectives.

Functionalist and Conflict Views. A basic difference in how law and society may be understood emerges from considering views drawn from *functionalist* and *conflict* perspectives. Popular in political science and sociology, these two theoretical perspectives offer contrasting assumptions of how society works. Sociolegal scholars have applied these two perspectives' general frameworks to law as a social institution. Before discussing functionalist and conflict views on law, it will be helpful to review the more general perspectives from which these views are drawn.

Functionalist theory—also known as functionalism, and better known as *pluralism* in political science—was developed as an intellectual reaction to the French Revolution of 1789 and to the Industrial Revolution of the nineteenth century. Although many, books have been written about both revolutions, for our purposes we simply need to know that both revolutions caused intellectuals to fear that social order was crumbling. The French Revolution, of course, involved a bloody revolt against the French aristocracy, while the Industrial Revolution led to the rise of large cities and their serious problems of crime and poverty. Concern about social order and stability in the wake of these two revolutions prompted European intellectuals to emphasize in their writings that societies need effective rules and socialization and strong social bonds to survive (Collins 1994). A key intellectual of this period was Emile Durkheim, whom we discussed earlier. Much of functionalist theory derives from his emphasis on the importance of social integration and socialization for social stability.

Functionalist theory today uses the human body as a model for understanding how society should and should not work. Just as the body has various organs and appendages that allow the body to function (our eyes allow us to see, our legs enable us to walk, and so forth), so does society have various parts, including the family, schools, and religion, that enable it to function. Because it emphasizes the importance of society's social institutions for social stability, functionalist theory is skeptical of sudden social change. Just as any sudden changes in our body's parts can endanger our health and general well-being, so can any sudden changes in society's various parts endanger its ability to function. A close offshoot of functionalist theory is *consensus theory*, which assumes that most people, regardless of their race and ethnicity, social class, gender, age, and so forth, generally hold similar views on important social and political issues.

Some of Durkheim's work is especially relevant for a functionalist understanding of law, crime, and society, as it involves the normality and functions of deviant behavior. If we define **deviance** as behavior that violates social norms and arouses negative social reactions, Durkheim (1895/1962) recognized that there will always be people who violate the many social norms that exist. Given this fact, Durkheim stressed that a society without deviance is impossible because the collective conscience is never strong enough to prevent all rule-breaking. Even in a "society of saints," he said, such as a monastery, rules will be broken and negative social reactions aroused. Because Durkheim thought deviance was inevitable, he considered it a *normal* part of every healthy society.

In a surprising and still controversial twist, Durkheim further argued that deviance serves several important functions for society. First, he said, deviance clarifies social norms and increases conformity. This happens because the discovery and punishment of deviance reminds people of the norms and reinforces the consequences of violating them. If your class were taking an exam and a student was caught cheating, the rest of the class would be instantly reminded of the rules about cheating and the punishment for it, and be less likely to cheat as a result.

A second function of deviance is that it strengthens social bonds among the people reacting to the deviant. If you are familiar with the famous story, *The Ox-Bow Incident*, from the original book (Clark 1940) or the classic film starring Henry Fonda, you would have seen this function in action. In the story, three innocent men are accused of cattle rustling and eventually lynched. The mob that does the lynching is very united in its frenzy against the men, and, at least at that moment, the bonds among the individuals in the mob are extremely strong. Another example of this function is a school pep rally. At these rallies, the "deviant" is the other school that your own school's team will soon play. As you and your schoolmates rally against this "deviant" school, you feel more united as a result.

A third function of deviance, said Durkheim, is that it can help lead to positive social change. Although some of the greatest figures in history—Socrates, Jesus, Joan of Arc, Mahatma Gandhi, and Martin Luther King, Jr., to name just a few—were considered the worst kind of deviants in their time, their heroic example contributed to the freedom of thought we now enjoy.

There is also a fourth function of deviance that Durkheim did not discuss. Deviance creates jobs, not only for some deviants themselves but, especially, for the segments of society—police, prisons, criminology professors, etc.—whose main focus is to deal with deviants in some manner. If deviance and crime did not exist, hundreds of thousands of law-abiding people in the United States would be out of work!

Conflict theory, derived from the work of Karl Marx and Friedrich Engels, who were discussed earlier, takes a different view in all these respects. Whereas functional theory emphasizes that social institutions contribute to social stability and in other respects a "healthy" society, conflict theory emphasizes that these institutions help perpetuate inequality by maintaining the control of the relatively few powerful individuals and groups. Whereas functionalist theory dislikes sudden social change, conflict theory asserts that such social change is needed to lessen and even eliminate social inequality. And whereas consensus theory assumes that people generally agree on important issues regardless of their social backgrounds, conflict theory assumes that people often disagree because of different needs and perceptions arising from race and ethnicity, social class, gender, and other aspects of their social backgrounds.

These general views of functionalist and conflict theory have been applied to law, and they offer a contrasting view partly reflected in the previous chapter's discussion of the functions and dysfunctions of law. The functionalist view emphasizes the importance of law for the proper workings of society. More specifically, law is said (to recall the functions of law outlined in the previous chapter) to help to preserve social order, to make social life possible, to settle disputes, to protect individual freedom, and to bring about needed social change. The functionalist view further assumes that the major sociodemographic groups generally agree on important legal issues such as appropriate sentencing for convicted criminals.

The pluralist variation of functionalist theory found in the field of political science adds an important assumption to these general views. Pluralist theory recognizes that society is filled with interest groups that compete for wealth, power, and influence. As they do so, they try to affect what legislation gets passed and what social policies get enacted. To cite just one example, environmental groups try to have legislation passed that would help protect the environment, while certain corporations try to prevent such legislation from being passed. Pluralist theory assumes that law serves as a neutral referee over the competition among interest groups like these. In this view, law does not take sides for one interest group or another but rather ensures that the competition is conducted fairly and that everyone "plays by the rules." Any legal decisions that help determine the outcome of the competition are thus based not on any bias for one side or the other but rather on dispassionate interpretations of law, evidence, and other considerations.

Once again, conflict theory differs from functional theory in all the beliefs just outlined. First, it assumes that law is dysfunctional for society; in particular, it assumes that law is characterized by inequality and also contributes to inequality in the larger society. Second, it assumes that major sociodemographic groups disagree on important legal issues in ways that reflect their life experiences and different positions on the socioeconomic ladder. Third, it assumes that law takes sides in the competition among interest groups and, more specifically, that it takes the side of powerful, established interests such as corporations.

Where does the truth lie? There is much evidence to support either the functionalist or the conflict perspective on law and society, and many scholars favor one perspective or the other. This book finds merit in both perspectives.

It is certainly true that law serves the several functions assumed by functional theory, but there is also compelling evidence that inequality sometimes characterizes the law and that law may reinforce social inequality (see Chapter 7). Similarly, there are certain beliefs on legal issues on which the major sociodemographic groups generally agree, but there are others on which they disagree. For example, a growing number of studies find that whites are more punitive than African Americans in regard to the death penalty and other types of legal punishment, and that racial prejudice among whites leads them to be more punitive (Barkan and Cohn 2005; Chiricos, Welch, and Gertz 2004; Soss, Langbein, and Metelko 2003; Unnever and Cullen 2007). Turning to the issue of interest group competition, the evidence is complex, but it does seem that the "haves" in society are often better able to take advantage of the law to win important advantages for themselves (Galanter 1974). Regardless of whether law does act as a neutral referee, it offers an arena for conflict in which individuals and groups with money and influence are better equipped to compete and win (see Chapter 7).

Marxist Perspectives. Recall that Karl Marx and his colleague Friedrich Engels thought that law helped the capitalist class maintain its position at the top of society by preserving private property and by convincing the working class that society was fair and just. Beginning in the 1970s, Marxist scholars drew on this basic view to develop at least three competing perspectives of the relationship between law and society. Although all these perspectives agree that law and the state serve the interests of the ruling class, they differ on some key elements (Balbus 1982; Gold, Lo, and Wright 1975; Hepburn 1977; Jessop 1977).

One perspective is called the **instrumentalist** view (and also *orthodox* Marxism). According to this view, law and the state are tools that can fairly easily be used by the ruling class to oppress the working class and in other respects to maintain its own superior position (Michalowski and Bohlander 1975; Miliband 1969). *Law Against the People*, the title of a 1971 collection of articles on the oppressive uses of the law (Lefcourt 1971), is a phrase that aptly summarizes this view. As a proponent summarized the instrumentalist outlook, "The state exists as a device for controlling the exploited class. . . . Contrary to conventional wisdom, law instead of representing community custom is an instrument of the state that serves the interests of the developing capitalist class" (Quinney 1977:45). Although this perspective was popular in the 1970s among many Marxists, other Marxists and, not surprisingly, non-Marxists criticized it for being much too simplistic and thus for being "one dimensional," as one critic put it (Chambliss and Seidman 1982; Greenberg 1976:610).

These critics took issue with instrumental Marxism for at least two reasons. First, they said, the ruling class does not always win in the legal and political arenas, contrary to what the instrumentalist view assumes. Second, individuals and groups comprising the ruling class sometimes disagree among themselves over important political, legal, and economic issues. Thus, the ruling class is not nearly as unified as instrumentalist Marxism believes. These criticisms were compelling in scholarly circles and led to the rise of the two remaining Marxist perspectives that take these criticisms into account.

The first of these is the **structuralist** view, which, contrary to the instrumentalist view, concedes that ruling-class individuals and groups sometimes disagree among themselves over important issues and compete with each other for short-term advantages. Such short-term victories may benefit one group at the expense of the others and may even lead to economic and political instability in the larger society. The state and its legal system prevent this from happening by being *relatively autonomous* from the ruling class (Beirne 1979). This relative autonomy helps convince the public that the economic and political systems are fair and impartial. As William Chambliss and Robert Seidman (1982:308) put it: "The state and the legal order best fulfill their function as legitimizers when they appear to function as value neutral organs fairly and impartially representing the interests of everyone." Because the state is relatively autonomous from the ruling class, the ruling class is not omnipotent, and sometimes the working class and other "have-nots" win short-term victories. According to the structuralist view, these victories are spurious because they do little to change the fundamental inequality in society and help blind the public to this inequality because the victories convince the public of society's fairness. In this way, the relative autonomy of the state and legal system helps preserve capitalism and serves the long-term interests of the ruling class (Poulantzas 1973). For this reason, the structuralist view dismisses attempts by the have-nots to use the law to better their position as short-sighted and self-defeating (H. Collins 1982).

Although the structuralist view has arguably become more popular than the instrumentalist view, some Marxists still criticize the former for refusing to admit that any legal or political victories by the have-nots are real and effective and not just short-sighted and self-defeating (Balbus 1982; Jacobs 1980). They also say that the instrumentalist view is guilty of tautological (or circular) reasoning by assuming that any state action helps support capitalism even if it does not appear to support capitalism: "One cannot disprove so tautological a theory. If the state acts in defense of capitalist interests, it acts because of capitalist control of the state. If it acts against capitalist interests, despite appearances, the state still acts in capital interests. A theory that data cannot conceivably contradict—that is, a nonfalsifiable theory—tells us very little" (Chambliss and Seidman 1982:314--315).

A third Marxist view attempts to address the problems of the instrumentalist and structuralist views. This view, called the **dialectical** or *class struggle* perspective, agrees with the structuralist view that the state and its legal system are relatively autonomous from the ruling class, but it also argues that legal and political victories by the have-nots can indeed be real and consequential and not just sham (Chambliss and Seidman 1982; Grau 1982). In this view, law can be an effective tool of the ruling class, but it also constrains the ruling class's ability to repress the poor and other have-nots.

Thus, whereas the structuralist view scoffs at civil liberties as a sham that helps obscure social and economic inequality, the dialectical view applauds the rule of law and civil liberties as an important benefit for have-nots (Gordon 1982). Commenting on crime in eighteenth-century England, legal historian, E. P. Thompson (1975:263–264) summarized this belief: "On the one hand, it is true that the law did mediate existent class relations to the advantage of the rulers. . . . On the other hand, the law mediated these class relations through legal forms, which imposed, again and again, inhibitions upon the actions of the rulers. For there is a very large difference . . . between arbitrary extra-legal power and the rule of law."

Critical Legal Studies. Critical legal studies (CLS) is a movement that formally began in 1977 with a conference at the University of Wisconsin-Madison attended mostly by law school students and young law professors. Active in the protest movements of the 1960s and 1970s, these individuals shared that era's distrust of authority and repulsion for poverty, racism, and militarism (Tushnet 2005). Given their legal interests, it is not surprising that they regarded law as a key contributor to social inequality and other social problems.

CLS incorporates many of the ideas we have already examined from legal realism and from Marxist views on law, but it also critically addresses the nature and quality of legal education. Although CLS proponents have diverse views, several general CLS beliefs may nonetheless be identified (Hutchinson 1989; Trubek 1989; Tushnet 1986; Unger 1986). First, law is *politics*: political views, biases, and considerations affect the creation of law, including legislative policymaking and judicial decisions, and also the operation of law. With this assumption, CLS joins legal realism in challenging the premise of legal formalism that law is merely logic. Second, law is *indeterminate*. This means that the evidence and law in any particular case do not necessarily lead logically to any particular conclusion (as legal formalism assumed); instead, the law is sufficiently vague and even contradictory that several outcomes are possible, opening the door for politics to affect the outcome.

Third, law is *ideology*. In the social sciences, "ideology" is a term that often means a set of beliefs and values that supports the *status quo* (Mannheim 1936). Recall that Marx and Engels thought that law created an appearance of justice that helped to placate the working class, and that today's structuralist Marxists make the same claim. By asserting that law is ideology, CLS proponents are saying that law comprises a complex set of beliefs, values, and symbolism that legitimizes the (unequal) *status quo* by suggesting the legal and political systems are fair, just, and impartial when in fact they are not. Fourth, *legal education as practiced in law school contributes to law student passivity and to social inequality in the larger society* when new law school graduates go out to practice. This critique of law school and education has aroused much controversy since it was first advanced about three decades ago, and Chapter 8 on the legal profession explores it at greater length.

CLS scholars have applied these beliefs in many interesting studies, and an example involving freedom of speech is illustrative. According to law professor David Kairys (1990:237), most people believe that freedom of speech has been a hallmark of American democracy since the Constitution and Bill of Rights were enacted more than two centuries ago. This belief is a myth, he argues, because "no right of free speech as we know it existed, either in law or practice" before the law changed between 1919 and 1940. Before this time, he observes, "one spoke publicly only at the discretion of local, and sometimes federal, authorities, who often prohibited what they, the local business establishment, or other powerful segments of the community did not want to hear." Thus, although the United States is about 230 years old, the freedom of speech that American enjoy today has in fact fully existed for less than one-third of the nation's history.

The judicial and other legal changes between 1919 and 1940, Kairys adds, primarily resulted from this period's massive labor movement. Because labor activists were often denied permission to hand out leaflets and to hold public meetings and rallies, they realized that free speech was essential to their efforts to organize unions. In the late 1930s, the Congress of Industrial Organizations (CIO) sued the major of Jersey City, NJ, after being denied permits for leaflets and public meetings and won an important free speech ruling from the U.S. Supreme Court in *Hague v. CIO* (307 U.S. 496 [1939]). Thus, says Kairys (1990:257), the Supreme Court's new rulings in favor of free speech during this time (as against its earlier rulings against free speech) are best understood "by examining the social and political contexts in which they were made, and by viewing legal decision-making and law as political processes."

CLS flourished during the late 1970s and through the 1980s and often aroused heated arguments in scholarly journals and inside law schools themselves. One law school dean accused CLS-oriented law school professors of being nihilists, or anarchists, and said they have "an ethical duty to depart the law school" (Carrington 1984:227). Several CLS professors at Harvard Law School had contentious tenure proceedings. A quieter period set in after this flurry of activity, and CLS continues to be influential today even if the tempers it once aroused are much less in evidence and CLS as a scholarly movement is said by many to no longer exist (Ellickson 2000). As law professor Mark Tushnet (2005) observes, CLS scholars who wrote influential articles two decades ago continue to do so. More important, he says, "[M]ajor components of critical legal studies have become the common sense of the legal academy, acknowledged to be accurate by many who would never think of identifying themselves as critical legal scholars" (Tushnet 2005:100).

Critical Race Theory. The critical legal studies movement was composed mostly of white males whose focus on law and inequality concerned social class and the economic structure. They devoted much less attention to issues of race and gender. Scholars interested in these issues faulted this neglect.

Many of these scholars were people of color or women, and they began developing legal theories and perspectives that took race and gender explicitly into account. Critical race theory and feminist legal theory were the two general views they developed, and ironically these views have outlived critical legal studies as active intellectual movements.

Critical race theory draws on sociology, philosophy, and several other disciplines to argue the racialized nature and impact of law in the United States and other modern societies. Although it has much to say about law, it also has much to say about race in the larger society. Its main argument is that the United States is a racist society and that the law both manifests this racism and contributes to it (Bell 2004; Brown 2003; Delgado and Stefancic 2001; Krenshaw et al. 1996). For this reason, law cannot be understood without taking race into account, and race cannot be understood without taking law into account. Critical race theory further questions the common view that racism and racial discrimination are atypical phenomena in American society that stem from prejudiced individuals and that are capable of being remedied through the wise application of legislative and constitutional law. Instead, say critical race theory scholars, these phenomena are rooted in the institutional fabric of our society and are not easily remedied via legal means (Gómez 2004).

Critical race theory scholars have written about many topics, but a particularly relevant line of work concerns the period of Reconstruction following the Civil War. This was an era dominated by racial issues, and critical race theory scholars have documented how these issues shaped the development of law during this time. For example, Pamela Brandwein (2000) discussed U.S. Supreme Court rulings relating to the Fourteenth Amendment to the U.S. Constitution. Whereas most legal historians have seen these rulings as arising from Supreme Court justices' views on the meaning of the Fourteenth Amendment and the powers of the Court vis-à-vis those of Congress, Brandwein argued that the Court's rulings were shaped by white supremacist beliefs that limited the help that the rulings gave to African Americans and other citizens. In another study, Michael Elliott (1999:613) discussed how racial ideology during Reconstruction centered on the need of whites to both assume and document that "each individual belongs to a race and to only one race." Whites felt they could not be superior to blacks if they were not actually "all white." For this reason, specific laws—for example, those banning racial intermarriage—arose that kept the races as separate as possible, and legal classifications regarding who by blood was "white" and who was "black" were developed. These classifications helped determine whether people were allowed to vote (if black, they were not allowed) and whether they were legally "Indians" for their land to receive federal protection.

Critical race theory has also made a methodological contribution to law and society scholarship. It is *pro-narrative*, meaning that it frequently relies for its evidence on personal accounts rather than on statistical studies to provide rich, often searing descriptions of racism in practice. Given that many critical race theorists are people of color, they have often experienced many types of discrimination and/or know people who have experienced discrimination themselves, and for either reason are well placed to write about these experiences. As Steven Vago (2006:71–72) observes, "Because of oppression, people of color perceive the world differently than those who have not had such experienced. Critical race theory scholars can thus bring to legal analyses perspectives that were previously excluded."

Feminist Legal Theory. Feminist legal theory, also known as *feminist jurisprudence*, is the other intellectual movement that arose out of concern that the critical legal studies movement had neglected issues of race and gender. Paralleling the argument that critical race theory makes for race, feminist legal theory argues that law both reflects and contributes to fundamental sexism in the larger society (Chamallas 2003; Dowd and Jacobs 2003; Lacey 2004). Again like critical race theory, feminist legal theory draws on several other disciplines and especially on the broader field of feminist intellectual thought to develop its general perspective on the gendered nature of law and the legal system. It recognizes that because most of the people who make and enforce laws are men, this gender imbalance "affects in major ways how women are thought of and treated by the legal system" (Sokoloff, Price, and Flavin 2004:19). The title of a recent book, *Women's Lives, Men's Laws*, by famed feminist legal scholar Catharine MacKinnon (2005), captures this dynamic nicely.

Feminist scholars differ in the extent to which they consider change possible within the existing political and economic system. *Liberal feminists* think change is very possible with the passage of legislation that effectively addresses discrimination against women, while *critical feminists* "do not seek to reform the existing system but rather aim to fundamentally restructure private and public life and to recast relations between women and men in political terms" (Sokoloff, Price, and Flavin 2004:21).

Much work in feminist legal theory is historical, as there is ample evidence of sexism in American legal history. In 1873, for example, the U.S. Supreme Court ruled in *Bradwell v. Illinois* (83 U.S. 130 [1873]) that women were not entitled to practice law. The plaintiff, Myra Bradwell, was denied permission to practice law in Illinois even though she had met all the necessary qualifications. The Court ruled 8-1 against Bradwell on the grounds that the Fourteenth Amendment's guarantee of certain rights did not extend to the practice of law. Joining the majority, Justice Joseph Bradley wrote a concurring opinion that women were naturally unsuited to practice law:

> The civil law, as well as nature herself, has always recognized a wide difference in the respective spheres and destinies of man and woman. Man is, or should be, woman's protector and defender. The natural and proper timidity and delicacy which belongs to the female sex evidently unfits it for many of the occupations of civil

> life. . . . The paramount destiny and mission of woman are to fulfill the noble and benign offices of wife and mother. This is the law of the Creator.

Reflecting the sentiments in Justice Bradley's concurring opinion, which was joined by two other justices, American law until very recently did not let women enter into contracts, serve on juries, and enjoy many other rights and responsibilities that men had long taken for granted (Baer 2002; Kuersten 2003). It was thought that women were simply not suited for these activities and that engaging in these activities might even upset their "delicate natures." Although we now regard these notions as outmoded, they served until the last two or three decades to reinforce women's inequality and are still accepted in some circles. For example, almost 24 percent of the respondents in the 2006 General Social Survey, a random sample of Americans of age 18 and above, agreed that "most men are better suited emotionally for politics than are most women." In the same survey, about 35 percent agreed that "it is much better for everyone involved if the man is the achiever outside the home and the woman takes care of the home and family." Traditional views of women's nature and roles are still popular and continue to affect many aspects of American society, including the law.

Since the 1970s, feminist legal scholarship and advocacy have made important contributions to several areas of the law, and perhaps most notably to the areas of family law and divorce, workplace discrimination and sexual harassment, and rape and domestic violence (Boyd 2004; Lipschultz 2003; Schneider 2000). However, much legal inequality for women continues to exist, and the legal profession and law school are not yet bastions of gender equality. Chapters 7 and 8 explore these issues further.

Summary

1. Law is very difficult to define precisely. A key problem lies in distinguishing law from custom. Another problem involves whether to extend the concept of law to encompass the rules and regulations that govern the operation of formal and voluntary organizations.
2. In classic social scientific definitions of law, several characteristics help to distinguish whether a rule is a law or only a custom: (a) the pressure to comply with the rule is external to the individual; (b) this pressure involves coercion and force and especially physical force; and (c) the party that applies such force is socially authorized to do so. This emphasis on physical force has been criticized for implying that physical force is the primary reason that people obey law. Although coercion is one reason for their obedience, many other reasons also exist.
3. These reasons include: (a) the belief that certain illegal actions are morally wrong; (b) a felt obligation to obey the law; (c) concern for peer

pressure or other informal sanctions; (d) fear that certain illegal behaviors might harm us; and (e) habit.

4. Positivist definitions of law describe what the law is but not what the law should be. In contrast, natural law definitions incorporate principles derived from nature and thought to be binding on human law. Sociologist Philip Selznick said that the essence of law is legality, or the rule of law. According to Lon Fuller, rules or commands constitute law only if they adhere to the inner morality of the law. To do so, they must satisfy several criteria, including: (a) rules must be general; (b) rules must be generally known; (c) rules must not be retroactive; (d) rules must be reasonably clear; (e) rules must not be contradictory; (f) rules must not require the impossible or the unreasonable; and (g) rules should be relatively stable through time.
5. The definition and conception of law we adopt affects decisions concerning the existence or nonexistence of law in any particular society. A natural law definition may lead to the conclusion that law does not exist in authoritarian regimes. Certain definitions of law would lead to the conclusion that law does not exist in some of the societies studied by anthropologists, while other definitions would lead to the conclusion that law does exist in these societies.
6. Legal formalism was popular in the nineteenth century and is still very influential today. Its central assumption was the belief that law is a self-contained system of logic that is independent of social and moral considerations.
7. Utilitarianism is a perspective that was developed during the eighteenth and nineteenth centuries by Cesasre Beccaria and Jeremy Bentham, who were critics of the harsh criminal justice practices of that era. They both thought that people act rationally and with free will and are chiefly concerned with maximizing their pleasure and reducing their pain. Given this belief, Beccara and Bentham reasoned that the criminal justice system certainly needs to be punitive only to the degree that it deters people from committing crime.
8. The historical school is the name given to the work of several nineteenth- and early twentieth-century scholars who discussed how law and legal systems changed over the centuries as societies developed from ancient times to the time the scholars were writing. Friedrich Karl von Savigny believed that law both reflects and expresses the culture, spirit, and values of a people. Sir Henry Maine wrote that a key change in law from the ancient world to the nineteenth century was the change from status to contract. Emile Durkheim believed that the type of punishment in society reflected its degree of heterogeneity of beliefs and values. Max Weber viewed the development of rationality as the key hallmark of modern society and thought that law became more rational as societies became more

modern. Karl Marx said that law reinforces elite dominance by helping to preserve private property and placate the working class.

9. Legal realism was a popular school of thought in the United States in the early to mid-twentieth century that greatly influenced the study of law and society then and now. It argued that how law really works can be very different from how it is supposed to work in theory. Key legal realists were Karl Llewellyn and Jerome Frank.
10. The school of sociological jurisprudence argued that law has social underpinnings and impact. Its two leading figures of were Eugen Ehrlich and Roscoe Pound. Ehrlich distinguished between positive law and living law, and Pound distinguished between law in action and law in books. He thought that the poor received little justice in the criminal justice system, and he called for a series of criminal justice reforms.
11. The early law and society movement generally began after World War II and involved work by anthropologists, political scientists, and sociologists. Bronislaw Malinowski wrote one of the earliest and still most influential anthropological legal studies based on his observations of the peoples of the Trobriand Islands near New Guinea in the South Pacific. A major emphasis in his book was that law existed among the Trobriand people even if they did not have the police, courts, and judges that typify modern societies.
12. Beginning in the 1940s and especially the 1950s, political scientists pioneered sophisticated empirical studies of the decisions of justices on the U.S. Supreme Court and on the federal appellate courts. This body of research examined how factors such as the socioeconomic backgrounds of justices, their religious upbringing, and their political leanings influenced their decisions on various types of cases. Beginning in the 1960s, sociologists published several accounts that emphasized the lack of justice in the criminal courts for the poor and people of color, and they later began to conduct quantitative analyses of the factors affecting whether criminal defendants were found guilty or not guilty and the harshness of the punishment they received if they were found guilty.
13. Two contemporary perspectives in the study of law and society are functionalist and conflict theories. Functionalist theory sees law as helping to preserve social order, to make social life possible, to settle disputes, to protect individual freedom, and to bring about needed social change. The functionalist view further assumes that the major sociodemographic groups generally agree on important legal issues such as appropriate sentencing for convicted criminals. Conflict theory assumes that (a) law is characterized by inequality and also contributes to inequality in the larger society; (b) the major sociodemographic groups disagree on important legal issues in ways that reflect their life experiences and different positions on the socioeconomic ladder; and (c) law takes sides in the competition among interest groups and, more

specifically, that it takes the side of powerful, established interests such as corporations.

14. Several contemporary Marxist perspectives on law exist. The instrumentalist view assumes that law and the state are tools that can fairly easily be used by the ruling class to oppress the working class and in other respects to maintain its own superior position. The structuralist view concedes that the state and its legal system are relatively autonomous from the ruling class but that any victories by the working class are sham victories that ultimately reinforce ruling-class dominance. The dialectical view also argues that legal and political victories by the have-nots can indeed be real and consequential and not just sham. Law can be an effective tool of the ruling class, this view believes, but it also constrains the ability of the ruling class to repress the poor and other have-nots.
15. The critical legal studies movement began in the 1970s and aroused much controversy for several years. Its basic beliefs are that law is political, that law is indeterminate, that law is ideological, and that legal education contributes to law students' passivity and to social inequality in the larger society when new law school graduates go out to practice.
16. Critical race theory arose as a reaction to the neglect in the critical legal studies movement of racial issues. Critical race theory draws on sociology, philosophy, and several other disciplines to argue the racialized nature and impact of law in the United States and other modern societies. Its main argument is that the United States is a racist society and that the law both manifests this racism and contributes to this racism.
17. Feminist legal theory arose as a reaction to the neglect in the critical legal studies movement of gender issues. It argues that law both reflects and contributes to fundamental sexism in the larger society. Although much feminist legal theory scholarship is historical and women's legal equality has greatly improved since the 1970s, much legal inequality for women continues to exist.

Key Terms

Conflict theory
Critical legal studies
Critical race theory
Deviance
Dialectical view
Feminist legal theory
Formal
Functionalist theory
Historical school
Informal
Instrumentalist view
Legal formalism
Legal realism
Legality
Natural law
Official law
Sociological jurisprudence
Structuralist view
Unofficial law
Utilitarianism

CHAPTER 3

Families and Types of Law

An implicit theme of the previous two chapters is that law is both diverse and complex. This chapter develops this theme further by examining the many families of law found across the world and the many types of law within the United States. Our look at families of law will reinforce a central idea of law and society: that law reflects the social context in which it is found. Meanwhile, our look at the types of law in the United States will reinforce the complexity of law and remind us of the many types of law beyond the criminal dimension with which Americans are familiar from TV shows, popular films, and, perhaps, their own experiences.

FAMILIES OF LAW

Although there are many legal systems around the world, they generally are grouped into a much smaller number of categories that are often called **families of law**. This situation is roughly similar to that found in the classification scheme used in biology, where *family* refers to a group of animals or plants sharing certain features that distinguish them from other animals or plants. Thus, dogs, coyotes, foxes, and wolves all belong to the family *Canidae*: although these animals obviously differ from each other, and although there are many types of dogs and of the other animals, as a whole they all share certain characteristics that distinguish them from other animal families. Similarly, the legal systems of the nations within each legal family share several important characteristics that justify calling them a family, even if these legal systems differ in other respects.

Five families of law are typically delineated: (1) common law, (2) civil law (also called code law), (3) theocratic law, (4) socialist law, and (5) traditional law. Common law systems are found in Great Britain and its former colonies, including the United States, Canada, Australia, Hong Kong, and India. Civil law systems are found in the nations of continental Europe and to some degree in these nations' former colonies in Africa and Central and South America; they are also found in Japan and South Korea, which long ago adopted the German civil code. As their name implies, theocratic legal systems are found in nations that are ruled or heavily influenced by religion and religious leaders. Islamic law is the dominant and most controversial example of theocratic law in the world today, but Israel displays some elements of theocratic law that will also be discussed. Socialist legal systems used to be dominant in the Soviet Union and China before the former dissolved and the latter began to move toward capitalism. Remnants of socialist law remain in China, however, and socialist law is still found in the Communist nation of Cuba. Finally, traditional law characterizes the premodern societies studied by anthropologists and reflects these societies' small size, homogeneity, and other traits.

Common Law

Common law legal systems originated in England almost one thousand years ago and now characterize Great Britain and its former colonies, the most notable of which, of course, is the United States. The U.S. legal system differs in important respects from those of Great Britain, Canada, and Australia, but all these legal systems share certain features and processes that reflect their origins in common law.

A key year for the development of common law was 1066, when, as you may have learned previously, William the Conqueror became King of England after the Battle of Hastings in October of that year. William was crowned king two months later on Christmas Day but did not secure his reign over the country for another six years. During that period and afterward, William faced the daunting task of bringing together a country with warring factions and more generally of establishing his authority. He took land away from the Saxons he had conquered and gave it to his Norman allies, and he instituted a local tax system to increase the royalty's wealth and power.

More important for our discussion, he initiated a new legal system that was national in scope and thus centralized his power. Before this time, disputes in England were decided either by the noble on whose land the dispute had occurred or in one of many shire (county) courts. This decentralized system meant that legal understandings and principles might vary from one shire to another. William instituted royal courts to handle disputes among nobles; these courts permitted a more uniform application of legal ideas but also, not incidentally, provided the king income from the fines that were collected. Perceiving these courts to be fairer than county courts, English commoners began to take their disputes to the new royal courts, which eventually superceded the local courts (Milsom 1981).

Over the next several decades, royal court judges began to travel around the country to hear and rule on cases. Because England had no written rules, the judges decided the cases according to principles and understandings—customs—that English people had long held. Their decisions in turn became applicable, or common, throughout England, yielding the term "common law." In the late 1180s, an advisor of King Henry II named Ranulf de Glanvill compiled laws as made and interpreted by the royal courts and published them in a collection called *Treatise on the Laws and Customs of the Realm of England*. This volume played an important role in nationalizing the common law of England. A key outcome as the common law developed was that private harms began to be treated as crimes against the state, thanks to a series of judges' rulings during the eleventh and twelfth centuries. Before this time, for example, if an assailant stabbed a victim, it was up to the victim and the victim's family and/or friends to even the score, an action we would now call vigilante justice. Judicial rulings during this period turned offenses like this into offenses against England itself, and it was now up to state agents—prosecutors—to deal with the suspected offender. This change helped to civilize England and both centralized and strengthened the king's authority over the land.

Political scientist Suzanne Samuels (2006) provides an interesting example of the uniformity that the centralization of common law during this time produced. Not surprisingly, different regions of England did not always rely on the same custom in settling disputes. In particular, when a father would die and leave his estate, different regions had different customs for deciding who would inherit the estate. Some followed the principle of *primogeniture*, under which the first-born son inherits the estate, while others followed the principle of *gravelkind*, under which all sons share the estate. Still other regions followed the principle of *borough-English*, under which the youngest son inherits the estate, and some areas even allowed daughters to inherit at least part of the estate. Faced with these competing principles, English judges settled on primogeniture, entitling first-born sons to inherit everything.

A key event in the development of the common law was the signing of the *Magna Carta* (Latin for *Great Charter*). For several years, King John (of Robin Hood fame) and English landowners had been feuding over taxes and land rights. Their struggles eventually led to the *Magna Carta*, which King John signed in 1215 and then generally ignored until he died a year later.

His son and successor signed three visions of the *Magna Carta* during the next nine years; the third one, signed in 1225, was finally accepted as official policy. The *Magna Carta* in its final form recognized the rights of English landowners, required the king to obey English law, and established several important due process procedures, including the right to be tried by one's peers. Although England was still not a democracy as we know it today, the *Magna Carta* was the formal beginning of the principle of the rule of law that is a key hallmark of modern democracies.

Glanvill's late 1180s collection of common law was supplemented in the middle of the thirteenth century by Henry of Bracton's *On the Laws and Customs of England*, which summarized judicial rulings on various matters. This volume also emphasized the need of judges to rely on past cases in deciding new cases and the need of the king to follow the law (Samuels 2006). During the 1760s, William Blackstone published his majestic, four-volume *Commentaries on the Laws of England*, which emphasized the need for the law to protect individual liberty and "was perhaps the most influential law book in Britain, her colonies, and her former colonies for more than a century" (Samuels 2006:61).

Because the common law originated in judges' decisions as described above, common law nations are ones where *case law*, or law made by judges through their rulings, is paramount. However, no nation is purely a common law nation, as no nation relies solely on case law to the total exclusion of statutory (legislative) law and other types of law. Thus, while the United States is a common law nation as discussed just below, it certainly also relies on these other types of law, which, as we shall see, have become more important over time. In saying that the United States, Great Britain, and other nations are common law nations, we simply mean that they rely much more on case law than do the nations grouped into the other legal families.

The reliance on case law makes *precedent*—the reliance on past judicial decisions in reaching a new decision—a hallmark of the common law tradition. Another hallmark of common law nations is the *adversary system*, in which attorneys for the opposing sides are said to rigorously contest the evidence in pretrial proceedings and in trials themselves. Throughout this process, judges typically play an important but fairly passive role as neutral referees whose main goal is to ensure that all parties to a case follow proper procedures and the rules of evidence. While scholars have in fact questioned the amount of adversariness in the U.S. legal system (see Chapter 9), common law legal systems are, overall, more adversarial than those in civil law systems. Another hallmark of common law systems is the jury. Although its importance differs among common law nations, overall the jury plays a more prominent role in common law nations than in civil law nations.

Common Law in the United States. The United States has a legal system that is in fact many legal systems. There is one legal system at the federal level, but there are fifty legal systems at the state level and obviously many, many more at smaller levels of government. Reflecting its British colonial origins, the United States follows the common law tradition with one exception. Reflecting its strong French influence, Louisiana follows the civil law tradition and is discussed in the next section.

The roots of the U.S. legal system lie, of course, in the colonial period (Cantor 1997). Most of the early colonists came from England, and, quite naturally, they relied on their knowledge of and experience with English law as they developed a legal system in the new land. Early colonial law was an interesting mixture of English common law and Puritan religious belief, and early Massachusetts and other New England colonies were essentially theocracies, as religious belief formed the basis for the substance of laws and for legal procedure. Puritan law in Massachusetts was very harsh in regard to certain behaviors. For example, Puritan belief banned swearing, sledding on the Sabbath, and many other behaviors that are obviously legal today. By the middle of the eighteenth century, however, Puritan law had given way to common law. Although the colonies' law was basically English, it was "fairly crude and stripped down" (Friedman 2004a:24), and never became as elaborate as the law in England. The colonists' society was much less complex than England's, and their law could also afford to be less complex. In developing their law, the colonists "took what they knew, what they needed, what they remembered" (Friedman 2004a:240). As the colonies grew throughout the eighteenth century and as the colonies' trade economy developed, colonial law became more complex but never as complex as England's law.

As noted earlier, an important development in colonial law was the publication of Blackstone's *Commentaries* in the 1760s. Blackstone's work, writes Friedman (2004a:31), "became a wild best-seller in legal circles on the American side of the Atlantic. Here, in limpid and elegant English, and in the short space of four volumes, was a skeleton key to the mysteries of English law: a guide to its basic substance." For this reason, the publication of *Commentaries* helped cement and expand the common law in the soon-to-be nation.

The revolution against England that began in 1775 and culminated in independence led to a more uniform legal system than had existed among the thirteen original colonies. Here the most important developments were undoubtedly the ratification in 1787 of the Constitution, which delineated the duties and powers of the federal government and of the states, and, four years later, of the Bill of Rights, which was intended to guarantee individual liberty by providing many types of legal protection against arbitrary governmental power. Both documents were heavily influenced by the colonial experience, in which England repeatedly used and abused the colonial legal system in its effort to quell dissent and maintain its own authority and wealth as a ruling power. The writers of the Constitution and Bill of Rights thus wanted to limit the potential abuse by the new government of individual freedom. For this reason, the Bill of Rights featured the right to freedom of speech and of the press, the right to trial by jury, protection from unreasonable searches, and other legal rights and protections.

Throughout most of the nineteenth century, case law continued to be the dominant source of law in the United States. The Congress and state legislatures certainly existed, but they were much less active than they are now in enacting statutes. That began to change toward the end of the century thanks to industrialization (Samuels 2006). The advent of an industrial economy after the Civil War meant that the U.S. economy was becoming much more complex than it had been when it was based primarily on agriculture and trade. New rules had to be devised to deal with the many kinds of matters that arise in an industrial economy, and legislatures were the bodies that devised these rules. Various groups wrote codes of conduct to guide the changing economy. One of the most important of these was the National Conference of Commissioners on Uniform State Laws (NCCUSL), which was established in 1892 to help make the states' laws more uniform. It initially comprised commissioners from seven states and by 1912 included commissioners from all the states. In 1940, it began to produce the Uniform Commercial Code to govern commercial relationships, and this code was adopted across the country by the mid-1960s.

Among common law nations, two features of the U.S. legal system stand out today. The first is the concept of **judicial review**, the principle that appellate courts may overturn legislative statutes and invalidate executive actions if they are deemed to violate Constitutional standards. As you may have learned previously, Chief Justice John Marshall established this principle in the famous 1803 case of *Marbury v. Madison*. However, it might surprise you to learn that American courts used judicial review only "rarely and gingerly" (Friedman 2004a:13) for another century; it did not become a common practice until the late nineteenth century and then accelerated through the next century. Today it is quite familiar to the average American and a hallmark of the American legal system, but it is less common and almost absent in other common law nations. Britain, for example, lacks a written constitution, and, because of this, never developed the tradition of judicial review that the United States now enjoys.

The other notable feature of the U.S. legal system among common law nations is the *jury*. As mentioned earlier, the jury is more important in common law nations generally than it is in civil law nations. Among common law nations, the jury has played a more important role and is more a part of the national culture in the United States than in its common law counterparts. Jury trials have been the subject of countless novels, films, and TV shows in the United States, and the jury as a legal institution continues to be the topic of much debate. At the same time, and somewhat surprisingly, jury trials are much rarer in the United States than the average citizen probably realizes. Chapter 9 discusses the jury at much greater length.

Civil Law

Civil law (also called **code law**) is the name given to the family of law found in the nations of continental Europe, in many of their former colonies in South America and elsewhere, and in other nations such as Japan and South Korea.

(Note that the term *civil law* also refers to the system of private law in the United States, discussed below. Thus "civil law" has two meanings in the study of law and society, and these meanings should not be confused with one another.) Civil law nations rely primarily on written, detailed codes, or collections, of law that have their origins in ancient Rome (Merryman and Pérez-Perdomo 2007). In the early sixth century, the Roman emperor Justinian had more than a dozen experts examine the many laws that were part of the vast Roman empire. He wanted to collect them in one volume that would eliminate laws that were unclear, misleading, or repetitive and that would be available in one source for lay people to read and understand. That publication, *Corpus Juris Civilis*, appeared in AD 533. During the Middle Ages, Roman law became an important topic in university classes in several nations in continental Europe and eventually was adapted to some extent by these nations' legal systems.

During the late eighteenth and early nineteenth centuries, continental nations increased their attempt at codification. This movement culminated with the adoption of the *Code Civil* in France in 1804, the year that Napoleon Bonaparte was crowned emperor of France; today, it is commonly referred to as the Napoleonic Code. Reflecting the move toward equality as expressed through French Revolution, this code among other provisions sought to limit large landowning by restricting inheritance and making it easier for common people to buy land (Bracey 2006). During the next few years, Napoleon conquered much of Europe in the Napoleonic Wars, and the nations he conquered, including Italy, Spain, and Portugal, adopted the *Code Civil*. The German Civil Code of 1896 was also influential, not only in Germany but also in Japan and South Korea.

Judges in civil law nations have less power to change the law than their common law counterparts, in large part because civil law codes are so detailed that they cover many circumstances and situations. As traditionally conceived, civil law judges' major role is merely to interpret the legal codes; the judges typically do not overturn laws, nor do they "make" law as common law judges sometimes do. At the same time, however, civil law judges play a more prominent role in directing cases than do common law judges. As noted earlier, common law judges act rather passively during trials as neutral referees who help ensure that all parties to a case follow proper judicial procedures. In contrast, civil law judges take a more active role in coordinating the investigation of the facts of a case, preparing it for trial, and questioning witnesses. In these respects, they perform the functions that attorneys fulfill in common law systems. Typically, a judge works with investigators to compile a written dossier on a case, and other judges then read the dossier and base their decision on its contents, questioning witnesses if necessary to clarify a point (Bracey 2006). The nature of this process is said to lead civil law systems to rely on an *inquisitorial model*, in contrast to the adversarial model characteristic of common law systems.

Whereas common law judges have typically been practicing attorneys who are then appointed or elected as judges, civil law judges are, in essence, civil servants who train to become judges by going to schools that are different from those intended for persons aspiring to be attorneys. In other differences between the two legal families, judicial review is unknown in civil law nations, and the jury plays a much more minor role in their legal systems than it does in common law systems and, in fact, is largely absent in most proceedings.

France's legal system reflects many of the civil law features just described (Dammer and Fairchild 2006; Reichel 2005). Not surprisingly, it is guided by the Napoleonic Code as revised since its introduction about two centuries ago. Its criminal court system follows the inquisitorial model; for felonies, this model involves extensive pretrial investigation conducted by judges called *examining magistrates*, and for misdemeanors it involves a similar investigation by civil servants called *procurators*. Examining magistrates conduct their pretrial investigation in secret and question witnesses and the defendant; the hope is that this type of investigation will prevent innocent people from having to be tried. At the end of the pretrial investigation, the examining magistrate either dismisses the charges or orders the case to be tried. Even defendants who wish to plead guilty are tried to ensure that they really are guilty and are not pleading guilty for some other reason, for example, because they just want to get things over with. At the trial itself, a group of three specially trained judges called *professorial magistrates* presides over the case. Juries sit only in very serious criminal cases and are not involved in civil cases. The standard of proof for a guilty verdict is that they have to be convinced of a defendant's guilt only by a preponderance of evidence, not, as in the United States, beyond a reasonable doubt.

Compared to the United States, France has a more restricted role for the jury and has a weaker standard of proof for a guilty verdict. These facts reflect France's lower commitment more generally to the rights of suspects and defendants. For example, French suspects can be held in jail after arrest for up to forty-eight hours without being charged; while they are detained, suspects may be subject to extensive police questioning without their attorney permitted to be present.

Despite the classic differences between civil law and common law systems, the two families of law seem to becoming more similar as the years go by. Friedman (2004a:10) attributes this convergence to the growing similarity of modern nations across the world thanks to globalization. As he puts it, "All modern, developed nations have income tax systems, stock exchanges, international airports, tall buildings, and traffic jams. They all face issues of copyright, pollution, air traffic control, and bank regulation. Similar problems tend to generate similar solutions; and similar problems and solutions means similar laws and legal systems." The growing number and use of statutes and codes in common law nations have also contributed to their growing similarity with civil law nations. Meanwhile, civil law judges have begun to take on the task of trying to change or make law and thus are beginning to resemble their common law counterparts. Thus, the distinction between civil law and common law is less clear than it once was, but significant differences still exist.

In the United States, Louisiana, as noted earlier, is the only state with a civil law tradition, though other states have developed their own codes. Reflecting its origins as a French colony, Louisiana adopted its code in the early nineteenth century and patterned it so closely on the Napoleonic Code that it was written in French. New codes in 1825 and 1870 (the latter written in English) supplemented and replaced the first civil code. Despite its identity as a civil law state, Louisiana's legal system incorporates many common law principles, which inform its state constitution and criminal and civil procedure. In addition, Louisiana judges act like common law judges in sitting as neutral referees during trial proceedings. Thus, while Louisiana is still considered a civil law state, much of its legal system resembles the common law model found in the other forty-nine states.

Quebec Province occupies the same place in Canada that Louisiana does in the United States: although Canada is a common law nation, Quebec, reflecting its French roots, follows the civil law tradition. A key development in Quebec civil law was the 1866 adoption of the *Civil Code of Lower Canada*; this code incorporated many of the elements of the Napoleonic Code. Revisions of the Civil Code began in 1955 and eventually led to the adoption of a new civil code, the *Code Civil du Quebec* (Civil Code of Quebec; CCQ), in 1991. The CCQ is very detailed and, like other civil codes, covers such topics as commercial law, contracts, evidence, family relations, property, and the rights of creditors. The code has a complex classification scheme and contains more than 3,000 articles, or clauses, grouped into its various sections. Article 368, for example, declares: "Before the solemnization of a marriage, publication shall be effected by means of a notice posted up, for 20 days before the date fixed for the marriage, at the place where the marriage is to be solemnized. No publication is required if the intended spouses are already in a civil union." Several other articles describe how other aspects of how marriages should occur and how they may be annulled if necessary. A number of articles even spell out the "rights and duties of spouses," as their section title is called, and include such proscriptions as "The spouses together take in hand the moral and material direction of the family, exercise parental authority and assume the tasks resulting therefrom" (Article 394), and "The spouses choose the family residence together" (Article 395). As these examples suggest, behaviors and expectations that are unwritten in common law nations are codified in the Quebec Civil Code.

Common law and civil law systems each have advantages and drawbacks (Samuels 2006). Because common law systems rely on precedent, they are thought to be more stable than civil law systems. However, because common law systems also rely on lawmaking by courts and judges, they are thought to be slower than civil law systems to respond to changing conditions. A key reason for this is that courts and judges play a passive role to the extent they must wait for cases to come to them. In contrast, legislatures, the primary lawmaking body in civil law systems, can play a much more active role in addressing new issues and changing conditions. Of course, legislatures obviously exist in a common law nation like the United States, and they have often taken a proactive role in passing legislation designed to deal with a controversial issue or changing condition. In recent years, for example, the U.S. Congress has tried to address such issues as immigration and same-sex marriages. Thus, even though common law systems may be less adaptable than civil law systems to changing conditions, they are certainly not unable to do so. Chapter 6 on law and social change explores further the use of judicial rulings and legislative statutes to bring about social change.

Theocratic Law

Theocracies are governments run according to religious rules and principles and ruled by religious leaders, and their law and legal systems reflect this religious dimension. **Theocratic law**, then, is *religious law* (also called *sacred law*), or, to be more precise, a system of law heavily dependent on religious belief. Islamic law is found in many nations and has become quite controversial in recent years, partly because of 9/11 and partly because of notorious cases involving the victimization of Muslim women. We will examine Islamic law in some detail because of this controversy and because many Americans are generally unfamiliar with its principles. We will also look at Israel's use of Jewish law in its public and private affairs; although Israel, unlike most Islamic nations, is a democracy, Jewish law nonetheless plays an important official role in Israeli life.

Islamic Law. Islam, the second largest religion in the world, has more than 1 billion adherents living primarily in the Middle East, northern Africa, and parts of Asia. The nations in which Islam is the dominant or major religion include Afghanistan, Egypt, Iran, Iraq, Jordan, Pakistan, and Syria. To a large degree, their governments are run according to Islamic law, as the separation of religion and state guaranteed in the First Amendment to the U.S. Constitution is not found in these nations. Instead, religion and state in Islamic nations are inseparable. Some nations, including Iran, interpret Islamic law very strictly, while other nations, such as Egypt, are guided by Islamic law but also include principles from Western law.

Islamic law, better known as the *Shariah*, includes rules about four areas of public and private life: (1) the individual's relationship to the state, (2) the individual's relationship to other individuals, (3) religious practices such as daily prayer and fasting, and (4) everyday practices such as parenting and personal hygiene. The principles underlying all these rules derive from the sacred book of Islam, the Koran, analogous to the Judeo-Christian Bible and considered the word of God as received by the Prophet Muhammad in the early seventh century. Although the Koran is centuries old, it "still wields tremendous influence in Muslim countries, since it largely determines what is sinful, and therefore illegal" (Holscher and Mahmood 2000:88). The behaviors specified in the Shariah derive from another Islamic holy text, the *Sunnah*, which is a collection of the Muhammad's words and religious practices.

The *Shariah* lists three types of crimes. *Hadd* offenses are the first type and are considered the most serious because both the Koran and the Sunnah prohibit them; examples include adultery involving sexual intercourse, drinking alcohol, and the abandonment of Islam. The Koran states that these offenses are to be punished by the community, not by any individual, and it also specifies the punishment that each hadd offense should receive. Theft, for example, results in the amputation of one's hand, while adultery results in being whipped one hundred times for an unmarried individual and death by stoning for a married individual. *Qisas* offenses are the second type and include homicide and assault; possible punishment for these offenses includes execution. *Tazir* offenses are the final type listed in the Shariah and are considered offenses against society, not against Allah; as such, they are found in neither the Koran nor the Sunnah and are considered less serious than the first two types of offenses. Tazir offenses include bribery, eating pork, and adultery not involving sexual intercourse.

Other elements of the *Shariah* are worth noting (Salama 1982). First, husbands are allowed to use corporal punishment to correct their wives' behavior as long as this punishment does not leave any bruises or scars. Second, criminal defendants are, like their Western counterparts, assumed to be innocent until proven guilty. Third, defendants have the right to counsel throughout the case against them. Fourth, offenders who are incarcerated should not be beaten or otherwise abused.

Most readers have probably read of harsh practices carried out in the name of Islamic law in nations such as Nigeria and Iran. Earlier in this decade, for example, civilian police in a region of Nigeria beat and whipped many people for offenses such as drinking, gambling, and premarital sex (Singer 2001). Iran sentenced a German businessman to death for having sex with a Muslim woman before allowing him to leave the nation after a two-year term of imprisonment. More recently, dozens of Pakistani women have been raped or murdered for "sinful" behavior that would be considered quite acceptable in Western society (Kristof 2005).

Although it might be tempting to blame the *Shariah* for all such punishment, many scholars of Islam, in fact, condemn these practices as violations of Islamic law. Even though the Koran does specify harsh punishment for many offenses, the Shariah provides so many protections for people accused of committing these offenses that harsh punishment should be relatively rare if the Shariah is followed as written. As two Islamic scholars note, "In practice, both in contemporary Muslim countries and historically, these traditional penalties have rarely been carried out" (Holscher and Mahmood 2000:88). For example, for theft punishable by hand amputation to be proven, testimony from at least two male eyewitnesses is required, and there are few thefts witnessed by at least two men who do not include the thief himself. Similarly, proof of adultery involving sexual intercourse requires testimony from four male eyewitnesses. Because most adultery obviously is not watched by four other men, conviction of adultery is and should be a rare event. As a result, say Holscher and Mahmood (2000:82), "By adopting a system of strict requirements of proof and testimony in criminal cases, Islamic law avoids despotic and arbitrary decisions and has limited the judges' discretion in the defendants' interest." Given this background, many Islamic scholars say that the harsh punishment for which some Islamic nations have become notorious in fact violates Islamic principles and should not be blamed on Islamic law.

Israeli Law. Israel is not nearly as theocratic as many Muslim nations are, but Jewish law still plays an important official role in Israeli public and private life.

Like the Shariah, Jewish law, or the *halakhah* (literally meaning "the path that one walks"), is complex and includes terminology unfamiliar to readers with little or no knowledge of Arabic or Hebrew; thus, a summary of the *halakhah*'s terms and principles will be helpful. The halakhah consists largely of 613 commandments, or *mitzvot*, included in the first five books (the Torah) of the Bible: Genesis, Exodus, Leviticus, Numbers, and Deuteronomy. The *Ten Commandments* are undoubtedly the most familiar of these mitzvot, but hundreds of others spell out, often in minute detail, how Jews should behave in many dimensions of their lives, including their interaction with Gentiles (non-Jews), their treatment of the poor, their learning of the Torah, their interaction with family members, work and employment, and dietary laws. (A list of all 613 commandments can be found at http://www.jewfaq.org/613.htm.) One mitzvah, for example, commands that a corner of fields and orchards should be left unreaped so that the poor can partake of the food left there, while another mitzvah forbids the eating of flesh (meat) and milk at the same meal. The hundreds of mitzvot in the Torah were the basis for a tradition of *oral law* that developed over many centuries during early Judaism to interpret the mitzvot, which often raised more questions than they answered. For example, although the Torah mentions marriage, it does not say how a marriage should be performed. The oral law addressed this and countless other matters and spelled out many behaviors that observant Jews practiced then and now (Elon 1994).

Beginning in about AD 200, a religious scholar named Rabbi Judah Ha-Nasi began to write down this body of oral law and eventually produced a volume called the *Mishnah* that contains six major sections and many subsections. For many decades thereafter, other rabbis studied and interpreted the Mishnah, and eventually a group of Palestinian rabbis published their commentaries around AD 400 in a volume now called the *Palestinian Talmud*. Babylonian rabbis began publishing their own version of the Talmud, now called the *Babylonian Talmud*, about a century later, and the Talmud that Jews study today is essentially this latter version. The Talmud consists of the Mishnah and the rabbis' commentaries, known as the *Gemara*.

When Israel became an independent nation in 1948, it faced the important task of adopting a legal system. What eventually evolved combines elements of both common law systems and civil law systems (Shapira and DeWitt-Arar 1995). As is true in other civil law nations, the jury system does not exist in Israel; instead, judges determine guilt and innocence in criminal cases and liability in civil cases. Before 2006, Israel had a codified legal system but no formal constitution, relying instead on the so-called Basic Laws, which the Knesset, Israel's parliament, adopted one at a time over more than thirty years. Taken together, the Basic Laws spell out the structure and function of the three branches of Israel's government and matters such as civil liberties.

Secular courts handle Israel's criminal cases almost all of its civil cases. Four types of secular courts exist: (1) magistrate courts, which handle the bulk of criminal and civil cases; (2) district courts, located in Israel's largest cities, which hear appeals of magistrate court decisions and handle more serious criminal and civil cases; (3) the Supreme Court, which hears appeals from district court decisions and hears direct challenges from individuals of the legality of actions by state officials (and in these ways has contributed to a body of case law characteristic of common law nations); and (4) special courts, which handle a variety of other matters, traffic violations, labor disputes, insurance cases, and juvenile misconduct.

Religious courts handle cases involving the following issues or matters: (1) dietary laws; (2) the Sabbath; (3) Jewish marriage, divorce, and burial; and (4) conversion to Judaism. Shariah courts and ecclesiastical courts handle matters of marriage and divorce for Muslims and Christians respectively.

Israel's decades-long attempt to blend democracy and Jewish law has resulted in some tension over the years, as a controversial Israeli Supreme Court ruling in 2006 illustrates. This ruling restricted the authority of the nation's rabbinic courts (the name given to religious courts that handle matters involving Jewish law) over financial disputes between Jewish spouses who have divorced. The decision involved a provision in a married couple's divorce agreement that a rabbinic court must handle any financial disputes between the couple that would arise after the divorce. The woman who filed the case said her ex-husband had refused to pay for her mortgage as the divorce agreement had required, but that a rabbinic court had failed to force him to do so. The Supreme Court's ruling declared that the provision restricting financial disputes to the rabbinic courts exceeded their authority over marriage and divorce. Civil libertarians said the ruling would strengthen the rights of women in the rabbinic courts, while one of Israel's chief rabbis said it would threaten Jewish family life (Associated Press 2006).

Israel's Supreme Court aroused more controversy a month later in a case involving the right of Palestinians married to Israeli citizens (most of them Arabs) to live in Israel. In 2002, Israel passed a law requiring Palestinian women married to Israelis to be at least 25-years-old to be eligible to live in Israel with their spouses; Palestinian men married to Israelis had to be at least thirty-five to enjoy the same right. During the next four years, the law forced hundreds of Palestinians married to Israelis to stay in the West Bank and Gaza Strip apart from their families in Israel. Although the Israeli government said the law was meant to enhance national security, critics said it was violated Arabs' civil rights and was designed to limit the size of Israel's Arab population. The Supreme Court upheld the law in a 6–5 decision in May 2006 (Myre 2006; Wilson 2006).

Socialist Law

Socialist law is a term used to describe the legal systems of Communist nations, primarily those comprising the former Union of Soviet Socialist Republics (USSR), the People's Republic of China, North Korea, Vietnam, and Cuba. With the USSR's demise almost two decades ago and China's gradual move toward capitalism since the early 1990s, socialist law is more a family of law of the past than it is one of the contemporary era.

Because Marx and Engels wrote that the state (and thus the legal system) would "wither away" after a communist revolution eliminated capitalism and class differences, the idea of law in a Communist society would seem to be a contradiction. Even so, every Communist nation has certainly had a legal system, and the proper role of law under Communism was a matter of great debate during the decades when Communism was much more influential than it is today (Tay 1990). This issue notwithstanding, historically Communist nations used their legal systems as convenient vehicles for the repression and elimination of their political opponents.

During the 1930s, for example, Joseph Stalin, the ruler of the Soviet Union, conducted a campaign of terror that came to be known as the Great Purge. This campaign involved the arrests, trials, imprisonment, and executions of hundreds of thousands of people. By the time Stalin died in 1953, the death toll from the Great Purge and later purges is estimated by some observers to have reached at least 20 million victims (T. Rosenberg 1995). Many of Stalin's political opponents sat through "show trials" in which their guilt was predetermined, with the trials designed to both publicize and legitimate the political repression they were experiencing (Hodos 1987).

The government of today's China wields a brutal criminal justice system to repress its opponents and to control crime. Its use of the death penalty amounts to government terror: at mid-decade, China was executing perhaps 10,000 persons every year, compared to only about 400 in the rest of the world. As one observer has noted, "The government's relentless death penalty machine has long been its harshest tool for maintaining political control and curbing crime and corruption" (Yardley 2005:A1). Many Chinese receiving a death sentence are thought to be innocent of murder or even of any crime at all, and the nation includes few, if any, legal safeguards for suspects accused of capital crimes. Although China theoretically guarantees its criminal suspects the rights to legal representation and a criminal trial, at least 300,000 people—primarily drug offenders, prostitutes, and political prisoners—are detained in special prisons without ever being tried or found guilty, and many of these inmates are thought to have not even committed any offenses. Known as labor reeducation camps, these prisons began in 1957 under the reign of Mao Zedong, China's Communist dictator, and have persisted to the present despite China's move toward a capitalist economy after Mao's death and its professed interest in political and social reform. Condemning the labor reeducation camps, a human rights advocate declared, "It's a violation of every due process right in every human rights law" (Yardley 2005:A1). Despite such criticism, a Beijing lawyer was skeptical that the government would soon abolish the camps: "It is important for the power holders that a system like labor re-education stay in place" (Yardley 2005:A1).

Despite the long history of legal repression in Communist nations, socialist law over the years has had several other goals that might sound more acceptable (Dammer and Fairchild 2006; Savelsberg 2000). A major goal is to implement the socialist distaste for private property. To achieve this aim, socialist legal systems have several provisions that support the socialist ideal of public ownership of land, real estates, factories, and other types of property that are typically owned in capitalist societies by private parties. Thus, even though China has been moving toward a capitalist economy, the government still owns all land if not the houses, stores, and other buildings on the land. In 2006, continuing antipathy in China for private property forced the cancellation of a draft law that would have provided additional protection for Chinese private property rights. Critics of the proposed law said it would undermine Chinese respect for socialist property and worsen an income gap and political unrest that had both been growing since China began moving toward capitalism. A critic of the proposed property rights law said sarcastically that it would equally protect "a rich man's car and a beggar man's stick" (Kahan 2006:A1). Defending the proposed law, a retired newspaper editor replied, "A widening gap between rich and poor is not the fault of market reforms." He wrote, "It's the natural result of them, which is neither good nor bad, but quite predictable" (Kahn 2006:A1).

Cuba's legal system has received considerable attention since Fidel Castro took power after Cuba's Communist revolution in 1959 (Salas 1983; Zatz 1994). Cuba enacted new laws that took property from large landowners and allocated it to the mass of less wealthy people. Perhaps most interestingly, the new Communist nation established *popular tribunals* to replace courts in handling many kinds of disputes. By 1969, more than 2,000 popular tribunals existed throughout the tiny island of Cuba. These tribunals were centered in neighborhoods, comprised ordinary citizens instead of legal professionals, and did without legal formalities and other legal trappings; this was an effort to make Cuba's law a true "people's law," in contrast to the system of formal law in Cuba under Castro's predecessor, Fulgencio Batista, that was widely seen as repressing Cuba's poor. Tribunals in rural areas tended, not surprisingly, to handle agricultural disputes and land redistribution, while urban tribunals tended to handle family disputes, disagreements among neighbors, and fighting. All such cases were typically heard by three lay judges and were usually held at night to encourage neighborhood residents to attend. Luis Salas (1983:590–91) notes the educational function of the tribunals' hearings: "The purpose of trials was not primarily to determine guilt or innocence. Rather, they were regarded as tools through which embarrassment and peer pressure helped rehabilitate offenders and deter others. Spectators played a vital role in the educational process, participating and expressing their opinions at will." A common penalty was fines, which were applied as quotas (e.g., three days' worth of wages) rather than as specific peso amounts and thus took into account the guilty party's financial status.

The popular tribunals were initially considered a success in implementing Cuba's socialist vision. Within a decade, however, their informality and wide discretion in administering sanctions became criticized as undermining respect for the law, and new state policy formalized their procedures. The absence of attorneys and professional judges in the popular tribunals was also criticized, with critics asserting that lay attorneys and judges lacked the necessary knowledge and experience to properly handle disputes. Various policy changes at the national level eventually reduced the scope of the popular tribunals by establishing a mufti-level system of conventional courts to handle many types of criminal and civil offenses. Thus, law in Cuba moved from a "people-based" system of informal justice to a more conventional, formal system (Salas 1983).

Traditional Law

Traditional law (also called *customary law*) refers to law that relies primarily on unwritten rules and customs. This family of law is found in the traditional or premodern societies that are most commonly studied by anthropologists. These include societies that stand alone, such as those found in the many islands of the South Pacific; those that exist in remote areas and rural villages of established nations like the Philippines and the many African nations; and native communities in the United States and Canada.

In general, these societies share certain related characteristics (Bracey 2006). First, they are relatively small. Second, because of their small size, people in a traditional society all tend to know each other very well and to value the close relationships they have. As a result, in these societies, "the rights and well-being of the group are more important than those of the individual" (Bracey 2006:71). Third, one's status in the society derives more from parents' status, birth order, gender, and other *ascribed* characteristics than from *achieved* characteristics such as talent and hard work. Fourth, most of these societies have preindustrial economies: horticultural, pastoral, or hunting and gathering. Fifth, the societies are fairly homogenous. People tend to look like each other in regard to skin and hair color and other physical characteristics and to share the same beliefs and values.

All these characteristics combine to produce a family of law that, compared to that found in most large, industrial societies, tends to be very group-oriented; to lack attorneys, formally trained judges, and the other trappings of modern legal systems; and to favor compromise solutions over win–lose outcomes in the settlement of disputes. When norm violation occurs, the close ties characteristic of societies with traditional law tend to make "reconciliation and restitution . . . more important than punishment and retribution" (Bracey 2006:71). Traditional law was featured in the previous chapter's discussion of the definition of law, and it will also be featured in the next chapter's discussion of dispute processing. In that chapter, we will see that modern societies have begun to rediscover traditional law's emphasis on reconciliation and restitution and to appreciate its aversion for win–lose outcomes and the costs they entail.

TYPES OF LAW IN THE UNITED STATES

Many types of law exist in the United States and other modern nations. We review the major types in the United States and devote most of our attention to the criminal law, the most familiar and interesting type to many Americans but also one that is very complex and the source of much controversy and debate.

Criminal Law

Criminal law has been defined as "the body of law that prohibits acts that are seen as so harmful to the public welfare that they deserve to be punished by the state, and that governs how these acts are handed by official state procedures" (Barkan and Bryjak 2003:G-2). U.S. criminal law generally distinguishes three types of crimes according to their seriousness: (1) *felonies*, serious crimes such as homicide and robbery that are punishable by incarceration of more than one year in state or federal prison; (2) *misdemeanors*, less serious crimes such as trespassing that are punishable by incarceration of up to one year in local jails; and (3) *violations* or *infractions*, minor offenses such as most traffic offenses that are punishable mainly by fines. In another distinction, U.S. law classifies crimes as either *mala in se* offenses or *mala prohibita* offenses. *Mala in se* offenses (evil in themselves) are behaviors considered so harmful or immoral that they would be wrong even if the criminal law did not prohibit them; examples include homicide, rape, robbery, burglary, and other crimes involving unwilling victims. *Mala prohibita* offenses (wrong only because prohibited by law) are behaviors that are crimes only because the law does prohibit them; examples include prostitution, illegal drug use, and other victimless crimes.

Elements of a Crime. For an act to be proven a crime beyond a reasonable doubt in court, several elements must be found to have existed at the time the crime was committed (Davenport 2006). The first is that the act *must* *be prohibited by the criminal law*. In a democracy, it is difficult to conceive that someone would be prosecuted for a behavior that does not, in fact, violate the criminal law, but that is the point: fundamental fairness in a democracy requires that no one should be arrested for a behavior that is not prohibited by the criminal law or for a behavior that becomes prohibited only after the person is arrested. A corollary of the first element is that the criminal law must also spell out the possible punishment for an act that violates it. In the absence of this stipulation, authorities would be free to mete out any punishment they desired. This may happen in an authoritarian regime, but it should certainly not occur in a democracy.

A second element, and perhaps the most important in terms of proving guilt, is *actus reus* (Latin for actual act). This means that an actual criminal act must have been committed, not just the thought of committing the act. For the criminal justice system, this element requires that a prosecutor prove beyond a reasonable doubt that a criminal act was actually committed. The third element, which is very important but sometimes difficult to prove in court, is *mens rea* (guilty mind), or *criminal intent*. For defendants to be convicted of a crime, the state must show that they intended to commit the criminal act for which they are accused. In practice, this means that defendants must have known and understood what they were doing and that they performed the act voluntarily. The major legal defenses to criminal responsibility (see below) all center on the issue of criminal intent. In a partial exception to the element of *mens rea*, the criminal law bans harmful behaviors that are committed out of gross negligence or recklessness even if the offender did not intend to commit the behavior. Thus, if a parent accidentally leaves an infant inside a locked car on a hot, sunny day and the infant dies, the parent can be prosecuted even if he or she obviously did not intend to harm or kill the infant in this manner. A tragic example of this offense occurred a few years ago, when a New Jersey father forgot that his two young sons, ages 1 and 2, were in his car when he drove to work after being distracted by a cell phone call. The boys died of heat exhaustion, and the father pleaded guilty the next year to involuntary manslaughter (Alaya and Kinney 2004).

The fourth element is *concurrence*. This is ordinarily not difficult to prove in court, but means that criminal intent and a criminal act must have occurred at about the same time and that the intent must have preceded the act. Suppose you are planning to poison your hated cousin who has taunted you relentlessly since your childhood. At a visit to your cousin's house, you hide the poison and make a date for a return visit, at which time you plan to stir the poison into your cousin's coffee. Your cousin then sees you off as you pull your car out of the driveway. However, your car fails to start the first time you try it, and then, when it does start, you begin backing out of the driveway only to hear and feel a loud thud. You rush out of your car and find your cousin dead under the back wheel. Unknown to you, your cousin had begun walking behind the car to come up to the driver's window to ask you if you needed any help with the car. Even though your cousin's death accomplished the goal of your dastardly plan to poison the coffee, your cousin's death is not a crime because you were not intending to kill your cousin with the car and the planned poisoning did not occur.

The fifth element is *causation*. This refers to the idea that the criminal act must have actually caused the harm that occurred as a result of the act. Suppose you get into a shoving match with a middle-aged man at a bar and some punches are thrown. Suddenly, the man has a fatal heart attack. Did the fight cause the heart attack? Should you be prosecuted for homicide (say involuntary manslaughter)? The prosecutor might argue that the fight caused the man's heart attack, while the defense would probably respond that a fight of this type normally does not cause a heart attack (thus not satisfying the element of causation), making a homicide prosecution inappropriate. The victim's health history would be a key piece of evidence in a case of this type: any evidence of heart problems would bolster the defense's argument.

The final element is *harm*. For an act to violate the criminal law, some harm to a person or property must ordinarily occur. This is usually what happens, of course, but whether harm occurs in so-called victimless crimes, such as illegal drug use and prostitution, is less clear. By definition, these crimes involve behavior in which people participate willingly and are thus not unwilling victims. Despite this fact, the criminal law assumes that the participants are indeed harmed and that their behavior harms the larger society. Victimless crimes raise important and fascinating questions of law and morality that Chapter 5 explores further.

Legal Defenses to Criminal Responsibility. Although defendants and their attorneys ordinarily challenge the weight and accuracy of the evidence of a crime, they may also claim that the defendant did not have the criminal intent that is required for a conviction. Several legal defenses to criminal responsibility challenge the existence of such intent.

Accident, Mistake, or Ignorance. A defendant may claim that he or she broke the law by accident, mistake, or ignorance and thus did not intent to commit a crime. If someone gives you drinks to hand out at a party and, unknown to you, one of the drinks contains the date rape drug Rohypnol, you should not be held criminally responsible for the consequences. If you are driving a car in the rain at a safe speed and your car hydroplanes anyway and hits a pedestrian, you should also not be held criminally responsible.

Duress. Because criminal intent requires that a defendant committed a crime willfully, a defense of duress may succeed if it can be shown that the defendant feared for her or his safety (or for that of a loved one) at the time of the crime and was thus forced to help a crime occur. Real-world examples of such cases are fairly rare, but TV shows and films sometimes feature someone who was forced to take part in a robbery or help an act of terrorism because the person's spouse and/or children were being held at gunpoint.

During the Vietnam War, some protesters who committed civil disobedience for protesting the war claimed that they did so because their consciences were putting them under duress. Judges almost always rejected this novel interpretation and expansion of the duress defense (Barkan 1985).

Self-Defense. Self-defense is probably one of the defenses to criminal responsibility with which you are most familiar from the popular media. If someone is about to rob, rape, or assault you or a loved one, the law allows you to use appropriate violence to prevent the crime from occurring or from being completed if it has already begun. The key issue is the level of violence you use and how much violence was needed to prevent the crime. If a young male demands your money but shows no weapon and you immediately fatally shoot him with a handgun, you may very well have succeeded the limits of self-defense. If someone is burglarizing your house and you again fire a fatal shot, it may again be difficult to show that you were acting in self-defense. Of course, the final determination of guilt is always up to a jury, and a prosecutor may decide that a jury may sympathize with your situation even if the law of self-defense does not support your fatal shooting. If so, the prosecutor may not prosecute the case at all or at least be quite willing to accept a plea bargain.

Entrapment. As a legal defense, a defendant is considered to have been entrapped if law enforcement authorities persuaded her or him to commit a crime that the defendant otherwise would have no intent of committing. Suppose you are living in a dormitory and one of the other dorm residents repeatedly offers you some marijuana to smoke. After you reject the offer each time, you finally relent and smoke a joint with the person, who turns out to be an undercover drug agent and then arrests you for using marijuana. Because it was apparent that you would not have smoked the joint if the narc had not induced you to do so, an entrapment defense might succeed. If, however, you had been looking to buy a joint and then buy one from the undercover officer, an entrapment defense will almost certainly not succeed, since you obviously needed little inducing from the officer to use marijuana.

Insanity and Other Diminished Capacity. Different states have different standards for an insanity defense, but this defense basically means that defendants must be able to appreciate the difference between right and wrong and/or to control their behavior. If they have a mental or emotional condition that prevents them from knowing right from wrong or that inexorably drives them to commit a criminal act, they should not be found guilty of committing this act. Perhaps the most notorious insanity defense involved John Hinckley, Jr., who shot President Ronald Reagan in 1981 and claimed later that he did so to help make actress Jodie Foster attracted to him. The jury's verdict of not guilty by reason of insanity was widely criticized (Clarke 1990). Despite the attention Hinckley's successful insanity defense received, this defense is, in fact, hardly ever used and succeeds only rarely when it is used and thus does not impede the administration of criminal justice (Walker 2006).

The insanity defense falls into a general category of *diminished capacity* defenses involving defendants' inability to understand or control their actions. Two other conditions that make up this category are developmental disability and youthfulness. If someone with a developmental disability commits an act of violence, the person ordinarily should not be criminally prosecuted if there is no question that the disability resulted in sufficiently diminished capacity. That said, many defendants with some degree of developmental disability have been found guilty of criminal offenses. For many years, developmentally disabled persons who committed murder were allowed to be sentenced to death, and some were executed. In June 2002, the U.S. Supreme Court finally banned such executions, with the majority opinion declaring that a national consensus had developed in the country against capital punishment for the developmentally disabled (Greenhouse 2002).

A final diminished capacity defense involves youthfulness. Historically, the common law has judged some individuals to be too young to be able to understand their actions and thus to possess criminal intent. Until recent decades, U.S. courts considered individuals below age 17 or 18 to fall into this category, and their cases were handed by the juvenile justice system rather than by the (adult) criminal justice system. Then, rising crime rates beginning in the 1960s prompted states to begin prosecuting youths who commit serious crime at age 13 or 14 as adults (Bishop 2000). Debate continues over whether such prosecutions will deter the juvenile crime rate or, as critics fear, reinforce the criminal tendencies of the juveniles who are prosecuted (Fass and Pi 2002). In March 2005, the Supreme Court banned capital punishment for youths who committed their crimes before age 18 (Greenhouse 2005). The Court's decision was by a narrow 5–4 margin, raising the possibility that changes in the Court's composition since 2005 could one day yield a different vote that would restore the death penalty for juveniles.

Civil Law

Civil law (also called *private law*) governs private transactions among individuals and organizations. (Recall that "civil law" also refers to the family of law found in the nations of continental Europe and their former colonies and that these two distinct means of "civil law" should not be confused.) Common examples of actions and statuses that fall under the civil law include marriage and divorce, contracts, wills, and real estate transactions. A violation of a civil law is called a **tort** (sometimes called a "civil wrong"), and a course called "Torts" or at least dealing with torts is a requirement in the first-year curriculum at most law schools.

Many kinds of torts exist. One type familiar to many readers is *slander*, a false spoken statement that defames the character or reputation of an individual. *Libel* is similar to slander but involves a written statement. For slander (or libel) to be proven in a court of law, the plaintiff must show that the statement was false and that it harmed the character or reputation of the plaintiff. When the plaintiff is a public figure like a politician or movie star, the plaintiff must also show that the defendant made the alleged defamatory statement with "actual malice." To satisfy this requirement, the plaintiff must generally show that the defendant knew the statement was false or should have known it was false.

One of the most important branches of civil law is *family law*, which governs marriage, divorce, adoption, and other aspects of family structure and relationships. As such, family law is a wonderful exemplar of the relationship between law and society and thus of the study of law and society itself. "Like everything else in the law," writes Lawrence Friedman (2004a: 59), family law "reflects what is happening in society, and what is happening in society has a profound impact on family life and family relations."

Changes in U.S. family law during the nineteenth century illustrate this dynamic. At the beginning of the century, the father literally "ruled the roost" as far as family law was concerned. As Friedman (2004a:59) bluntly observes, "Husband and wife were, as the saying went, one flesh; but the husband was definitely in charge of that flesh. And more than the flesh. The wife had, in many ways, as few rights as a newborn baby or a lunatic." For example, land or any other valuable property that a woman might have owned when she married automatically became her new husband's property as she lost ownership altogether. Once she was married, a woman could not buy or sell property unless her husband gave his approval, and she also was unable to enter or contract or leave a will. As the nineteenth century moved along, however, so did family law, and women gained more rights in all these respects by the end of the century. Women's growing equality in family law thus reflected their growing equality in American society in general over the course of the nineteenth century. Chapter 6 further discusses the evolution of family law during this period.

The civil law differs from the criminal law in several ways. As we have seen, when a crime occurs, it is considered an offense against the state, not just against an individual, and the state is the party that handles the case by considering whether to prosecute someone for committing the crime. In contrast, when a tort occurs, it is up to the aggrieved party to seek legal redress by filing a *civil suit*, more popularly called just a *lawsuit*. When a crime is prosecuted, possible punishment for a defendant who is found guilty includes imprisonment; when a civil suit occurs, the penalty for a defendant who is found liable excludes imprisonment and instead involves *compensatory damages* (e.g., payment for medical expenses or lost wages), taking some action to remove the grievance (e.g., dismantling a fence that was on a neighbor's property line), or other actions designed to compensate the plaintiff (the party bringing the lawsuit) for any harm the plaintiff experienced. *Punitive damages* may also be awarded; as their name implies, punitive damages are intended to punish a defendant whose behavior is especially malicious, reckless, and/or injurious.

An important difference between the criminal law and civil law is the standard of proof that governs whether a defendant can be found guilty (criminal law) or liable (civil law) for committing a crime or tort, respectively. In the criminal law, the standard of proof is reflected by the very familiar phrase, *beyond a reasonable doubt*. As this phrase implies, the judge or jury does not have to be 100 percent certain of a defendant's guilt, but they have to be almost this certain. In the civil law, the standard of proof is weaker and involves a *preponderance of evidence*. In practice, this means that a judge or jury must believe only that there is more than a 50 percent chance that the defendant is liable for any harm done to the plaintiff. Another important difference between criminal and civil law is that many of the due process rights and protections given to criminal defendants are unavailable for civil defendants. For example, whereas criminal defendants have the right to legal representation by an attorney provided by the state if the defendant is too poor to afford a lawyer, civil plaintiffs and defendants must pay for their attorneys themselves. Although legal services attorneys for the poor do exist, in practice this requirement means that many poor or near-poor defendants are unable to bring lawsuits or to defend themselves against a suit. (Chapter 7 explores this issue further.)

A telling illustration of the different standard of proof in criminal and civil proceedings involved football star and celebrity O. J. Simpson a decade ago. As you are no doubt aware, Simpson was arrested in 1994 for allegedly murdering his ex-wife, Nicole Brown, and one of her friends, Ron Goldman. In 1995, a jury found him not guilty after a trial that presented both strong and weak evidence of his guilt but also accusations of police racism. The families of Brown and Goldman later brought a wrongful death suit against Simpson, and in 1997, a civil jury found Simpson liable for the death of Goldman and for battery against Brown. Although many factors undoubtedly explain the very different outcomes of Simpson's two trials, the different standard of proof was very likely one such factor: Given the same amount of evidence, it is easier for a civil jury to find a defendant liable than for a criminal jury to find the defendant guilty.

When Americans attempt to settle disputes, they often turn to the legal system as the civil law comes into play. The use of the civil law for this purpose has several advantages, but it also has several disadvantages. The next chapter examines the civil law and dispute settlement at much greater length.

Executive Orders

As their name implies, **executive orders** refer to laws passed and actions undertaken by the executive branch (the president or the fifty state governors) of the government. Several types of executive orders exist (Samuels 2006). A first type is the very familiar *proclamation* that sets aside a specific day, week, or month to honor an individual or group or to give attention to a worthy cause. In 2007, for example, President George W. Bush proclaimed

May 19 through May 25 as National Safe Boating Week. A list of presidential proclamations from the last few years appears at http://www.whitehouse.gov/news/proclamations/. Proclamations may attract some attention to their subject, but their value is primarily symbolic.

A second type of executive order involves the *establishment of task forces*, groups of experts on a particular issue who eventually write a report outlining the dimensions of the issue, its causes, and possible solutions or strategies to undertake. Many times these reports receive little or no publicity, and their recommendations are ignored, but some task force reports do win heavy attention and make quite an impact. One such task force was the so-called Kerner Commission, appointed by President Lyndon Johnson in July 1967 to address the urban riots that beset the nation's major cities during that summer and preceding summers. The Kerner Commission's report a year later in many ways blamed the riots on fundamental racial inequality in American society and famously warned that the United States was "moving toward two societies, one black, one white—separate and unequal." The report also included a series of recommendations to reduce racial inequality, including better urban housing and improved educational and employment opportunities.

A third type of executive order involves the *organization of the executive branch*. The president or a governor may create or eliminate executive branch departments and administrative agencies and combine other departments or agencies. President Bush's creation in October 2001 of the Department of Homeland Security is a recent notable example. As with new laws (see Chapter 6), executive orders that create or reorganize executive branch departments may not always achieve their intended goal of greater effectiveness and efficiency. For example, at mid-decade the Department of Homeland Security came under criticism for not taking sufficient measures to protect national security and in particular for not protecting the nation's chemical plants, railways, and other critical terrorist targets (Mintz 2005).

A final type of executive order involves *policymaking*: the president or a governor may issue an executive order that changes governmental policy or undertakes governmental action to address an issue or problem. Two of the most famous if controversial executive orders in U.S. history were Present Abraham Lincoln's Emancipation Proclamation and President Franklin Delano Roosevelt's order during World War II to place Japanese Americans in internment camps (Samuels 2006). Certain examples of this final type of executive order have been criticized for exceeding the executive branch's proper powers under the nation's system of checks and balances, and courts have deemed some executive orders of this type to be unconstitutional for this reason.

Administrative Law

The United States has a countless number of *administrative* (also called *regulatory) agencies* at the federal, state, and local levels. At the federal level, some of the best-known administrative agencies include the Internal

Revenue Service, the Food and Drug Administration, and the Federal Communication Commission. All administrative agencies have a countless number of rules and regulations that govern the behavior of the organizations falling under their jurisdiction and specify procedures for the agencies to follow in their dealings with these organizations. **Administrative law** refers to these rules and regulations and affects Americans in all walks of life (Stewart 2003). Many Americans work in jobs that are governed by administrative agency rules or at least have employers who must conform to the requirements of the Occupational Safety and Health Administration, and we all use products manufactured by companies regulated by the Food and Drug Administration, National Highway Traffic Safety Administration, Environmental Protection Agency, or other agencies. Administrative law is obviously important in all these ways.

An interesting aspect of administrative law is that regulatory agencies perform both an enforcement and a judicial function in investigating possible violations of their rules and then holding hearings to determine whether violations have indeed occurred and, if so, then imposing appropriate penalties or taking other actions. Although jury trials do not occur, defendants in these hearings have the right to be represented by legal counsel. Partly for this reason, many law firms throughout the country, but particularly in the Washington, DC area where the federal agencies are located, have attorneys who specialize in the body of administrative law that a particular agency follows. For example, attorneys with expertise in environmental law represent clients who appear before the Environmental Protection Agency.

Before the late nineteenth century, administrative agencies and thus administrative law hardly existed, as the federal government was much weaker than it is today (Friedman 2004a). Industrialization changed this situation. The new railroads, factories, and other components of industrialization led to the rise of a national economy and national problems. In particular, the new railroad, oil, steel, and other large corporations engaged in a variety of unsavory practices such as price-fixing, bribery, and kickbacks to maximize their profits, and they operated workplaces that endangered the health and lives of their employees. To control the behavior of these "robber barons," as they have been called (Josephson 1962), Congress established agencies such as the Interstate Commerce Commission in 1887 and the Food and Drug Administration in 1906, and it enacted new laws, perhaps most notably the Sherman Anti-Trust Act of 1890. These developments led, in the twentieth century, to the rise of the so-called administrative-welfare state and an explosion in administrative law. New Deal legislation during the 1930s under President Franklin Roosevelt and "Great Society" initiatives during the 1960s under President Lyndon Johnson expanded the importance of regulatory agencies and the amount and influence of administrative law.

Administrative agencies differ regarding whether their rules and decision making are more formal or more informal (Kagan 2006). Some agencies have rules that are very detailed and have decision-making processes that enforce these rules rigidly, while other agencies have rules that are less detailed and allow for more discretion in their enforcement. In the latter agencies, regulatory violations are typically handled through negotiation and mediation, while in the former agencies, regulatory violations are typically handled through legalistic hearings involving the presentation of evidence and arguments by attorneys. In some agencies, the style of regulatory enforcement has changed over time. An example of this pattern is the California Industrial Accident Commission, which was established in the early twentieth century to handle workers' compensation complaints; its goal was to provide compensation to injured workers in a fairly automatic manner that would avoid the time and expense of legal proceedings. Dissatisfied with their compensation, however, workers hired attorneys to represent them at administrative hearings, and the process eventually became very legalistic and sometimes contentious (Nonet 1969). In general, the more formal, legalistic style of administrative decision-making is more likely to be found in agencies with heavy caseloads and thus the need to resolve their cases as quickly as possible (Kagan 2006).

Other Distinctions

Several other types of law exist in the United States that can be discussed more briefly. [For a fuller treatment see Hames and Ekern (2006) and Samuels (2006).]

Statutory Law and Case Law. **Statutory law** refers to the body of law passed and enacted by Congress at the federal level and legislatures at the state level. Many but certainly not all statutes outline the elements of specific crimes and are thus part of the criminal law. In the American system of checks and balances, the legality of specific statutes is subject to review by appellate courts. Most statutory law before the early twentieth century came from state legislatures, not from Congress (Samuels 2006). Beginning in the 1930s, Congressional lawmaking accelerated, especially in the area of interstate commerce. During the last two decades, however, the Supreme Court has ruled against several Congressional bills that the Court interpreted as affecting only intrastate (within a single state) commerce, not interstate commerce. One ruling here was the Court's 1995 5–4 decision in *United States v Lopez* (514 U.S. 549), which invalidated a Congressional law that made it a federal crime to have a gun near a school. Because this activity did not involve interstate commerce, the Court said, Congress could not ban it. The minority opinion held that the law should have been upheld because education has an effect on interstate commerce and the presence of guns at or near schools may seriously impair the quality of schooling received by students.

As noted in the earlier discussion of common law systems, **case law** refers to the decisions of appellate court judges. As such, case law is a prime topic of study in law school curriculums everywhere, with most law schools following the "case study" method in which a class is taught via lecture and discussion of collections of appellate judges' opinions on cases dealing with the class's subject matter (see Chapter 8). Through their decisions, appellate court judges make law, and their decisions are binding on the jurisdiction over which the court presides. The Supreme Court is, of course, the most powerful and influential court in the United States, and its decisions become the law of the land. Debate began in the 1960s and continues today over whether the Supreme Court and certain federal appellate courts have become too involved in the nation's political and social affairs, with their policymaking in these areas condemned as "judicial activism" (Powers and Rothman 2002). Supporters of such activism say it has been necessary to address the nation's many social problems.

Public Law and Private Law. Another basic distinction is between **public law** and **private law**. Public law is the body of law that outlines the structure of government, the powers and responsibilities of public officials, and the government's relationship to private citizens. Criminal law and administrative law both fall into the larger category of public law. Private law, as we have seen, is another term for civil law and, as such, is the body of law that governs relationships among individuals and organizations.

Substantive Law and Procedural Law. Yet another important distinction is between **substantive law** and **procedural law**. Substantive law refers, as the name implies, to the substance of law: which behaviors are permitted and which behaviors are prohibited. Procedural law, on the other hand, refers to how the substantive law is implemented and enforced. Procedural law is complex but includes such components as the rules of evidence that govern what kind of evidence is admissible or inadmissible during a trial and more mundane matters such as the deadline for filing an appeal after a criminal conviction.

Substantive Justice and Procedural Justice. A similar distinction lies between **substantive justice** and **procedural** (also called *formal*) **justice**. Substantive justice is achieved if the outcome of a case is "right" according to prevailing standards of morality and fairness. Procedural justice is achieved if a case during its various stages follows the many procedural rules that govern the judicial process. Although the ideal situation is to have a case in which both substantive justice and procedural justice occur, that does not always happen. In particular, the "wrong" result may occur, with substantive justice not being achieved, even if all procedural rules are followed (thus achieving procedural justice). For example, an innocent defendant may still be found guilty even if all of her or his legal rights have been observed. Chapter 1 discussed the case of a man who was left in a radiation-induced vegetative state after mistakenly being diagnosed with brain cancer. Because his family's malpractice lawsuit was filed after the statute of limitations (which began after the misdiagnosis and not after the discovery of the misdiagnosis) had expired, the plaintiff's lawsuit was dismissed. Most readers will probably agree that the wrong moral outcome occurred in this case even though (and, in fact, because) all procedural rules were followed. Thus procedural justice was achieved, but not substantive justice.

SPECIAL TOPICS

Certain types of law do not fit neatly into the preceding categories and are discussed here separately.

Military Law and Military Justice

The armed forces in the United States have their own legal system (Shanor and Hogue 1996). The military justice system handles violations of military discipline and other aspects of military life and also charges of conventional criminal behavior such as physical assault, rape, and theft (which can also be handled by the civilian criminal justice system). The content of military law may be found in the Uniform Code of Military Justice (UCMJ) at http://www.au.af.mil/au/awc/awcgate/ucmj.htm. The UCMJ, adopted during the Korean War and revised since, specifies the persons who are subject to military law, the procedures to be followed and the defendant's rights to be observed when violations are alleged, and the rules of military discipline. For example, Section 886, Article 86 defines *absence without leave* as follows:

> Any member of the armed forces who, without authority,
>
> 1. fails to go to his appointed place of duty at the time prescribed;
> 2. goes from that place; or
> 3. absents himself or remains absent from his unit, organization, or place of duty at which he is required to be at the time prescribed; shall be punished as a court-martial may direct.

The military maintains its own system of law and justice because historically it has been considered better able than civilian courts to appreciate and enforce its need for strict obedience to military discipline and hierarchy. Despite this tradition, military law has been criticized in recent decades for at least two reasons. The first reason is summarized by the wry title of a 1970 book, *Military Justice Is to Justice as Military Music Is to Music* (Sherrill 1970). As this title implies, military justice does not produce much justice, say its critics. More specifically, say the critics, military law lacks some of the rights and protections afforded civilian defendants and sometimes "stacks the deck" against military defendants in ways that are arbitrary and unfair.

As a news report summarized this critique, "Compared with the accepted legal practices of the civilian world, the military's justice system is neither particularly free nor particularly democratic. Justice is not secondary, . . . but it is subject to other considerations, not least of which is accomplishing the military's mission, often in the middle of war" (Myers 2004:Section 4, p. 3).

A key problem, said the report, stems from the huge role that military commanders play at all stages of the military justice system, as they in effect serve as prosecutor, judge, and jury. They decide what charges should be brought against a defendant, they help decide the defendant's guilt or innocence, and they determine the defendant's punishment if convicted. The power given to commanders in these respects means that military justice "ultimately depends almost entirely on the judgment of commanders," said the report. Because different commanders may exercise their judgment in different ways, cases are handled quite differently from one military base or unit to another. In a related problem, a military base's commander selects the officers who will reside as the judge and jury in a courts-martial. Because the commander may select officers who share the commander's opinion on a case and because the officers may feel beholden to the commander, critics question whether the officers can exercise independent judgment.

The second reason for recent criticism of military justice is summarized in the familiar phrase, "the fox guarding the chicken coop." Because the military handles its own cases, say the critics of military justice, too often it has covered up crimes and other transgressions or has meted out only minimal punishment. In this regard, the military's investigation and prosecution of crimes stemming from torture and abuse by U.S. personnel of Iraqi prisoners at the Abu Ghraib prison camp in Iraq in 2003 was criticized by many observers for the weak punishment that was handed out to low-ranking personnel and for generally ignoring possible crimes committed by personnel higher in the chain of command. As the title of one newspaper critique put it, "On Abu Ghraib, the Big Shots Walk" (Herbert 2005:A25).

This problem is not confined to notorious cases like Abu Ghraib. Critics say that because military commanders' careers often depend on their unit's reputation, they have a vested interest in not investigating and prosecuting ordinary military offenses and may be tempted to hush up those they do investigate (Myers 2004). A shocking example of this problem came to light in late 2003 when the *Denver Post* published a series of articles about thousands of rapes and physical assaults (domestic violence) committed against women on U.S. military bases by their fellow members of the armed forces (Herdy and Moffeit 2004). The *Post* found that when women reported their victimization, their complaints were often ignored; when an offender was occasionally punished, the punishment usually was fairly minor. In 2000, for example, women in the military reported more than 12,000 cases of domestic violence, but only 26 led to courts-martial. In an older case, a drill sergeant and four other soldiers gang-raped a woman who had just completed basic training and also broke several of her bones, urinated on her, and burned her with cigarettes. The five men were never punished, and years later the woman was still having serious health problems because of the gang rape and beating she suffered.

Native American Law

Native American law was mentioned earlier as an example of traditional or customary law. Because Native American reservations have their own legal systems, they in effect are legal enclaves within a much larger nation, and these enclaves' understanding of law differs in important ways from that characterizing the United States as a whole. A brief history of Native American law is relevant for understanding it today.

When white people conquered and relocated Native Americans during the nineteenth century, treaties were signed that confined the latter to reservations, where they were allowed to govern themselves and have their own law (Bracey 2006). Then, in the Dakota Territory in 1881, a Sioux named Crow Dog fatally shot Chief Spotted Tail in an apparent act of self-defense. As tribal law prescribed, the families of the two men negotiated a settlement, with the family of the victim, Spotted Tail, receiving restitution in the form of $600, eight horses, and one blanket. The two families were satisfied that the case was over, but non-Indian acquaintances of Spotted Tail, including an agent of the Bureau of Indian Affairs (BIA), had the case brought into a U.S. territorial court. They thought the defendant needed much harsher punishment, and they also wanted to use this case to further the dominance of U.S. government law over tribal law. As a result, Crow Dog was tried, convicted by an all-white jury, and sentenced to death (Harring 1994).

Hoping for a ruling from the U.S. Supreme Court that would invalidate tribal law, the BIA paid for Crow Dog's legal appeals. In 1883, the U.S. Supreme Court disappointed the BIA by overturning the conviction on the grounds that the Sioux tribe's sovereignty prohibited legal intervention by the U.S. government. At the same time, the Court added that the government had the right to eliminate tribal sovereignty but that its failure to have already done so meant that Sioux law must prevail in the Crow Dog case. Given this opening by the Court, Congress passed the Major Crimes Act two years later that gave the federal government jurisdiction over serious violent and property crimes committed on Indian reservations but continued to permit Native Americans to handle minor offenses and civil cases. According to anthropologist and criminal justice scholar Dorothy H. Bracey (2006:78), this bill "undermined the ability of the Indians to govern themselves" and reflected Congress's low opinion of Indian law and its belief that Western law would help civilize Native Americans.

Today there are more than 500 Native American tribes in the United States, and it is fair to say that no one type of Native American law exists. Just as the different tribes have their own histories and cultures, so do they each have their own system of law to deal with the minor criminal offenses and civil cases that the U.S. government has left under their jurisdiction. That said, their various understandings and application of law do share certain characteristics that justifies their grouping into the traditional family of law: they base their law on unwritten rules and customs, and, when disputes arise, they tend to favor compromise solutions and reconciliation and restitution (as illustrated by the Crow Dog case) over win–lose outcomes and harsh punishment (Nielsen and Zion 2005; Richland and Deer 2004). As such, Native American law is a type of law that transcends the individual differences among the hundreds of Native American tribal communities. Yet, thanks to decades of forced assimilation, law in these communities resembles white man's law much more than it did two centuries ago.

An interesting development in tribal law surfaced a few years ago, when some Native American communities revived an old penalty, banishment, in an effort to deal with drug and alcohol abuse, gambling, and related violence. Within the Native American context, banishment is often regarded as much worse than imprisonment. Tribal members who are banished may lose health and education benefits, burial rights, and other entitlements. Banishment can also have a strong spiritual impact. As one member of the Lummi reservation in Washington State said, "Spiritually, it's going to take your insides and turn them inside out" (Kershaw and Davey 2004:Section 1, p. 1). Even so, banishment was being seen as necessary to deal with the growing problem of drug abuse, gambling, and violence. One tribal official said, "We need to go back to our old ways. We had to say enough is enough." An Iroquois journalist added, "It's out of desperation. They could either reinforce the ancestral discipline, or they go the American route, which has proven to be a failure" (Kershaw and Davey 2004:Section 1, p. 1). Ironically, it is precisely because the Native American sense of community is so strong that banishment is seen as such a serious punishment.

International Law

International law refers to the body of law that governs the relationships among the nations of the world and among organizations with international dealings (Janis 2003; Joyner 2005). It consists of treaties and agreements signed by various nations, acts passed by the United Nations and other international bodies, and disputes between parties from different nations. Specific matters covered by international law include international commerce, the conduct of war and the treatment of war prisoners, and maritime issues. Unlike the law found in individual nations, international law has no enforcement power in terms of police. Economic sanctions may be used, but the primary force behind international law is diplomacy, moral persuasion, and a nation's concern for its international reputation. Nations must consent to be bound by international law for international law to be effective, and if an individual nation refuses to abide by a particular law, the international community may not be able to compel it to do so. In this sense, international law is something of a paradox: It is most effective when it is needed least, when a nation already feels it should obey international law, and least effective when it is needed most, when a nation wants to disobey international law.

Not surprisingly, the key body in international law today is the United Nations (UN), whose 1945 Charter highlighted the need for international law to resolve worldwide issues without resort to armed conflict. Toward this end, the UN Charter established the International Court of Justice (also called the World Court), which is centered in The Hague, Netherlands. The Court hears legal disputes submitted to it by members of the UN and also renders advisory opinions when asked to do so by UN agencies and other international organizations. It consists of fifteen judges, each one from a different nation, who serve a nine-year term and are appointed by the General Assembly and Security Council.

When the Court hears a dispute, the disputing parties submit legal briefs and other documents and then argue their cases in a public hearing. The Court's two official languages are English and French; all written documents are prepared in both languages, and anything said in a public hearing is translated into the other language. If the parties do not agree to a settlement before the oral hearing concludes, the Court will eventually issue a judgment that is published in both English and French. The judgment is considered binding on the disputing parties and may not be appealed (International Court of Justice 2005), but a nation is still free to ignore the Court's decision with relative impunity. In fact, several major nations, including the United States, do not accept the Court's jurisdiction in principle and instead agree to its jurisdiction only on a case-by-case basis. Because the Court's legitimacy is enhanced when the disputing parties abide by its decisions, observers say it shies away from controversial cases and instead takes on cases in which the stakes are not that great and in which the disputing parties can be counted on to accept the Court's judgment.

The International Court of Justice hears the equivalent of civil disputes, but until recently no international body existed to prosecute international crime, such as war crimes or other crimes against humanity. After World War II, the victorious allies established the Nuremberg and Tokyo tribunals to prosecute German and Japanese political and military personnel accused of these crimes, but this development did not lead to a permanent structure and process for the prosecution of similar crimes that have occurred since, including genocide and other war crimes committed in Rwanda, Croatia, and elsewhere in the early 1990s. These and other events spurred the UN to intensify its efforts to establish a criminal court equivalent of the International Court of Justice, and a major conference took place in Rome in July 1998. With 160 nations participating, 120 of these nations endorsed the Rome Statute that outlined an International Criminal Court. After several more years of negotiations on the structure and function of this court, it was established in July 2002 and, like the International Court of Justice, is located in The Hague, Netherlands. The Court hears cases pertaining to genocide, war crimes, and other crimes against humanity committed by members of nations that have ratified the Rome Statute.

Although the International Criminal Court was an important development in international law, two major problems threaten to limit its effectiveness (Rothe and Mullins 2006). First, as is true of international law generally, the Court has no effective way of enforcing its verdicts and other decisions and of implementing any punishment it may mandate. For example, when individuals are indicted by the Court, it is up to the nation in which these individuals live to ensure that they appear before the Court. If the nation refuses, there is little the Court can do. Second, many nations have refused so far to ratify the Rome Statute and thus have not agreed to be bound by the Court's verdicts and other demands. One of these nations is the United States. Although President Bill Clinton signed the Rome Statute in December 2000, his successor, President George W. Bush, annulled the U.S. ratification in May 2002, with the Bush Administration saying the Court would weaken the fight against terrorism (Lewis 2002). The Administration later announced an end to military aid to nations that refused to pledge they would not extradite Americans to the Court (Brinkley 2006). The lack of ratification by the United States and several other major nations may well impair the Court's scope and effectiveness.

International law existed long before the United Nations, of course. It developed rapidly during the nineteenth century thanks to a series of international conferences. One of the first and most significant conferences was held in 1815 after the defeat of Napoleon's attempt to dominate Europe. This conference, the Congress of Vienna, addressed national boundaries, denounced slave trade, and discussed navigation rights on rivers that flowed across national borders.

The founder of the International Red Cross played a major role in a conference that met in Geneva, Switzerland, in 1864 and led to the signing of the first Geneva Convention that addressed the conduct of war. Jean Henri Dunant, a Swiss philanthropist, came up with the idea for the conference after an 1859 visit to a battlefield of the Austro-Sardinian War that had been the site of 40,000 casualties (Moorehead 1999). He spearheaded two historic international events: a meeting in 1863 that established the International Red Cross, and the 1864 conference, attended by twelve nations, that established the first Geneva Convention. This convention addressed the care of battlefield casualties, including the safety of medical personnel helping the casualties. Subsequent Geneva Conventions in 1906, 1929, 1949, and 1977 built on the first Convention to govern the conduct of war, the treatment of prisoners of war, and other war-related matters. Almost all nations are Geneva Convention signatories.

Although the Geneva Conventions, taken together, are considered one of the most important aspects of international law, nations have been accused of violating them over the years. During the Vietnam War, for example, critics accused the United States of violating Geneva Convention prohibitions against the killing of innocent civilians and the use of chemical and biological warfare (Neilands 1972), and North Vietnam was accused of violating Geneva Convention rules for the treatment of prisoners of war. Similar accusations emerged after 9/11, when the United States overthrew the Taliban government in Afghanistan and imprisoned several hundred suspected members of Al Qaeda and Taliban forces in the U.S. military base on Guantanamo Bay, Cuba. The United States announced that the prisoners were terrorists and not soldiers of any one nation; for this reason, the government added, they were not prisoners of war, and their treatment thus not subject to the Geneva Conventions (Bravin 2002). Civil liberties advocates in the United States and several European nations criticized this narrow interpretation of the Geneva Conventions and worried it could increase the possibility that future American prisoners of war would be treated harshly. The treatment of detainees at Abu Ghraib prison involved torture and other abuse, and these actions, too, were criticized as violations of the Geneva Conventions.

Summary

1. The many legal systems around the world are generally grouped into five families. The legal systems of the nations within each legal family share several important characteristics that justify calling them a family, even if these legal systems differ in other respects.
2. Common law systems are found in Great Britain and its former colonies, including the United States, Canada, Australia, Hong Kong, and India. The common law originated in Britain about 1,000 years ago and emphasizes case law, or law made by judges through their rulings. However, no nation is purely a common law nation, as no nation relies solely on case law to the total exclusion of statutory (legislative) law and other types of law. The reliance on case law makes *precedent*—the reliance on past judicial decisions in reaching a new decision—a hallmark of the common law tradition. Another hallmark of common law nations is the *adversary system*, in which attorneys for the opposing sides are said to rigorously contest the evidence in pretrial proceedings and in trials themselves. Reflecting its British colonial origins, the United States follows the common law tradition with the exception of Louisiana.
3. Civil law is the name given to the family of law founding the nations of continental Europe and in many of their former colonies in South America and elsewhere. Civil law nations rely primarily on written, detailed codes, or collections, of law that have their origins in ancient Rome and the 1804 Napoleonic Code. Judges in civil law nations have less power than their common law counterparts to change the law, in large part because civil law codes are so detailed that they are thought

to cover a wide variety of conceivable circumstances and situations. As traditionally conceived, the major role of civil law judges is merely to interpret the legal codes; the judges typically do not overturn laws, nor do they "make" law as common law judges sometimes do.

4. Theocratic law is religious law, a system of law heavily dependent on religious belief. Islamic law, or the Shariah, is the major type of theocratic law in the world today. Its principles underlying all these rules derive from the sacred book of Islam, the Koran. Although harsh practices have been carried out in the name of Islamic law, many scholars of Islam in fact condemn these practices as violations of the Shariah. Jewish law, or the halakhah, plays an important official role in Israeli public and private life, and consists of 613 commandments, or mitzvot, included in the first five books of the Bible. Written about 1,500 years ago, the Talmud consists of rabbinical commentaries that interpret and explain Jewish law. In Israel today, rabbinic courts handle cases involving the following issues or matters: (1) dietary laws; (2) the Sabbath; (3) Jewish marriage, divorce, and burial; and (4) conversion to Judaism.

5. Socialist law refers to the legal systems of Communist nations, primarily those comprising the former Union of Soviet Socialist Republics (USSR), the People's Republic of China, North Korea, Vietnam, and Cuba. With the USSR's demise almost two decades ago and China's gradual move toward capitalism since the early 1990s, socialist law is more a family of law of the past than it is one of the contemporary era. Historically, Communist nations used their legal systems as convenient vehicles for the repression and elimination of their political opponents. China's government today wields a brutal criminal justice system to repress its opponents and to control crime. Cuba's legal system has received considerable attention since Fidel Castro took power and early on established popular tribunals nation established popular tribunals to handle many kinds of disputes.

6. Traditional law refers to law that relies primarily on unwritten rules and customs. This family of law is found in the traditional societies most commonly studied by anthropologists. All these characteristics combine to produce a family of law that, compared to that found in most large, industrial societies, tends to be very group-oriented; to lack attorneys, formally trained judges, and the other trappings of modern legal systems; and to favor compromise solutions over win–lose outcomes in the settlement of disputes.

7. U.S. criminal law distinguishes felonies, misdemeanors, and violations or infractions. For an act to be proven a crime, several elements must exist, one of the most important of which is criminal intent. In practice, this means that defendants must have known and understood what they were doing and that they performed the act voluntarily. The major legal defenses to criminal responsibility all center on the issue of criminal intent. These defenses include: (1) accident, mistake, or ignorance; (2) duress; (3) self-defense; (4) entrapment; and (5) insanity or other diminished capacity.

8. The civil law governs private transactions among individuals and organizations, including marriage and divorce, contracts, wills, and real estate transactions. A violation of a civil law is called a tort. The civil law differs from the criminal law in several respects, for example, when a tort occurs, it is up to the aggrieved party rather than the state to seek legal redress.
9. Executive orders refer to laws passed and actions undertaken by the executive branch of the government. Several types of executive orders exist: (1) proclamations, (2) the establishment of task forces, (3) orders to create or reorganize executive branch departments and agencies, and (4) orders that change governmental policy or undertake governmental action to address an issue or problem.
10. Administrative law refers to the rules and regulations of administrative and regulatory agencies. Regulatory agencies perform both an enforcement and judicial function. Administrative is far more extensive and complex now than it was during a century ago, thanks to industrialization and the development of the administrative-welfare state.
11. Other distinctions in types of law include statutory vs. case law, public law vs. private law, and substantive law vs. procedural law. A related distinction is substantive justice vs. procedural justice.
12. The U.S. military has its own legal system as outlined in the Uniform Code of Military Justice. Critics say that military law can be arbitrary and unfair because of the power given to military commanders and because military defendants lack some of the rights and protections their civilian counterparts enjoy. Critics also say that the military sometimes hides or minimizes the importance of crimes and other transgressions committed by military personnel.
13. Although more than 500 Native American tribes exist, they share an understanding of law that helps distinguish Native American law from the law practiced by the federal and state governments. In general, Native American law favors compromise solutions and reconciliation and restitution. A traditional penalty, banishment, has been revived in recent years as Native American communities have tried to deal with growing problems of drug and alcohol abuse, gambling, and violence.
14. International law governs the relationships among the nations of the world and among organizations with international dealings. It consists of treaties and agreements signed by various nations and acts passed by the United Nations and other international bodies. Specific matters covered by international law include international commerce, the conduct of war and the treatment of war prisoners, and maritime issues. Unlike the law found in individual nations, international law has no enforcement power in terms of police. Economic sanctions may be used, but the primary force behind international law is diplomacy, moral

persuasion, and a nation's concern for its international reputation. The International Court of Justice of the UN handles many kinds of international disputes but lacks effective enforcement powers. One of the most important developments in international law has been the Geneva Conventions, which govern the conduct of war, the treatment of prisoners of war, and other wartime matters.

Key Terms

Administrative law
Case law
Civil law
Code law
Common law
Criminal law
Executive orders
Families of law
International law
Judicial review
Private law
Procedural justice
Procedural law
Public law
Socialist law
Statutory law
Substantive justice
Substantive law
Theocratic law
Tort
Traditional law

CHAPTER 4

Law and Dispute Processing

Chapter Outline

Have you ever had a quarrel with your parents? Did your parents ever have a problem with a neighbor, or perhaps with a plumber or auto mechanic who did shoddy work? Have you ever had a difference of opinion with a friend, partner (boyfriend or girlfriend), roommate, or anyone else you know? Have you ever had an argument in a bar or at a party? Have you had a problem with a landlord? When any of these events occurred, what did you do? Did you just disagree and let it go at that? Did you try to work it out? Did you seek the help of a third party? Did anyone go to court?

Many ways of addressing disputes exist, and the use of the legal system is one of these ways. As noted in Chapter 1, an important function of law is, ideally, to help settle the many problems that may occur in our society or those elsewhere. Some problems may be too trivial for an attempt at a legal solution to make sense, but others may be sufficiently serious for one or both parties to a dispute to look to the law for relief. Such use of the law—adjudication—may indeed help settle a dispute, but it may also ironically aggravate the problem rather than relieving it. This chapter examines adjudication and other possible ways of dealing with disputes, discusses the factors that prompt societies and individuals to favor one method of dispute settlement over another, and then focuses on law and dispute processing in the United States.

THE DISPUTING PROCESS

Every society has disputes, and everyone in every society becomes involved in many disputes, some very minor and others much more serious, over their lifetimes. Many disputes involve two or more individuals: the students in the dorm room next to yours may play their music too loudly late at night; a married couple may fight a lot and begin thinking about divorce; siblings may contest a will after their remaining parent dies; in one of the many premodern societies studied by anthropologists, someone may steal sheep, crops, or other important goods. Other disputes involve various kinds of groups and organizations: a citizen group may challenge plans to build a "big box" store like Wal-Mart or seek to end mercury pollution of a nearby river; a department store may take action against someone with a long-standing unpaid bill; one corporation may accuse another corporation with copyright infringement.

As these examples indicate, disputes occur in many ways and for many reasons. They are found in even the most harmonious societies and the best of relationships. Whether we prefer the term "conflict," "contention," "dispute," "disagreement," "friction," "quarrel," "squabble," or any other synonym, disputes arising from different views, perceptions, and conclusions are universal. When they do arise, they must be addressed, even if this means just ignoring or glossing over the problem. Just as every society has disputes, so does every society have one or more customary ways of dealing with disputes. Just as individuals and groups become involved in many disputes over time, so must they decide what to do when a dispute arises. This process of dealing with disputes goes by many names in the literature on this subject, including "handling," "management," "processing," "resolution," and "settlement." Whatever we call it, this process is an essential part of any society.

As Chapter 2 discussed, many anthropologists have stressed that law, or at least the equivalent of law, may be found in the ways that premodern societies process disputes. Even if these societies are too premodern to have courts, judges, and other legal trappings familiar to the Western mind, they nonetheless have regular procedures for handling disputes, and these procedures help them maintain social order just as law ideally helps maintain social order in modern Western societies.

This interest in disputes has led legal anthropologists to focus much of their attention on the disputing process in premodern societies. In many ways, this process looks very different from that found in modern societies, but in other ways, as we shall see, it may seem familiar. Perhaps most important for our purposes, an understanding of the disputing process in premodern societies provides an illuminating contrast to the use of law to settle disputes in modern societies. As people in the United States and elsewhere seem to becoming increasingly frustrated with the cost, slow pace, and other problems of using the law to address disputes, perhaps we have much to learn from the societies that anthropologists have studied. Accordingly, the first part of this chapter examines what anthropologists have said about the disputing process in tribal societies. One goal here will be to see what lessons might be learned for appreciating the advantages and disadvantages of using the law when disputes arise in the United States and other modern societies.

Another goal will be to appreciate how a society's disputing process reflects the culture and social structure of the society itself and, in so doing, to reinforce this book's theme that law is a social phenomenon. As historian Jerold Auerbach (1983:3–4) has written,

> The varieties of dispute settlement, and the socially sanctioned choices in any culture, communicate the ideals people cherish, their perceptions of themselves, and the quality of their relationships with others. They indicate whether people wish to avoid or encourage conflict, suppress it, or resolve it amicably. Ultimately the most basic values of society are revealed in its dispute-settlement procedures.

By understanding the disputing process, then, we better understand the society in which the disputing process occurs.

Stages in the Disputing Process

Following common practice, this book uses the term "dispute" to refer to a problem or disagreement between two (or more) parties that is handled by one or more of the methods of dispute processing discussed in the next section. However, some scholars have developed typologies that describe stages in the disputing process and in particular reserve the term "dispute" for a problem that escalates beyond an initial disagreement. Two popular typologies are worth describing here.

A typology by anthropologists Laura Nader and Harry F. Todd, Jr. (1978a) is the shorter (though no less influential) of the two and outlines three stages of the disputing process. The first stage involves a situation or disagreement that one party regards as unjust or unfair. Nader and Todd call this the *grievance* stage. The grievance may be real or only imagined, they note, but add that the "important thing" is that the party feels "wronged or injured" (p. 14). Once a grievance arises, it may fade away or stay under the surface if the aggrieved party does nothing about it, or it will escalate if the aggrieved party chooses to confront the offending party by indicating "resentment or feeling of injustice" (p. 15). If so, the grievance has now entered the *conflict* stage. If the two parties cannot resolve their conflict by themselves (through the methods of coercion or negotiation; see below), the disagreement then enters the *dispute* stage by becoming public with the involvement of one or more third parties. Thus, a grievance becomes a dispute if and only if a third party knows about the grievance and becomes involved in trying to resolve the grievance. Nader and Todd note that some grievances skip the conflict stage by proceeding directly to the dispute stage (as when a person files a lawsuit without ever telling the offending party about a perceived grievance), and they add that the process may reverse with a dispute deescalating to a grievance if the aggrieved party decides to give up. A dispute may also disappear altogether if the offending party makes amends to the satisfaction of the aggrieved party.

The other popular typology comes from sociolegal scholars William L. F. Felstiner, Richard L. Abel, and Austin Sarat (1980–1981). They begin with the concept of an *injurious experience*. This is a situation in which people are being hurt (in terms of their health, finances, or some other way) whether they realize it or not. Initially they might not realize they are being hurt, and this is the first stage, or the *unperceived injurious experience* stage. As an example, the authors cite higher cancer rates among people living downwind from a nuclear test site. These people are certainly being harmed long before their cancers emerge, yet this harm remains unperceived. When they do learn of their cancer, they now enter the *perceived injurious experience* stage by *naming* the problem. However, not all these cancer victims will necessarily connect their cancer to the nuclear test site. Those who do make this connection now enter the *grievance* stage by *blaming* the problem on the nuclear testing. More generally, the grievance stage occurs when a person decides that another party (individual or organization) is responsible for the injury the person has suffered. The person must feel both that a wrong has been committed and that the wrong should be remedied. If and when the aggrieved party communicates the grievance to the offending party, the problem then enters the *dispute* stage through a process of *claiming*.

Despite the differences between them (for example, the dispute stage in the latter typology is equivalent to the conflict stage in the first typology), the two typologies both remind us that an aggrieved party must make three key decisions: (1) whether a situation is unjust, (2) whether to confront the offending party with the problem, and (3) whether to take the problem to a third party if the two disputing parties cannot resolve the problem. As we will see in the next section, many aggrieved parties decide not to confront the offending party at all, or, if they do, to proceed no further if a resolution cannot be reached.

Methods of Dispute Processing

Although there are innumerable kinds of disputes, there is a smaller, finite number of procedures for dealing with disputes once they arise. "Sue the bastard" is a term often heard in the United States, but litigation is in fact just one of seven ways of handling a dispute. As we shall see, the method a society tends to favor depends on several factors, including the society's legal culture and structure of relationships. Each method has certain advantages and disadvantages, and whether a society favors one method or another depends on its assessment of whether the method's advantages outweigh the disadvantages or vice versa. Within a given society, the method an individual favors also depends on several factors and the individual's assessment of the method's advantages and disadvantages. With this brief foundation in mind, we now turn to the methods of dispute settlement as identified and discussed by anthropologists and other scholars (Felstiner 1974; Gulliver 1979; Nader and Todd 1978a). In order of presentation, they are (1) lumping it, (2) avoidance, (3) coercion, (4) negotiation, (5) mediation, (6) arbitration, and (7) adjudication.

Lumping It. One way of dealing with a dispute is simply to ignore it by taking your lumps, or **lumping it**. If there was a one-time matter that led to your grievance, you just let it go; if there is a continuing problem, you just decide to live with it the best you can. If the students in the dorm room next to yours are too noisy one night, you might decide to lump it by putting on your headphones or using ear plugs. If your partner ticks you off one day with an insensitive comment, you might decide to let it go rather than respond in kind. If you just spent $85 at the grocery store and find when you get home that they forgot to pack the onion you paid for, you might decide to just forget about the onion rather than taking the time and spending the gas to drive back to the store to claim your onion. No doubt you can come up with many examples in your own life where you decided to "lump it" when some problem occurred and you felt you had a legitimate grievance.

Several factors affect whether or not we decide to lump it when a dispute occurs. One factor, which might be evident from the example of the onion, is that problem is simply not important or serious enough to address. If the noise in the dorm room next door is a one-time event as the example above suggested, you might lump it, but if it becomes a continuing problem, then your grievance has become more serious and you will probably decide that you can no longer ignore it. If the grocery store clerk forgot to pack your onion, that cost you very little money and, after all, it is only an onion. But what if the item the clerk forgot to pack was an expensive bottle of champagne that you had intended to use at a romantic dinner that very night? Between the expense of the champagne and your urgent need for it, lumping it will probably make no sense, and, assuming you have the time, you may well drive back to the grocery store to reclaim your bubbly.

Another factor that affects whether we decide to lump it when a problem occurs is the extent of the power difference between the two disputants. If one of the two disputants has much less power than the other and thus may suffer various consequences if he or she does not lump it, then lumping is more likely to occur. Say, for example, that you are spending your summer serving food at an upscale restaurant. You are making very good tips, but the restaurant manager is obnoxious and condescending without doing anything, like sexual harassment, that is illegal. You realize that if you complain to the manager or to the restaurant owner about the manager's conduct, you risk losing your job. Under these circumstances, you may well decide to lump it by choosing to live with the manager's boorish behavior. Differences in physical prowess may also matter: a woman who is being beaten by her husband or boyfriend may decide not to seek help because she fears being beaten yet again if she tries to do anything.

Avoidance. A second way of dealing with a problem is to practice **avoidance** by ending the relationship that produced the problem or by physically removing yourself from the situation or location in which the problem is located. Whereas lumping it allows a dispute to continue by basically ignoring the dispute, avoidance deals with a dispute by ending it, or, more precisely, by ending the relationship or situation producing the dispute. In the noisy dormitory room example discussed just above, avoidance would involve your moving to another room on the same floor, on a different floor, in a different dormitory, or off campus. In the grocery store example, avoidance would involve never shopping again at the store where the clerk forgot to pack your onion or champagne. In the restaurant example, avoidance would, of course, involve quitting your job. Separation and divorce are avoidance to deal with a bad marriage, and "breaking up" is the analogous avoidance for a nonmarital romantic relationship that is no longer working.

Several factors similar to those affecting lumping-it decisions again affect whether we decide to practice avoidance when a dispute occurs. If the problem is not very serious, we are not likely to undertake avoidance. In the dorm room example, you probably would not consider moving away unless the noise continued for many nights and no other method of dispute processing (see below) offered any hope. In the grocery store example, an unpacked onion is less likely than an unpacked bottle of champagne to prompt you to decide never to shop at the store again. One minor tiff is not apt to end a marriage or romantic relationship, but a more serious or continuing problem, such as a spouse having an affair, may well do so.

Another factor affecting avoidance decisions is the need for a relationship or situation to continue. In the restaurant example where you are earning a lot of money but labor under a boorish manager, you are less likely to practice avoidance by quitting your job than if you had the same problem at a fast-food restaurant where you were making minimum wage. If you are deeply in love, you are less apt to end a romantic relationship once some problems arise than if your feelings are less intense. If the grocery store that caused your problem is the nearest store or offers consistently lower prices than other nearby stores, you are less likely to refuse to shop there again than if other stores were reasonable options. For several reasons, then, avoidance is often not a practical option.

Note that some societies deal with a troublesome individual by forcibly removing the individual from the society. The extreme form of such avoidance is, of course, execution, a common practice in China (see Chapter 3) and one that is found only in the United States in the Western world. Many societies have practiced forms of avoidance by banishing or shunning a perceived offender. As the previous chapter discussed, a few Native American communities began banishing drug abusers and other offenders earlier this decade in an effort to deal with rising problems of drug use and crime. The classic literary example of shunning is Nathaniel Hawthorne's *The Scarlet Letter*, in which young Hester Prynne in seventeenth-century Puritan society had to wear the scarlet letter 'A' and was treated like a pariah for having committed adultery.

Coercion. Lumping it and avoidance are common ways of dealing with various disputes, but neither method attempts to resolve the dispute to the satisfaction of the aggrieved party. In this sense, the other party "wins" the dispute because the aggrieved party either lives with the dispute by lumping it or leaves the relationship or situation spawning the dispute. Although disputants may choose lumping it or avoidance for the reasons we have seen, many quite naturally do not desire either outcome and will prefer to actually address the dispute with one or more of the remaining methods of dispute processing.

For better or worse, a common such method is **coercion**. As its name implies, coercion (also called *self-help*) involves the use of threats or pressure to compel a change in someone's behavior or thinking. The use or threat of physical force (*interpersonal violence*) comes most readily to mind as an example of coercion, but other kinds of coercion exist and include ridicule, divulging a secret (as through gossip or "ratting out" somebody), and blackmail. Coercion may be an attempt to deal with a dispute, but by trying to force an outcome it does not address the underlying reasons for a dispute and may well aggravate the problem rather than alleviating it (Nader and Todd 1978a).

The likelihood of engaging in coercion depends on several factors; some relate to the circumstances of a dispute and others relate to characteristics of the disputants. One circumstantial factor is the seriousness of the grievance underlying the dispute. Just as a less serious grievance may lead a disputant to lump it, a more serious grievance may prompt an attempt at coercion, as it is more likely to produce anger, annoyance, and other emotions that underlie coercion. To return to our noisy dorm room example, one night of noise might be easy to ignore, but repeated nights are more difficult to ignore. As we saw, avoidance is a more likely outcome than lumping it under this circumstance, but so is coercion. You might turn up the volume on your own music and have a bout of "speaker wars," but you might also threaten the offending residents next door with physical harm to their persons, their music equipment, or some other possession. It is difficult to imagine someone resorting to violence or other coercion just because a grocery store forgot to pack your champagne, but in theory, at least, such a response is more likely when an expensive bottle of champagne is at issue than a mere onion. It is also difficult to imagine a restaurant server with an obnoxious manager resorting to coercion, but it at least stands to reason that coercion is more likely if the manger's behavior is so obnoxious that your temper might flare.

Because interpersonal violence is a key topic in the field of criminology, sociologists, psychologists, and other scholars have tried to understand why such violence occurs. The literature on this subject is voluminous, and details are beyond the scope of this discussion. Generally, sociologists cite the importance of social background factors such as geographical location, gender, social class, age, and race for the use of interpersonal violence (Eller 2006; Holmes and Holmes 2004; Zahn, Brownstein, and Jackson 2004), while psychologists cite authoritarian personalities and other personality traits that make some individuals more likely than others to be aggressive (Carrasco et al. 2006; Ehrensaft, Cohen, and Johnson 2006). These personality problems begin because of inadequate parenting and perhaps genetic and other biological factors and can persist into adolescence and beyond. In some individuals, aggressive tendencies are thought to be so extreme that they suffer from *intermittent explosive disorder*, which causes them to lash out in an uncontrollable rage. A recent study estimated that 7.3 percent of American adults have experienced this disorder at least once in their lifetimes (Kessler et al. 2006).

Negotiation. Like coercion, **negotiation** is another method of trying to resolve disputes. Unlike coercion, negotiation does not seek to force an outcome. Instead, it involves discussions between the two disputing parties in which they try to persuade one another of their way of thinking about the grievance underlying the dispute. Sometimes, one disputant will persuade the other disputant to completely change the behavior or situation that led to the grievance, but often negotiation succeeds in resolving a dispute because the disputants are able to forge a compromise outcome in which both parties win a little and lose a little but can each be sufficiently happy with the result. Although we discussed lumping it, avoidance, and coercion before turning to negotiation, negotiation may well be the first method of dispute processing that a disputant attempts. If it fails to resolve the dispute, then one or more of the other methods may be the next recourse.

Negotiation is easier to contemplate and more likely to succeed in some situations than in others. In the noisy dormitory room example, many students bothered by the loud music would probably simply ask the music players to turn down the volume. The latter may do so immediately or ask whether it would be okay to play the music for another ten or fifteen minutes until a set of songs ends, and the aggrieved students may then say that would be okay. In this ideal situation, the disputing parties have quickly negotiated an outcome that apparently satisfies both parties. Obviously, the loud music players may resist any such solution; if so, the aggrieved student(s) would then have to decide which method of dispute processing would next make sense. In the grocery store champagne example, negotiation would involve simply returning to the store, receipt in hand, and asking for your bottle of champagne. A problem might arise, however, if the grocery store does not believe that you failed to receive your champagne or if you no longer have your receipt. In either case, the negotiation is not so simple, and you may have to use your best persuasive powers to finally obtain your purchase. In the obnoxious restaurant manager example, negotiation would involve your talking directly with the manager and asking for a change of behavior. As noted earlier, many servers in this situation will be very reluctant to undertake this step, either because they dislike confrontations of this type or because they fear being fired. They may also feel that the obnoxious manager is unlikely to change her or his personality and thus that any such negotiation will prove futile.

As these examples suggest, at least three factors affect whether we are willing to try to negotiate a settlement of a dispute. First, some situations have certain circumstances that lend themselves more or less readily to negotiation. Second, and related, power differences between the disputants may matter: a disputant with less power may fear retaliation by the disputant with more power and thus be less willing to negotiate (and more likely to lump it). Third, some people are bolder or have other personality traits that make them more willing to try to negotiate (which, even if it goes well, does involve an element of confrontation).

Mediation. **Mediation** is the first of the methods discussed so far to involve a *third party*, a significant feature to scholars of dispute processing. Both disputants must agree to mediation in advance, and the mediator tries to help the disputants reach some resolution, usually through a compromise that both parties will regard as a win–win outcome. However, either party is free to decline any solution or change in thinking or behaving a mediator might suggest, and mediation often does not succeed in resolving a dispute. Mediation is a common and even predominant practice in many premodern societies but is also found in modern ones. The mediator may be someone that both parties know and trust, a stranger who is considered particularly wise or perceptive and therefore likely to be a good mediator, or someone in a position of authority. Regardless of the mediator's relationship, or lack of same, to the disputants and any good qualities the mediator may have, it is essential that the disputants consider the mediator impartial. The mediator's typical search for a compromise solution in turn reinforces the parties' belief in the mediator's impartiality.

In an influential essay, Torstein Eckhoff (1969), a Norwegian law professor, wrote that a mediator may use several strategies to help disputants reach an agreement. One strategy is to have the parties recognize their common interests and to deemphasize their competing interests. Another strategy is to come up with possible solutions that the disputants may not have previously considered. Because the mediator is considered impartial, these possible solutions may carry some weight with the disputants. A third strategy, which is possible only in certain circumstances, is to promise to help the parties in the future if they can resolve their dispute or threaten to take the side of one of the disputants if the other does not agree to a resolution. (Here one might think of the many movies or TV shows about organized crime in which a "boss" calls together the heads of two warring factions and simultaneously encourages them to set aside their differences while threatening retaliation to any faction that does not do so.) However, this strategy is risky, Eckhoff added, because a party who is pressed to agree to the mediator's proposed solution may begin to regard the mediator as biased toward the other party. If the mediator's impartiality is called into question for this reason, her or his effectiveness in helping the parties achieve a compromise solution is weakened.

We mentioned earlier that scholars regard the presence of a third party to be a significant dimension. That is because the involvement of a third party takes the dispute beyond the two disputing parties and, therefore, brings it to the attention of people other than the disputants themselves. In this way, the dispute becomes more public than it was before. The need to enlist the help of a third party also signifies that the disputing parties are incapable of resolving the dispute by themselves, and they may not wish other parties to hold this perception. For these reasons, some disputants may prefer not to use mediation despite its possible help in resolving their dispute.

Like negotiation, mediation makes more sense and is more possible under some circumstances than under others. A married couple that is having problems may turn to a friend, relative, member of the clergy, or marriage counselor for help. The noisy dorm room example is another one in which mediation is possible. If the two parties, the student playing the loud music and the student finding it too loud, fail to negotiate a mutually satisfactory outcome, or if the aggrieved party simply does not want to bother with negotiation, they may turn for assistance to a dormitory officer, often called a resident assistant, who will then try to help the parties work things out. Mediation by an individual is less conceivable in the grocery example. If the store does not provide the customer with a bottle of champagne, it is unlikely that any individual could be called in to mediate. Instead, the customer may contact the local Better Business Bureau or Chamber of Commerce, with either organization then serving as a mediator. In the restaurant manager example, a likely mediator would be the restaurant owner if (and this might be a "big if") the server were convinced of the owner's impartiality.

Arbitration. **Arbitration** is a common method of dispute processing in many premodern societies but is also present in their modern counterparts. In the United States, arbitration is a common method for addressing disputes between labor and management in business, industry, and academia

(O'Meara 2002). As a method of dispute processing, arbitration is similar to mediation in two ways. First, it involves a third party with whom the disputants consent to meet in an attempt to resolve their dispute. Second, this third party, an arbitrator, is expected to be impartial and also to consider all the facts and other circumstances of the dispute. However, arbitration differs from mediation in three other respects. First, an arbitrator devises her or his own resolution after considering the disputants' claims, whereas a mediator works with the disputants to come up with a resolution. Second, the disputants agree beforehand to abide by the arbitrator's decision (hence the popular term "binding arbitration"), whereas either disputant in mediation is free to refuse a solution a mediator may propose. Third, the arbitrator's resolution is often a win–lose outcome, whereas the resolution reached via mediation is, as noted earlier, typically a win–win compromise solution. Thus, an arbitrator decides that one party is right and the other party is wrong, with the former winning the dispute and the latter obviously losing it.

Because arbitration involves a third party, it has the same implications for a dispute as discussed earlier for mediation. In particular, the dispute becomes more public than it was before, with the resulting perception that the disputing parties are incapable of resolving the dispute by themselves. As with mediation, these consequences may lead some parties to reject arbitration. An additional consequence may also lead to rejection of arbitration: the win–lose outcome that arbitration entails. Because the disputing parties realize that one of them will lose if their dispute becomes arbitrated, one or both may wish to avoid arbitration and instead try to work out their differences via negotiation or mediation or force an outcome with coercion.

Adjudication. The final method of dispute processing is **adjudication**, in which the third party, the judge, "has the authority to intervene in a dispute whether or not the principals wish it" (Nader and Todd 1978b: 11). This is the key feature that distinguishes adjudication from arbitration. In other respects, adjudication is similar to arbitration in that the third party is expected to be impartial and, more important, imposes a win–lose outcome on the disputants. But whereas one or both disputants may decline arbitration (or mediation) for the reasons discussed earlier, neither is free to decline adjudication once the adjudicative process has begun. Thus, if you are involved in a dispute with someone, and your antagonist brings a lawsuit against you, you are not free to decline to be sued. Instead, you have to respond with appropriate legal actions. A judge may eventually dismiss the suit, but only after you (or, assuming you have legal representation, your attorney) file the necessary papers and makes the necessary arguments to convince the judge that the suit should be dismissed.

Not every dispute qualifies for adjudication, and certain requirements must be satisfied for a case to be adjudicated. First, a case can be heard by a court only if the court has *jurisdiction* over the case because the case both occurs in a geographical location and involves a subject matter that a particular court is authorized by law to hear. For example, a dispute must involve a matter of federal law for a federal court to hear it; otherwise it must be heard by a state or local court. Second, the case must involve an issue or matter that is justiciable. *Justiciability* is a complex concept but basically means that the case involves an issue that is appropriate for the courts to hear. To take a silly example, if a professional baseball player sued an umpire for calling him out on a close play at home plate, a judge would almost certainly dismiss the suit on the grounds that it did not involve a justiciable matter. To be justiciable, a case must involve an actual dispute between two parties rather than a hypothetical dispute and a dispute that is "ripe" rather than "moot." In effect, this latter element means that a controversy must have actually developed and still be ongoing for it to be justiciable. If you sue a neighbor for building a fence on your lawn but then move away while the suit languishes in the courts, a judge may decide to dismiss your suit because the issue is no longer ripe.

The third requirement for adjudication is *standing* (sometimes called *legal standing*). For a party to bring a dispute to a court, that party must have standing to do so. This means that the party must be actually involved in the dispute or have a right or interest directly affected by the dispute. The requirement of standing has played a key role in environmental law, because endangered species, for example, obviously cannot sue on their own behalf. It might seem to make sense that environmental groups should be allowed to sue on the behalf of endangered species, but the standing requirement has sometimes made it difficult for the groups to do so (Echeverria 2003).

Adjudication is an important feature of modern societies, but it is also found in premodern societies that obviously lack courthouses and other legal trappings but still have procedures in which one or more people serve as judges to hear disputes and impose a decision. However, it is in modern societies and especially in the United States that adjudication has aroused controversy because of its perceived costs in time and money. We return to this issue below in our focus on the United States, but consider, for example, a situation in which you are sued. In our complex legal system, you almost certainly will feel it necessary to hire an attorney, but legal representation can obviously be quite expensive. This fact may prevent you from hiring an attorney and force you to lose the suit, or may at least limit the fees you can afford to pay the attorney and thus the extent and quality of the legal help you get. For these reasons, adjudication can favor the party with greater financial resources, even though adjudication is often thought to level the playing field, as the familiar blindfolded statue of Lady Justice signifies (see http://www.statue.com/lady-justice-statues.html), by offering an impartial forum for dispute processing. As you probably realize, adjudication can also be very time consuming, and victory may well go to the party—again often the one with greater financial resources—who can simply afford to outwait the other party.

Adjudication has at least one other cost that is also present in arbitration and that is captured in Eckhoff's (1969:175) observation that the judge's "task is not to try to reconcile the parties but to reach a decision about which of them is right." Thus, adjudication results in a win–lose outcome in contrast to the win–win outcome ideally achieved by mediation. As a result, adjudication does not really address the causes of the dispute and may well worsen the relationship between the parties. In particular, say anthropologists, the losing party in adjudication is more likely than parties that undergo negotiation or mediation to dislike the outcome and to bear a grudge against the other disputant. Thus, whereas negotiation and mediation can help reconcile the disputants, adjudication may further estrange them and may worsen the situation that led to the adjudication. Recall from the previous chapter that premodern societies often favor reconciliation over punishment. When they process disputes, many premodern societies thus tend to favor negotiation and mediation over adjudication because the former methods are more likely than the latter to produce the reconciliation these societies favor.

The expense, time, and estrangement that adjudication often involves has led various groups in the United States to promote alternative dispute resolution (ADR), which attempts to resolve disputes without resort to adjudication. We discuss ADR below, but note here that it tries to achieve in American society what premodern societies ideally have achieved through their reliance on negotiation and mediation.

EXPLAINING DISPUTE PROCESSING DECISIONS

Sociolegal scholars have identified several sets of factors that help explain preferences for one or more of the methods of dispute processing just examined over other methods. In doing so, they try to answer two related questions: (1) Why do some societies favor mediation (or some other method) more than other societies? and (2) Why do some individuals favor one particular method more than other individuals? Societal factors help answer the first question, while individual factors help answer the second question. We now turn to these two kinds of factors.

Societal Factors

Societies differ in certain aspects of their structure and culture, and these differences help explain differences among societies in their preferred methods of dispute processing.

The Structure of Social Relationships. Anthropologists stress the importance of the nature of social relationships in a society for its dispute processing preferences (Gluckman 1955; Nader and Todd 1978a). As Chapter 3 discussed, people in the traditional societies studied by anthropologists tend to know each other very well and to value their close relationships. They are very group-oriented and thus tend to favor reconciliation and compromise outcomes in the settlement of disputes. This means that negotiation and mediation are more likely than adjudication to be the preferred methods of dispute processing.

As Nader and Todd (1978b:12) observe, "[T]he nature of the relationships in which litigants or disputants are involved will affect the manner in which they attempt to manage the problem." Following this logic, a major reason for the preference of traditional societies for negotiation and mediation rests on the nature of their social relationships. In these societies, one individual will know another individual in multiple ways: he or she will be the individual's relative by birth or marriage, a neighbor, a member of a hunting or gathering party, and so forth. Such societies are said to be characterized by *multiplex* relationships. Because any two individuals in such societies literally see one another everyday and interact in many roles, it is important that their relationship continue beyond any particular dispute that might develop between them. This emphasis on enduring relationships in turn promotes compromise outcomes and thus negotiation and mediation as dispute processing preferences. Nader and Todd (1978b:13) summarize this dynamic as follows: "Relationships that are multiplex and involve many interests demand certain kinds of settlement, such as compromise, which will allow the relations to continue."

In contrast, large societies such as the United States are filled with people who do not know each other at all or know each other in only a relatively superficial context: an individual is your cashier at the local supermarket, your plumber, your server at a restaurant, and so forth, but nothing else. These societies are said to be characterized by *simplex* relationships. If you have a dispute with these individuals, you probably do not care whether you continue to have any relationship with them afterward. Because these societies thus lack an emphasis on enduring relationships, they care less about compromise outcomes (and thus about negotiation and mediation) and are more likely to favor win–lose outcomes (and thus arbitration or adjudication).

Although multiplex relationships make negotiation and mediation more likely, they do not guarantee them for at least three reasons (Nader and Todd 1978b). First, the very intensity of multiplex relationships may give rise to especially intense disputes that cause the parties to lose any interest in continuing their multiple relationships and thus any interest in a negotiated or mediated outcome. Second, and on the other hand, a multiplex relationship may help keep an initial disagreement from rising to the level of a dispute that needs negotiation or mediation. Finally, when scarce resources (such as land or livestock in a premodern society) are at issue, disputants may decide that the importance of the scarce resource outweighs the importance of the continuing relationship and thus seek a win–lose outcome through coercion, adjudication, or arbitration rather than a comprise outcome through negotiation or mediation (Starr and Yngvesson 1975).

Power and Inequality. In many small, homogeneous societies, some individuals have more power and influence than others, but most members are fairly equal in this regard, and most members live in the same economic circumstances, that is, poverty or near-poverty, or perhaps even on the edge of starvation. Other small societies have less equality of this type, with some lineages, or family groupings, ranking higher than others in terms of wealth, prestige, and influence. Overall, though, small societies are relatively unstratified, to use a sociological term, compared to large, heterogeneous societies like the United States. In these societies, many individuals have much wealth, power, influence, and other advantages, while the mass of the citizenry enjoy fewer resources of this type. Larger societies, then, tend to have more wealth and power inequality than small societies.

How and why does the level of stratification affect a society's dispute processing preferences? As the earlier discussion of adjudication indicated, wealth and other resources can make a significant difference in the outcomes of adjudication. Individuals and groups with such resources can use the courts to their advantage and, more specifically, use the courts to the disadvantage of parties without wealth and power. Thus, even though the law may sometimes act to equalize power differences as described earlier, Nader and Todd (1978b:21) observe that "the weight of law as equalizer appears light in comparison to the power derived by the already powerful from routine actions of law." Recognizing this dynamic, the many more numerous powerful individuals and groups in large societies will be inclined to "manipulate legal means for their exclusive advantage" (Nader and Todd 1978b:21) and to use adjudication. In contrast, because power is more evenly distributed in small societies, there is overall less inclination to use adjudication because there are fewer individuals with significantly greater power. Adjudication is less common in small societies, then, not only because they have multiplex relationships but also because they have relatively little stratification.

Legal Culture. A final societal factor is **legal culture**, which refers to a society's general views about the law and specific dimensions of the legal system and its perceptions about using the law to address disputes (L. M. Friedman 1969; Nelken 2004). A key perception here is whether it is appropriate and even desirable according to a society's cultural values to use adjudication when a dispute arises, or whether it is wrong to do so. The idea that a culture may consider it wrong for moral or other reasons to use the courts may be incomprehensible to many Americans, accustomed as we are to lawsuits and other kinds of litigation. However, the view that litigation is indeed wrong for moral or other reasons characterizes many traditional societies, partly because of their emphasis, discussed above, on enduring relationships. This view also characterized many communities earlier in American history, a point to which we return later in this chapter. Here, we briefly discuss one traditional society, a Sunni Muslim village in Lebanon, and one larger society, Japan, to illustrate how their legal cultures often prompt them to reject adjudication as a method of dispute processing.

Anthropologist John E. Rothenberger (1978) conducted field work in Qarya, a remote Sunni Muslim village in the Northern Lebanon mountains, during 1966 and 1967. The village had about 1,100 residents and an agricultural economy. It had a mayor, the *mukhtaar*, but no other form of government; the nearest national courts were almost an hour away by motor vehicle and the nearest police also a distant drive. Rothenberger (1978:164) found that the villagers strongly preferred negotiation and mediation, either by a family member, the *mukhtaar*, or another respected villager, over adjudication by a national court, with 98 of the 105 disputes he studied processed in this fashion.

This strong preference derived from two beliefs. First, the villagers felt that their community would have little say in the resolution of a dispute if it ended up in a national court and that the Lebanese police and judges were too unpredictable. Second, they placed great emphasis on enduring relationships: "There is a clear recognition of the importance and necessity of ongoing relationships within the community. The villagers of Qarya have known each other all their lives and will continue to deal with each other for the rest of their lives. All but a very few of the villagers handle their affairs in a way that indicates recognition of this principle" (Rothenberger 1978:165). These two beliefs in Qarya's legal culture, then, help explain why this small society favored mediation and negotiation over adjudication.

Note that Qarya's emphasis on enduring relationships and distaste for adjudication derived from its small population size and resulting multiplex structure. This process raises the question of which factor, structure or culture, is more important in explaining Qarya's dislike of adjudication and, more generally, the similar distaste found in other traditional societies. Some scholars discount the importance of culture in this regard and say that the structure of relationships is the key factor (Kidder 1983), while other scholars highlight culture as the key factor (Bracey 2006). This is an important theoretical debate for which there is no clear answer, and many scholars take an eclectic view in saying that *both* structure and culture matter for a society's preference for mediation and negotiation versus that for adjudication.

Japan is certainly not a small society, but some scholars say it also has a legal culture that leads its residents to shun adjudication in favor of other methods of dispute processing. In a classic article, Takeyoshi Kawashima (1969) observed that Japan has relatively little litigation and a relatively low number of attorneys even though it is an industrial society. As an example, he cited two Japanese railroads that were involved in more than 500 traffic accidents in 1960. None of these accidents resulted in any litigation, and in only three of these accidents did the injured party hire an attorney. Kawashima argued that the time and cost of litigation could not explain Japan's low rate of litigation, since other industrial societies have more litigation despite its time and expense. Although he acknowledged that Japan's low monetary damages for traffic accident injuries may play a role in its lack of accident-related lawsuits, he wrote that a "more decisive factor" (p. 184) lies in Japan's legal culture: "Traditionally, the Japanese people prefer extrajudicial, informal means of settling a controversy." They do so because, as we have seen, litigation "admits the existence" (p. 185) of a dispute, highlights conflict between the disputing parties, potentially jeopardizes a relationship, and involves a right/wrong decision in which the disputants have no say.

These characteristics of litigation apply in other societies, but the Japanese culture, wrote Kawashima, prompts them to dislike adjudication so much that the Japanese tend to avoid the courts in situations where more Americans would be ready to litigate. Kawashima cited certain aspects of Japanese society and culture that underlie this distaste. First, the Japanese culture emphasizes respect for and deference to authority, and such respect and deference characterize most relationships; for example, a renter is expected to defer to a landlord and a seller is expected to defer to a buyer. At the same time, the higher-ranking party is expected to be gracious toward the lower-ranking party and to respond to any legitimate concerns. Because adjudication theoretically puts both disputing parties on the same footing and because it challenges the views or interests of the defendant party, litigation is incompatible with the norm of deference. Second, the Japanese culture emphasizes relationship *harmony*. As Kawashima (1969:186) put it, "There is a strong expectation that a dispute should not and will not arise; even when one does occur it is to be solved by mutual understanding." For this reason, when disputes do arise, many Japanese will be rather quick either to apologize for or to forgive a perceived wrong.

This twin emphasis on deference and harmony leads the Japanese to think litigation is "morally wrong, subversive, and rebellious," wrote Kawashima (1969:186). He added that the Japanese culture similarly affects contractual relationships, as a strong expectation exists that parties to a contract not become involved in any serious disputes, let alone a lawsuit. The Japanese are often even reluctant to insist on a contract before entering into a formal arrangement because doing so would imply a lack of trust and a contentious personality.

A vivid example of the Japanese aversion to litigation was seen in the aftermath of mercury poisoning by factories of some Japanese villages' waterways during the 1950s and 1960s. Although many villagers eventually became sick and deformed and some died from eating contaminated fish, many victims refused to sue the companies and instead either did nothing (lumped it), accepted mediation by the national government, or relied on direct negotiations with the companies that polluted their waterways. A study of this experience attributed the reluctance to sue to several factors, including a fatalistic attitude and shame over the disease and physical deformities caused by the mercury poisoning. But another factor was the traditional Japanese reluctance to litigate and the perception that anyone who did sue was acting selfishly and too assertively (Upham 1976).

Some scholars minimize the importance of Japanese culture in explaining its lack of lawsuits and instead point to other factors. Calling Kawashima's thesis the "myth of the reluctant litigant," one scholar attributed Japan's low litigation rate to a lack of attorneys and judges and not to the cultural factors Kawashima highlighted (Haley 1978). The contrasting experience of the contaminated Japanese villages in the example just discussed is instructive. Although the members of one village, Minamata, were very reluctant to sue, the residents of the other, Niigata, filed lawsuits in great numbers. This contrast, says sociologist Robert L. Kidder (1983), indicates that the Japanese as a whole are not reluctant to sue when the need arises and that Kawashima's cultural argument is off target. How then to explain the difference in the experience of the two villages? Kidder notes that Minamata was a "company town" in that the factory that polluted its waterway provided many jobs for the villages' members, while Niigata lived some distance from the factory that polluted its river and otherwise had little connection with the factory. If so, he adds, the reluctance of Minamata victims to sue stemmed from two factors: (1) their fear that the factory's corporate owner would fire employees or take other punitive actions (which it had already threatened to do) if the victims did sue and (2) the multiplex relationships that Minamata residents enjoyed with factory officials since they all lived in the same small area.

In further discounting the importance of culture, Kidder (1983) discussed examples of the reluctance of Americans to sue or otherwise use the law when we might expect them to do so. One poignant example involved a flood of the Buffalo Creek, West Virginia mining community in 1972 that killed 125 people and left more than 2,500 others homeless (Erikson 1976). The flood was caused by the bursting of an artificial dam composed of mine waste that violated federal safety regulations. Although the mine company's gross negligence led to much death and tragedy, many Buffalo Creek survivors were very reluctant to sue the company. Many believed the company's claim that the flood was an unavoidable accident and trusted it to take care of them. When the company failed to come to their aid, some but by no means all of the survivors finally joined a lawsuit against the company. The remaining survivors accepted their fate and thought those who joined the lawsuit went too far. Citing this example, Kidder (1983:47) observed that Americans are not "trigger-happy about invoking the law" and thus are not so different from the Japanese after all. We return to this critique below in our discussion of the American legal experience but reiterate that many scholars think culture does matter in explaining societal differences in the willingness to litigate.

In an interesting sidelight, Japan's litigation rate increased rapidly during the 1990s, some three decades after Kawashima (1969) wrote his classic article, even though it still remains relatively low in international comparisons. It is not clear whether this increase reflects an ebbing of Japanese respect for authority and emphasis on harmony, or instead, as some scholars believe (Ginsburg and Hoetker 2006), an increase in the number of lawyers and, especially, procedural reforms that made it easier and more attractive to litigate.

Individual Factors

Even if a society favors or opposes a method of dispute processing for the many reasons just discussed, it is also true that people within that society will differ in the methods they like or dislike. That is because people within a given society differ from each other in ways that affect their willingness to use one method or another. Our earlier discussion of dispute processing methods outlined some of these differences; in this section, we discuss this issue more systematically, if briefly, by focusing on individual-level factors that correspond to the societal ones just examined. For the sake of argument, we acknowledge that the importance and seriousness of a dispute obviously matter, but in the following discussion hold these dimensions constant to answer the following question: Given disputes that are equally important and serious, why are some individuals in a particular society more or less willing than other individuals to use certain methods of dispute processing?

The Nature of the Relationship. Our earlier discussion highlighted the importance of the relationship between two disputants for the methods of dispute resolution they favor. If they are involved in a close or multiplex relationship where they will ordinarily want that relationship to continue or at least not want to cause difficulties in future interaction, the aggrieved party will be less likely to adjudicate and more likely to try negotiation or mediation but also more likely just to lump it (Black 1976). If you have a problem with your plumber but the plumber is also your brother-in-law or sister-in-law, taking your plumber to court could cause all kinds of family problems that you would rather avoid. On the other hand, if the two disputants have only an impersonal, simplex relationship, the aggrieved party will be more likely to adjudicate or perhaps to practice avoidance. If your plumber is just your plumber, you do not care about an enduring relationship; accordingly, you might not call the plumber again, and you may even go to small claims court.

Note that a close or multiplex relationship might also prevent other uses of the law, since an individual's expressed desire to use the law may indicate distrust in another individual. For example, what would happen if you propose marriage to someone (or are proposed to), and the next words out of your mouth are that you want a prenuptial contract? What would your intended spouse think of you at that moment? What does your wish for a "prenup" imply? Why would your desire for a prenuptial contract lead to more hurt feelings in this regard than your desire, say, for example, for a contract with a carpenter you want to hire to work on your home? How does this scenario help you understand the importance of the nature of the relationship?

Power Differences in the Relationship. Our earlier discussion also highlighted the importance of power differences in a relationship. If the aggrieved party has less wealth, power, influence, eloquence and the power of persuasion, or some other important resource than the offending party, the aggrieved party will be more likely to lump it or to practice avoidance than to address the dispute more proactively. Aggrieved parties in this position will do so either because they fear retaliation (e.g., being fired, to recall the boorish restaurant manager example) or because they perceive that the offending party will win the dispute if the aggrieved party does proceed proactively.

By the same token and for corresponding reasons, an aggrieved party with significant advantages in a relationship will be less likely to lump it or practice avoidance and more likely to use adjudication. As noted earlier, even though the legal system ideally puts disputants on an equal playing field before an impartial judge, in practice it favors the party with greater wealth and other advantages. Such parties are, moreover, likely to be "repeat players" in the legal system and thus to have more knowledge of how the law works and other legal advantages (Galanter 1974), increasing their willingness to use adjudication. (We return to this point in Chapter 7's discussion of law and inequality.)

Finally, when the disputing parties are of roughly the same status and have equal resources, they are more likely to rely on negotiation or mediation. Neither party has sufficient advantages to make adjudication an attractive method; by the same token, neither party has sufficient disadvantages to make lumping it or avoidance the only viable options. In general, the level of resources a disputing party enjoys may make a difference from the very beginning of the disputing process. If, as discussed earlier, significant steps involve the decisions to blame an offending party and to confront the party, some people are more likely than other people to make these decisions because they have higher levels of education, greater wealth, more self-confidence, and other resources conducive to more proactive approaches in dispute processing.

Personality. A third individual factor affecting dispute processing preferences is personality, or temperament, consisting of an individual's cognitive and emotional traits (Heen and Richardson 2005). Some people tend to be rather meek and submissive, others tend to be rather assertive and even overbearing, and many of us fall in between these two extremes. When a dispute arises, those on the meek end of this spectrum are, all things equal, more likely than those on the assertive end to lump it or practice avoidance. By the same token, those on the assertive end will be more likely to practice coercion or to use adjudication.

Sociodemographic Factors. Social class and race/ethnicity have also been examined as possible factors that affect the willingness to litigate (Seron and Munger 1996). Early research relying on survey evidence found that the poor were less likely than wealthier people to use the courts when problems arose and attributed this difference to the poor's lack of knowledge about the law and legal rights, their lower economic resources, and their sense of hopelessness induced by their poverty (Abel 1973; Carlin, Howard, and Messenger 1966; Levine and Preston 1970; Mayhew and Reiss 1969). Because people of color are disproportionately poor, they, too, were considered less likely to use the courts for similar reasons (Merry 1979; Moulton 1969).

However, recent research yields a more complex picture (Silbey 2005). One branch of this research confirms that the poor and people of color are indeed less likely to use lawyers and the courts but attributed this more to the kinds of problems these citizens experience rather than to the reasons just described (Engel 1984; Miller and Sarat 1980–1981; Silberman 1985). Relying on ethnographic fieldwork and intensive interviewing, a second branch of this research found low court use overall and few, if any, differences by social class and race/ethnicity (Bumiller 1988; Greenhouse 1986; Merry and Silbey 1984). In view of these conflicting findings, class and race differences in litigation remain unclear; a cautious conclusion would be that relatively small differences do in fact exist for certain kinds of grievances and for reasons that remain to be fully determined.

The influence of gender on litigation has also been considered. Although many scholars think that until recently men "have dominated the civil litigation arena" (Morgan 1999:68), the reasons for the earlier gender gap remain unclear. Some researchers argued that women were less likely than men to litigate because they had fewer resources such as time, money, and legal expertise (Gleason 1981). Others argued that women litigated less than men primarily because they were not working in the business and professional worlds that give rise to so many issues that end up in litigation (Morgan 1999:191).

Whatever the reason, the traditional gender gap in litigation seems to have narrowed and even disappeared for many types of grievances. Scholars attribute these changes to the increased education and other resources that women have gained in recent decades and to their greater involvement in the business and professional worlds (Morgan 1999). Moreover, the nation's expansion of the legal rights of women through such measures as the Violence Against Women Act and Title IX, which prohibited gender discrimination in educational institutions, created litigation opportunities regarding denial of these rights that women have pursued many times since (Hoyman and Stallworth 1986). Finally, studies of ordinary citizens' litigation in the nation's lower courts find that women are in fact more likely than men to seek court redress for family and neighborhood problems (see below). Thus, the question of gender differences in litigation in the contemporary era seems to hinge on the type of issue or problem rather than on broad gender differences *per se*.

DISPUTE PROCESSING IN THE UNITED STATES

Anthropological studies of the disputing process helped spawn a more general interest in the disputing process in modern societies. The United States has been the subject of many disputing studies during the past few decades by law and society scholars in several disciplines. A central task has been to explain Americans' apparent interest in law, to understand the amount of litigation the United States has, to explain how Americans go about using or not using the law, and, perhaps especially, to explore the popular view that the United States is beset by a so-called "litigation crisis" involving a glut of frivolous lawsuits and runaway juries.

Dispute Processing and Litigiousness: Then and Now

In June 2006, actress Reese Witherspoon sued the tabloid magazine *Star* for writing that she was expecting her third child and that she did not want her new films' producers to know about her pregnancy. Both allegations were false, Witherspoon said in the suit, which added that the second allegation harmed her reputation by making her sound deceitful (Reuters 2006). At about the same time, a public health interest group sued KFC (aka. Kentucky Fried Chicken) for using partially hydrogenated oils containing trans fats, which clog arteries and contribute to heart disease. The plaintiff, the Center for Science in the Public Interest, asked the District of Columbia Superior Court to order KFC either to stop using trans fat oils or to tell its customers that it uses the fats. KFC replied that its food was safe and that the lawsuit had no merit (Burros 2006). Meanwhile, environmental groups in Arkansas sued to stop a $300 million irrigation project that, they said, would impair the habitat of an endangered woodpecker, and they were heartened when a federal judge suspended the project pending further review (Demillo 2006).

You may or may not think these were the most important or serious lawsuits that were ever filed, but it probably does not surprise you that they were filed. If so, your lack of surprise probably stems from your perception that Americans sue all the time for all kinds of things, trivial and important alike. Whatever you might think of Reese Witherspoon as an actress and about the substance of her lawsuit, you probably do not think that the act of filing a lawsuit made her a bad, selfish person, and you probably think that she had every right to sue the tabloid. Even if you have never heard of the Center for Science in the Public Interest and think that people should be allowed to eat what they want, you probably also think that it had every right to sue KFC. And litigation has certainly been an important strategy for the environmental movement.

Your perception that these three lawsuits were unsurprising and even normal reflects the widespread belief that suing someone is as American as apple pie. According to this belief, the United States is said to be "a society profoundly rooted in law," to cite political scientist Seymour Martin Lipset (1996:270) and, in particular, an especially litigious society in which all kinds of matters become legal issues. A famous observation from Alexis de Tocqueville (1994 [1835]:280), the perceptive nineteenth-century French observer of American democracy, captures this sentiment: "Scarcely any political question arises in the United States that is not resolved, sooner or later, into a judicial question. Hence all parties are obliged to borrow, in their daily controversies, the ideas, and even the language, peculiar to judicial proceedings."

Many scholars since have echoed this belief. "The notion of justice without law seems preposterous, if not terrifying," writes legal historian Jerold Auerbach (1983:3). "A legal void is especially alarming to Americans, who belong to the most legalistic and litigious society in the world." Adds law professor Jethro K. Lieberman (1981:3), "Judicial decrees have changed the face of the social order, and Americans seemingly take to the courtroom at the merest whisper of an insult." To the extent the United States is an especially litigious society, Americans' litigiousness is attributed to several factors: (1) the nation's simplex structure; (2) its cultural emphasis on individualism and success, both of which contribute to an affinity for the win–lose outcomes that adjudication achieves; (3) the distrust of government and respect for law as a protector of individual freedom that grew out of the colonial experience and was written into the Constitution and Bill of Rights (see Chapter 3); (4) a decentralized political structure in which power is fragmented among many political and legal bodies, thus encouraging litigation and other efforts to affect public policy; and (5) the granting of rights by the U.S. Supreme Court to African Americans and other disadvantaged groups during the 1960s and early 1970s, which furthered Americans' perceptions that the courts are a source of justice (Kagan 2001; Lieberman 1981; Lipset 1996).

Whether or not Americans are especially litigious is a debate discussed later when we examine the litigation crisis controversy. Whatever the situation today, many American communities earlier in U.S. history were definitely *not* litigious. Historian Jerold S. Auerbach (1983:4–5) writes,

> In many and varied communities, over the entire sweep of American history, the rule of law was explicitly rejected in favor of alternative means for ordering human relations and for resolving the inevitable disputes that arose between individuals. . . . Historically, arbitration and mediation were the preferred alternatives. . . . Sharing a suspicion of law and lawyers, [Americans] developed patterns of conflict resolution that reflected their common striving for social harmony beyond individual conflict, for justice without law.

Wishing to preserve their own traditions, fearing that litigation would disrupt their harmony by exacerbating conflict and by implying mutual mistrust, and mistrusting lawyers, these communities developed dispute processing methods that kept "courts and lawyers as remote as possible" (p. 5).

These methods were perhaps most salient in the early American colonies, whose residents, writes Auerbach (1983:20), "understood that legal disputation, with its adversarial imperatives, was destructive of the group solidarity upon which they depended for the fulfillment of their mission in the New World." Much of these colonists' aversion to "legal disputation" derived from their religious beliefs. In Puritan Massachusetts, religion was certainly all-important. Law was considered an "alien value system, antithetical to Christianity itself," and litigation was considered "a form of self-aggrandizement contrary to the best interests of the community" (p. 22). Accordingly, "legal dispute settlement was explicitly discouraged" (p. 23), and the Puritans relied largely on arbitration, mediation, and negotiation, and, failing those, church intervention.

Outside New England, the Quakers of early Pennsylvania also shunned legal methods of dispute processing. Like Puritans, Quakers felt that conflict and contentiousness would threaten the harmony of their Christian society and favored dispute processing that would preserve social harmony. They thus developed a process that first involved negotiation, then mediation by one or two other Quakers, and then arbitration by other Quakers. If these methods all failed, the dispute then went to the local monthly meeting, or church, which assigned arbitrators. A disputant who refused to honor the arbitrators' judgment could then be disowned by the monthly meeting. This entire process, writes Auerbach (1983:30), "was designed at every stage to suppress conflict." Quakers' disputes with non-Quakers did enter the courts, but even here "the pressure for harmonious resolution was evident" (p. 30), with arbitration always available as an alternative.

The aversion to litigation also characterized the early colonies' mercantile economy. When disputes involving merchants needed resolution, merchants preferred commercial arbitration over litigation because it was faster and less expensive and more attuned to their needs and interests. They also feared that litigation would make their disputes more public and open mercantile trade to outside intervention.

Eventually, however, courts and litigation became more popular in colonial communities and among colonial merchants alike by the beginning of the early eighteenth century. As the colonies grew, the sense of community weakened and relationships became more simplex, to use our earlier term, with strangers moving into colonial towns and merchants and customers no longer knowing each other outside their economic relationship. The law thus became a common meeting ground for the resolution of disputes as the colonists became more willing to embrace a "legal process [that] encouraged the clash of individual differences amid constant jostling for private advantage" (p. 34). This development had an important consequence, writes Auerbach (1983:35): "Paradoxically, law encouraged contentiousness while channeling it. The mixed benefits provided by litigation made colonists uneasy even as they went to court more frequently." As the colonies grew, then, they became less multiplex and more simplex, and their dispute processing changed accordingly as the importance of enduring relationships lessened. As Auerbach (1983:41) notes, "Arbitration and mediation had been appropriate for neighbors and parishioners, but the disagreements of strangers, who lacked any basis for mutual trust, were for lawyers and judges to resolve."

Although courts and adjudication began to take hold in the new nation, they did not become dominant everywhere and in every era. In particular, utopian communities that arose during the early eighteenth century in New England and the Midwest disliked litigation for reasons similar to those held by the early colonial communities just discussed. Some of these utopian communities gathered because of their religious beliefs and others because of their political beliefs, but both types embraced communal harmony as an underlying principle and feared that state law would undermine this harmony. Disputes were addressed in these communities by peer pressure and moral persuasion and the threat of expulsion. If these did not work, disputes would be heard by the entire society. Although almost all of these utopian communities lasted at most a few decades, their aversion to legalism again reminds us that the United States has not always been a litigious society.

Another reminder occurred about a century after the utopian communities first appeared when a new wave of immigrants came to the United States in the early nineteenth century. Although they quickly settled into the various neighborhoods of large cities on the eastern seaboard and elsewhere, they strove to hold onto their old ways even as they slowly but surely began to adopt the culture of their new nation. Not surprisingly, dispute processing and the law became one venue in which this struggle became manifest. As Auerbach (1983:69–70) writes,

> New immigrants had good reasons to resist, at least temporarily, litigation and the judicial process. They often dwelled in communities where personal relationships were intricately social and enduring, not impersonally contractual and transitory. . . . The newcomers understood that the delicate equilibrium of continuing community relationships would be upset by the narrowing of issues, the designation of winner and loser, and the abrupt, abrasive finality of a legal verdict. Control over conflict was crucial for preserving communal values from the corrosive effects of assimilation.

Their aversion to litigation led several of these immigrant communities—those from Bulgaria, Greece, Italy, and Turkey—to rely on arbitration by a *padrone*, a wealthy, powerful individual in their neighborhoods who ran many businesses and provided many jobs and much housing. His influence over their lives meant that his decision in the disputes he arbitrated would be followed. Eventually, however, padrones began to lose their power, and their arbitration of disputes gave way to litigation in American courts. Other immigrant communities that never had padrones also began to turn to litigation in place of internal methods of dispute processing, in part because they wanted to act like Americans.

Three immigrant groups resisted this trend. One group was the Scandinavian communities in Minnesota and North Dakota; for these immigrants, reconciliation was an important cultural norm, and they feared that litigation and the involvement of attorneys would aggravate disputes rather than reconcile them. Chinese and Jewish immigrants were two other groups that resisted the turn to litigation. The former favored mediation by elders and other revered individuals, while the latter favored arbitration; both groups disliked litigation because they valued harmony, because they wanted to keep their cultures free of American influence, and because they feared that the legal system was racially biased against them.

Evidence of such bias readily existed. In 1854, for example, the California Supreme Court ruled that Chinese witnesses could not testify against white defendants because the Chinese were "a race of people whom nature has marked as inferior" (quoted in Auerbach 1983:74). Although all three immigrant groups resisted litigation for all these reasons, they, too, eventually began to turn to the law for dispute resolution. As they became more assimilated over time and as their values changed, they became more willing to litigate: "The Americanization process, with its overriding emphasis individual achievement, encouraged adversarial competitiveness and communal fragmentation" (Auerbach 1983:93). As this happened, immigrants' informal methods of dispute processing gave way to American legal formalism.

Legal Consciousness and Going (or Not Going) to Court

Earlier, we mentioned a recent line of research involving ethnographic fieldwork and intensive interviewing. Some of these studies have taken place in small towns across the country and thus may be considered community studies, while other studies have involved selected groups of individuals, such as welfare recipients and people who have experienced sexual harassment or racial discrimination in the workplace or other venues. Regardless of the scope of the research, studies of this type have tried to understand how Americans' **legal consciousness**—their everyday understandings of and experiences with the law—have shaped their willingness to use lawyers and courts and also how law and other aspects of society have shaped legal consciousness. As such, legal consciousness also encompasses people's ideas about justice, power, and rights and how these ideas affect their everyday lives and, especially, their reactions to grievances and disputes (Engel and Munger 2003; Ewick and Silbey 1998; Marshall 2005b; Merry 1990; Nielsen 2004a; Silbey 2005).

An early study of legal consciousness that never used the term was Stewart Macaulay's (1963) classic discussion of the reluctance of businesses to use formal written contracts when they initiate agreements with each other. Although you might be surprised by this finding, Macaulay wrote that many businesses actually prefer "handshake" agreements because an insistence on a written contract would imply a lack of trust and threaten the enduring relationships that businesses have with one another. They also feel that formal contracts would make their arrangements less flexible and more complex. When disputes arise between businesses even when there is a contract, many prefer to handle the dispute without involving lawyers and without threatening to sue, again because they wish to preserve their long-term business relationship. As one purchasing agent put it, "If something comes up, you get the other man on the telephone and deal with the problem. You don't read legalistic contract clauses at each other if you ever want to do business again. One doesn't run to lawyers if he wants to stay in business because one must behave decently" (p. 61). Thus, the related desires for a good reputation and for continuing relationships with other companies help explain why law is often avoided in the business world.

Legal Consciousness and Everyday Life. One of the most notable recent studies of legal consciousness is Patricia Ewick and Susan S. Silbey's (1998) *The Common Place of Law: Stories from Everyday Life*. They interviewed a random sample of 430 New Jersey adults about their thoughts and reactions to law and other authority (e.g., hospital administration) in their everyday lives. The subjects' rich descriptions of their lives enabled the authors to develop a typology of three types of legal consciousness. Some people, they found, have a *before the law* consciousness because they respect the law and even hold it in awe. These people tend to turn to the law only when especially severe problems arise and can become frustrated by its complexity when they do so. Other people are *with the law* because they are quite ready and willing to use the law to achieve their self-interests. They view law as a game in which the rules may be manipulated if necessary to win and in which the most skilled players have the greatest chances of winning. The third type of legal consciousness is *against the law*. People with this consciousness distrust the law and other authority and react with various violations of norms that the authors call "daily acts of individual resistance" (p. 183). Examples of such acts include pilferage, the use or threat of violence, delays, and small acts of deceit. For example, one elderly man in a high-crime Newark, NJ, neighborhood told the authors that the police always ignored his calls to them until one day he used a higher-pitched voice to sound like a woman when he called them. This time they responded quickly. Another interviewee, 17, said she falsely reported being 18 in order to receive emergency room treatment at a hospital without her abusive parents having to be contacted.

Legal Consciousness Among the Working Class. Sally Engle Merry (1990) studied legal consciousness among largely working-class residents of two Massachusetts towns during the early 1980s. Like many other people, these residents had various problems with family members and other relatives, lovers and other close friends, and neighbors: a neighbor would be noisy, a husband would be violent, a teenaged child would be disobedient. They sometimes decided to seek legal redress for these problems either by calling the police or by going directly to court and talking with a court clerk or other official they encountered. Typically they had tried to resolve the problem in other ways that proved futile and thus went to court "reluctantly and only as a last resort" (Merry 1990:3). Once they did so, they found that the dispute escalated for the reasons discussed in our earlier section on methods of dispute processing.

They also found another harsh reality: legal officials typically considered their interpersonal problems to be "garbage cases" that did not belong in court. This result, wrote Merry (1990:2), surprised and disappointed these citizens, who went to court "because they see legal institutions as helpful and themselves as entitled to that help. They see the court as an institution which has a responsibility to protect their fundamental rights to property and safety, rights they acquire as members of American society. Moreover, they think that settling differences by legal rules and authorities is more civilized and reasonable than violence." Given their strong belief that their issues belonged in court and their expectation that the law would recognize their rights, these litigants were taken aback when they did not receive the legal help they anticipated.

Another unexpected outcome of going to court related to the issue of power. On the one hand, going to court empowered the litigants in relation to the person with whom they were having a dispute, however petty or serious. On the other hand, going to court meant that the court would now control what happened with the dispute. As Merry (1990:2) put it, "People who take personal problems to court become more dependent on the state to manage their private lives. Recourse to court strengthens the hand of the plaintiff against his or her neighbor, relative, or friend, but at the same time it leaves her dependent on the court for support." Thus, going to court was ultimately a disempowering experience for these plaintiffs.

Contributing to this disempowerment were several other realities. First, they found legal proceedings difficult to comprehend. Second, they often were unable to meet with a judge and instead had to content themselves with a court clerk or other court personnel. As Merry (1990:3) observed, "It is easy to get into the door of the courthouse but far more difficult to arrange a hearing in front of a person in a black robe." Third, any penalties handed down against defendants tended to be relatively light. In general, Merry found that women litigants who brought charges of violent abuse against their husbands or boyfriends were particularly disadvantaged, as courts refused to take their claims seriously and as the act of going to court antagonized the men who were beating them. For these and other reasons, going to court changed the legal consciousness of the working-class plaintiffs in Merry's ethnographic study. In particular, they began to think that the courts were "ineffective, unwilling to help in these personal crises, and indifferent to the ordinary person's problem" (Merry 1990:70).

In other respects, Merry's book-length study shed important light on some of the litigation issues discussed earlier in this chapter. First, she found that the less educated, working-class people she studied had a strong sense of their legal rights and of their entitlement as American citizens to seek redress in the courts. Second, she found that few of these citizens actually used the courts despite the many problems they had, as most sought to resolve the dispute in nonlegal ways and turned to the courts only reluctantly. Third, she found that working-class citizens were more likely than wealthier ones to use the courts for the family and neighborhood problems, while wealthier people use the courts for other types of problems such as real estate and consumer problems. This last point exemplifies the finding in recent research, discussed earlier, that the involvement of the various social classes in courts depends more on the types of problems they experience than on differences in legal consciousness, material resources, and other such dimensions.

A final set of findings in Merry's study concerned gender. Women in her study were much more likely than men to bring family and neighborhood problems to court. The reason for this, she wrote, was that women have less power than men in the troubled relationships that lead them to go to court: "In this social world, relative power depends to a large extent on strength, willingness to use violence, and economic resources. Women are usually less well endowed with these qualities than are men. They turn to court because they feel vulnerable and because they hope it will provide a powerful ally, but it is not a first choice" (Merry 1990:4).

Legal Consciousness and Offensive Speech. Another notable study of legal consciousness and litigation is Laura Beth Nielsen's (2004a) book about the reactions of people in three Northern California cities to offensive public speech, including racist and sexually suggestive speech (aimed at women and gays and lesbians) but also begging. Public comments that are sexually suggestive or racist are quite common in American society and no doubt elsewhere as well. One woman reported to Nielsen what a man once said to her: "I love that smile. I would have liked to have been there this morning when your man put that smile on your face. What did he do to put that smile on your face? I'll bet he (expletive) you so long you'll be smiling all day." A lesbian reported, "When I am walking down the street with my girlfriend we get lots of comments like, 'Try me and you'll never go back' or 'I can show you things that she can't.' " An African American woman reported a man's shout to her: "Monkey for a dollar!" (Nielsen 2004a:1).

Most people in today's society, and especially the recipients, would consider comments like these to be highly offensive. But they are uttered in an American society that ordinarily values and protects freedom of speech, including speech that many find offensive. This protection provides people uttering such speech a "license to harass" that is the title of Nielsen's book and also the point of departure for her study. Among other topics, her book examined the willingness of three groups of people—white women, white men, and people of color—to have the law limit or ban offensive public speech. This examination thus touched on one aspect of their legal consciousness.

Not surprisingly, Nielsen found that the white women and people of color in her study were much more likely than white men to experience offensive public speech. All three groups agreed that such speech is indeed offensive, but they also agreed that it should be permitted in American society. However, they differed in the reasons they gave for believing it should be permitted: Whereas white men opposed offensive speech laws because they favored the principle of freedom of speech, white women and people of color opposed these laws for other reasons, including the belief that the laws would be ineffective and difficult to enforce and the fear that any such laws would be used to restrict their own speech in different contexts. In this way, wrote Nielsen, the study of offensive public speech illustrates the different types of legal consciousness that these groups hold: "It is clear that the relative social status of the target of offensive public speech makes a difference in how different types of offensive public speech are legally managed" (p. 12).

Nielsen drew two other conclusions that are relevant for some of the litigation issues discussed earlier. First, because the people she interviewed were generally opposed to legal restrictions on offensive public speech, she joined Merry (1990) in concluding that Americans are far less litigious than commonly depicted. Second, she noted that the law generally does not limit the type of offensive public speech, sexually suggestive and racist comments, aimed at women and people of color, respectively, but it does limit the type of public speech, begging, aimed most often at white men and also opposed by merchants and other businesses. In this way, Nielsen concluded, the law reflects and reinforces social class, racial, and gender bias.

Legal Consciousness and Sexual Harassment. Some of the themes of the books by Merry and Nielsen are echoed in Phoebe A. Morgan's (1999) study of sexually harassed women. Today's legal system theoretically provides extensive legal redress to women who suffer sexual harassment in the workplace and elsewhere. The rate of sexual harassment claims filed with the federal Equal Employment Opportunity Commission rose by about 12 percent annually from 1980 to 1994, and sexual harassment certainly is an issue with which many people are familiar from news coverage and popular media depictions on TV and in film. Despite this new awareness and the opportunity for legal action, however, in practice, only about one-sixth of sexually harassed women take such action (Marshall 2005b; Morgan 1999).

Why do so many decline to file suit? Morgan interviewed thirty-one sexually harassed women. All the women had considered filing a lawsuit, but only four eventually did so. A major factor in their decisions regarding litigation was their family relationships as wives and/or mothers, specifically "how litigation might affect those to whom they were most closely tied" (Morgan 1999:86). Some filed suit because they thought it would bring their families needed financial resources, but others declined to file suit because they feared that legal action would be too stressful for their families. Women with children and unsympathetic husbands especially held this fear. "For such women," Morgan wrote, "litigation pits the need to meet familial responsibilities against personal longings for formal justice" (Morgan 1999:87).

Thus, these women's family relationships and responsibilities played a major role in their decisions to seek or not seek legal redress for the sexual harassment they had experienced.

Legal Consciousness and Disability. A final study illustrates how the idea of legal rights can affect individuals' legal consciousness. David M. Engel and Frank W. Munger (2003) interviewed sixty people with disabilities (either people with learning disabilities or those confined to a wheelchair) about their views of their legal rights and about their everyday behaviors and experiences after the passage of the Americans with Disabilities Act (ADA) in 1990. Everyone in their study had suffered disability-based discrimination, but no one had ever sued. Even so, the authors found that the ADA had helped improve the self-image of many of their interviewees while also giving them a sense of entitlement to equal treatment. As a result, some pursued career paths that they otherwise might not have pursued if the ADA had not been enacted. The authors found that their interviewees received better treatment in their workplaces after the ADA was passed. To the extent the ADA had these benefits, they accrued more to the interviewees who were white and middle class than to those who were African American or lower class. Gender affected the choice of careers but not the extent of benefits that otherwise occurred.

Rights Consciousness. The Engel and Munger study is part of a larger body of research and theory on Americans' perceptions of their legal and civil rights, or on their *rights consciousness* (Nielsen 2004b; Scheingold 1974). In general, this research finds that Americans have a strong sense of their rights as reflected in the Declaration of Independence's assertion of the rights to "live, liberty, and the pursuit of happiness" and in claims of the Southern civil rights movement and other social movements of the last several decades. It also finds that Americans believe aggrieved groups are entitled to pursue their rights through the legal and political process.

At the same time, however, this research also finds that Americans with perceived grievances do not usually pursue their rights through the legal and political process, as most such people practice lumping it or avoidance for reasons discussed in the preceding pages: they may believe that pursuit of their rights will be futile, expensive, or time-consuming; they may lack the self-confidence or material resources to pursue their rights; they may not trust the legal and political systems to hear their claims impartially; or they may fear that litigation will aggravate a situation and make them look like selfish individuals (Bumiller 1988; Galanter 1983). As Engel and Munger (2003:3) summarize this research, it finds that

> Americans usually deal with legal problems by absorbing perceived wrongs without overt response. Americans seldom consult lawyers when they believe themselves to be the victims of

> rights violations, and lawyers seldom bring lawsuits on behalf of those who consult them. . . . From this perspective, America is a nation of "law-avoiders." Potential claims of rights tend to be repressed, wrongdoers are often free to repeat their transgressions without fear of legal reprisal, and relatively powerless individuals suffer the consequences of an inability or unwillingness to invoke the law to protect their interests.

Thus, a large body of research finds a large divide between what Americans believe about rights and what they actually do, or fail to do, about their rights (Trubek et al. 1983). In related areas, this research also finds that when people do end up in court, they often care more about being treated impartially and fairly, with dignity and respect, and about being allowed to have their say, than about the actual outcome of the case (MacCoun 2005; Tyler and Huo 2002). In short, they care about whether they have received *procedural justice* (see Chapter 3) (Tyler 2004). When they perceive they have received procedural justice in all these respects, they are more satisfied with their legal involvements.

The Litigation Crisis Controversy

Since the 1970s, many observers have said that the United States is suffering a litigation crisis characterized by a markedly increasing amount of litigation, much of it involving frivolous cases; clogged courts that make the processing of cases extremely expensive and time-consuming; and, perhaps especially, runaway juries that render verdicts in the millions of dollars for claims that did not merit such huge sums (Huber 1988; Kagan 2001; Olson 2002; Sykes 1992). A *Newsweek* article as far back as 1977 led with the provocative headline, TOO MUCH LAW? and featured a large drawing of a swarm of people trying to cram into, and at the same time falling out of, a courthouse (Footlick 1977). The article began by summarizing two cases: a frivolous lawsuit in which frustrated fans of a professional football team filed a lawsuit to try to overturn a referee's decision that had cost their team the game, and a much more noteworthy case involving Karen Ann Quinlan, a comatose woman on life support whose parents asked a court to let her die. The article then went on to say,

> For good or ill, Americans have come to rely on the courts to solve their problems to an unprecedented degree. . . . The mounting influence of law and lawyers on modern life constitutes one of the great unnoticed revolutions in U.S. history: the ever-increasing willingness, even eagerness, on the part of elected officials and private citizens to let the courts settle matters that were once settled by legislatures, executives, parents, teachers—or chance. (p. 41)

A list of actual cases, some very important and some less important, then followed, involving such issues as: whether an employer can refuse to hire anyone the employer does not want to hire; whether a school may restrict the hair length of its male students; whether a factory should be permitted to dump waste in a lake; whether a hockey player should be subject to criminal prosecution for hitting an opponent; and whether a woman has a legal right to an abortion. The article then noted a sharp increase in the number of federal and state civil cases since 1960, much of this stemming from the fact that "events that would have seemed inconceivable as legal matters a few years ago are finding their way into the courts" (Footlick 1977:45). Reflecting the article's concern, a case in San Francisco Small Claims Court a year later involved a man who wanted compensation from a woman for standing him up on a date. After he drove fifty miles to pick her up, she told him she did not want to go out with him after all. The plaintiff wanted reimbursement of $38 (worth about $120 today) for his expenses (*San Francisco Chronicle* 1978).

A quarter-century later this purported litigation crisis again won headlines as political conservatives and business leaders led a call earlier this decade for "tort reform" that would limit the kinds of issues juries would be allowed to hear and, especially, the financial amount of the awards and damages they could give to defendants they found liable. Much of this campaign focused on medical malpractice, as President George W. Bush gave a series of speeches in early 2005 that called for Congress to limit malpractice awards to reduce what he called "junk lawsuits" (Pear 2005:A1). Although the Congress never did pass the malpractice legislation the President wanted (Stolberg 2006), it did pass legislation that limited the ability of state courts to hear class action lawsuits. Although the votes in both houses were not close, they were nonetheless criticized by civil rights, consumer, environmental, and labor groups who feared the legislation would "provide new protections for unscrupulous companies," according to a news report (Labaton 2005:A1).

These efforts came in the wake of successful tort reform efforts at the state level during the 1990s, when about two-thirds of the states enacted limits on awards in malpractice or other tort cases. Much of the motivation for these new limits came from a few cases in which juries awarded millions of dollars for trivial injuries. In Alabama, for example, a jury in 1999 awarded $581 million to a family who claimed that a company had overcharged them $1,200 for two satellite dishes. Five years earlier, a Georgia jury awarded $50 million to a plaintiff who claimed that he had been overcharged $1,000 on a car loan, and two years before that a plaintiff won $4 million after suing the auto company BMW for secretly repainting the car he had bought; the U.S. Supreme Court later reduced this award to $50,000. The satellite dish case prompted the Georgia legislature to sharply limit punitive damages by juries (Firestone 1999).

Legal Legends and News Media Coverage. These cases from Alabama and Georgia were just three of many examples of "bizarre jury verdicts and huge damage awards" (Glaberson 1999a:D1) that have won news headlines during the last few decades. However, many of these cases have proven to be urban legends (or, to be more precise, *legal legends*) that either never happened or were distorted in their retelling (Galanter 1998; Glaberson 1999a; Haltom and McCann 2004).

One of the most notorious such cases involved a plaintiff who spilled McDonald's coffee on herself and won $2.9 million from a jury in 1994. Although this case received much publicity as a frivolous lawsuit with an absurd jury award, the news coverage generally omitted some important facts that might have cast the case in a different light. First, the plaintiff suffered third-degree burns over 6 percent of her body that required skin grafts and kept her in a hospital for more than a week and partially disabled for two years. Second, McDonald's coffee was at least 20 degrees hotter than other restaurants' coffee, and more than 700 people had told McDonald's in the previous decade that they had been burned by its coffee. When a judge reduced the plaintiff's jury award to about one-fifth of its original amount, this event received much less news coverage than did the original award itself.

This example reflects a more general problem in news media coverage of lawsuits (Bailis and MacCoun 1996; Haltom and McCann 2004). The legal legends typically receive heavy coverage, as do the relatively few cases, such as medical malpractice awards, that end in multimillion-dollar awards. This coverage paints a grim but *false* picture of a nation besieged by law and lawsuits and runaway juries. For example, in 1989 the *New York Times* reported on many cases in and around New York City, and the average award in the cases that won the *Times'* attention was $20.5 million. However, many more cases did not win the Times' coverage, and their average verdict was $1.1 million (Glaberson 1999a). Thus, media coverage of lawsuits gives the public and policymakers an exaggerated idea of the damages that the bulk of lawsuits really provide. Media coverage also exaggerates the degree to which plaintiffs win tort suits. Political scientists William Haltom and Michael McCann (2004) studied 3,500 articles in the nation's leading newspapers of tort cases from 1980 to 1999. In these cases, plaintiffs won 78 percent of the time, even though plaintiffs win only about 50 percent of all cases (see below). Thus, media coverage suggests that plaintiffs win much more often than they actually do win.

Social Science Evidence on the Litigation "Crisis." Aided by media coverage, the idea of a litigation crisis characterized by runaway juries and soaring litigation has become accepted wisdom among citizens and public officials alike. However, social science research finds that this problem is greatly exaggerated (Daniels and Martin 1995; Galanter 1983; Haltom and McCann 2004; Kritzer 2004). Cases like the satellite dish suit in Alabama do occur and gain much publicity, but they are the exceptions rather than the rule. Punitive damages by juries are, in fact, rare in comparison to the total amount of litigation, and fairly small when they are awarded (Glaberson 1999a).

According to a federal report, for example, of the 12,000 civil cases disposed of by trial in the nation's seventy-five largest counties in 2001, plaintiffs won only 55 percent of the time (including only one-fourth of the time in medical malpractice cases), and they won less often in jury trials than in bench trials (when only the judge decides the verdict). When they did win, their median award was only $37,000 from juries and $28,000 from judges. Only 6 percent of victorious plaintiffs won punitive damages (meaning that punitive damages were awarded in only about 3 percent of all trials), and their median punitive award was only $50,000. Indicating that juries are not overly generous, they awarded punitive damages only about as often as did judges, and their punitive awards were *not* significantly higher than those from judges. Only 8.4 percent of victorious plaintiffs won more than $1 million in total damages, and only 18 percent won more than $250,000. In related findings, the percentage of plaintiffs who won punitive damages in 2001 was the same as in 1992, and the amount of punitive damages was lower in 2001 than in 1992 (Cohen 2005; Cohen and Smith 2004).

Another study also found that juries and judges are equally likely to award punitive damages (each in about 4 percent of all trials) and to award similar amounts (Eisenberg et al. 2002). These findings led the study's lead author to observe, "Policy is being determined on the notion that there are these crazy jurors out there that need to be reined in by legislatures and courts. The evidence is that juries are not out of control" (Glaberson 2001:A9). A later study by the lead author and colleagues of 11,610 civil cases won by plaintiffs again found that juries and judges awarded punitive damages at similar rates (juries, 5 percent; judges, 4 percent); juries awarded punitive damages more often than judges in nonbodily injury cases and less often in bodily injury cases. This last finding ran counter to "conventional wisdom . . . that juries, not judges, should be the relative pushovers for injured plaintiffs" (Eisenberg et al. 2006:291).

Haltom and McCann (2004) point to additional evidence against the litigation crisis thesis. First, the number of liability suits has not been soaring, as tort reform critics maintain. Although such suits rose by 58 percent from 1975 to 1997, they actually declined by 9 percent after 1986. Thus, they rose for about a decade and then declined for about the next decade. Second, most people who suffer disabling injuries in their workplaces or elsewhere never sue; a large study found that only 4 percent of these potential litigants hired an attorney and only 2 percent filed a lawsuit. Third, most people who suffer injuries from medical malpractice also do not sue, with one study finding that only 16 percent of patients who suffered serious, permanent injuries took legal action. Thus, the authors conclude, "Americans' modal response to injury is to do little or nothing" (p. 82). In this regard, the reluctance of injured people to sue reflects Americans' general reluctance to go to court as discussed in the previous section on legal consciousness.

A recent study of medical malpractice cases from 1988 to 2002 in Texas, one of the few states for which appropriate data are available, also provides evidence against the litigation crisis thesis (Black et al. 2005). The authors found that "malpractice claims and payments were stable over the period for which we have data" (p. 209) and that jury awards did not rise during this time. These and other findings led the authors to further conclude that "no crisis involving malpractice claim outcomes occurred" (p. 210) and that only a "weak connection" (p. 210) exists between malpractice claims costs and malpractice insurance costs.

In sum, the body of evidence on jury awards and the amount of litigation suggests that the litigation "crisis" is not really a crisis after all (Galanter 1993). Moreover, the evidence discussed in this and the previous section on legal consciousness also suggests that the litigiousness of Americans has been exaggerated, as they only rarely litigate even when they have sufficient grievances. Comparative data support this conclusion. Although accurate international data on litigation rates are difficult to obtain (Blankenburg 1994; Boyle 2000), they indicate that U.S. litigation rates are not dramatically higher than those in other democracies (Galanter 1983; Kritzer 1991). Moreover, historical studies of U.S. litigation rates indicate that these rates have periodically risen and fallen and that current rates in the state courts are not higher than earlier rates (Friedman 1989; McIntosh 1990). At the state level, then, where more than 90 percent of all cases occur, it does not appear that the litigation "explosion" cited by tort reform advocates has actually occurred.

Another recent study again supports this conclusion (Seabury, Pace, and Reville 2004). The authors examined forty years of jury verdicts in tort cases in San Francisco County and Cook County, IL (greater Chicago). During this period, the number of jury verdicts in San Francisco actually declined by 70 percent and remained stable in Cook County. The average jury award did rise during this period in both locations but at a lower rate than real income. Moreover, the increase that did occur stemmed primarily from a decrease in automobile cases and an increase in medical malpractice cases and in patients' medical expenses. The authors concluded, "Our results suggest little evidence to support the hypothesis that juries are awarding substantially higher awards on average, though they may be doing so for certain kinds of cases" (p. 23).

If there is a litigation crisis because juries are "out of control," we would expect that judges would attest to this in surveys of their opinions. To the contrary, surveys of judges find that they are generally satisfied with the performance of juries in civil trials (Dwyer 2002). A 2000 survey of Texas state trial judges and all federal trial judges found that most had a very favorable opinion of civil juries (Pusey 2000). More than 80 percent said that juries were impartial as they decided their verdict; 96 percent said that they agreed with jury verdicts most or all of the time; and 90 percent said that juries are able to understand the various issues that arise in cases.

Although a legal crisis thus does not appear to exist, two-thirds of the states, as noted earlier, passed tort reform measures during the 1990s to limit the ability to sue and/or the size of punitive damage awards. For better or worse, these measures have had their effect, but they have also prevented plaintiffs who suffered serious injuries because of gross negligence from receiving adequate compensation. According to a 1999 news report, this result has led critics to charge that "the 'reform' label is being used to force the most extensive cutback in the legal protections for citizens in this century" (Glaberson 1999b:A1).

In Texas, a leader in the tort reform movement, a jury awarded $42.5 million in a case in which it found an oil refinery negligent in the death of a worker who died in an explosion; citing a new state law that limited punitive damages, the judge reduced the award to $200,000. The judge's decision prompted a juror in the case to remark, "$200,000 is just pocket change. They'll just write this off" (Glaberson 1999b:A1). Texas's tort reform effort was aided by its state Supreme Court in several rulings. In one case decided by the court, a father won $250,000 from a jury after his daughter was murdered by a psychotic patient, the daughter's husband, who was released from a hospital without proper medication. The court invalidated the award, and the father received no compensation. He later said that the court's decision sent the wrong message to psychiatric hospitals: "If they can get by with that, they can get by with anything. If that can happen, what good is your court system?" (Glaberson 1999b:A1).

To try to get around the new limits on punitive damages, some plaintiffs' attorneys developed a strategy of convincing juries to provide pain-and-suffering awards for emotional and psychological harm in addition to compensation for lost wages, healthcare costs, and other expenses. In a sexual harassment case, a Michigan woman won a $20 million pain-and-suffering award that was widely interpreted as the equivalent of a punitive award. The woman had worked in a Detroit auto assembly plant and for years had been subject to pornographic materials and vulgar comments. In one incident, a photo of a penis was taped to her toolbox. Her attorney said the harassment led to a suicide attempt and hospitalization (Liptak 2002).

Some evidence also suggests that in states that have limited pain-and-suffering awards, juries have responded by providing greater economic damages, as their total awards are roughly equal to those in states without such limits. However, it is possible that attorneys in these states are simply choosing to represent clients with high-paying jobs and thus significant loss income, while those in states without pain-and-suffering limits are taking on clients with lower-paying jobs (Liptak 2005). Research that controls for the career path of the client would thus be needed to determine whether juries in states with pain-and-suffering limits are indeed responding with higher economic damages.

Alternative Dispute Resolution

Dissatisfied with the cost, slow pace, and other problems of adjudication, the United States began two or three decades ago to adopt arbitration, mediation, negotiation, and other **alternative dispute resolution** (ADR) measures in cases involving divorce, small claims, and other disputes. Divorce attracted much early attention as a dispute that would benefit from ADR (Mansnerus 1994), and several states now require that spouses wishing to divorce receive mediation to try to save the marriage. Arbitration is also a common procedure used to process labor-management disputes, and mediation and arbitration are increasingly being used in small-claims courts around the country (Wissler 1995). In New York City, about 85 percent of small-claims court cases are resolved by these two methods (Zane 2005). ADR's development has been so widespread that it is often referred to as the ADR *movement*.

A rapidly growing literature addresses the features and techniques of ADR and discusses its advantages and disadvantages (Grenig 2005; Marshall, Picou, and Schlichtmann 2004; Nolan-Haley 2001; Ware 2001). Proponents say ADR has several advantages over adjudication. First, it saves time and money. Second, it helps avoid the escalation of conflict that often accompanies adjudication and may even aid in reconciling the disputing parties. Third, because ADR is more informal and less rigid than adjudication, it is better able to consider the disputants' personal needs and other nonlegal and nonfactual aspects of a case. Fourth, this informality also leads disputants to be less stressed and more satisfied with the procedures than they would be with adjudication.

Critics dispute some of these points and also indicate certain problems in the use of ADR. While generally acknowledging the first three points on time and money, escalation, and informality, they question whether ADR necessarily results in more satisfied disputants than adjudication. Their skepticism on this issue rests on research findings that disputants who go to trial in fact "tend to view the trial process favorably" (MacCoun 2005:177) whether or not they win their cases. Thus, litigants who use adjudication appear to be as satisfied with the processing of their disputes as those who use ADR.

In a related point, some scholars also question whether ADR can succeed in resolving disputes as well as it does in the traditional societies studied by anthropologists, whose research on dispute processing helped inspire ADR's growth. Anthropologist Sally Engle Merry (1982), whose study of legal consciousness was discussed earlier, notes several features of mediation in small societies that contribute to its effectiveness. First, it usually occurs soon after a dispute arises, before the disputants' positions have had time to harden. Second, it occurs in public, allowing onlookers to voice their views of the disputants' conduct. Third, many hours or days of mediation may be needed before a settlement is reached. Fourth, after a settlement is reached, the outcome, including compensation, is put into effect as soon as possible, again often in front of onlookers. Fifth, mediators are usually respected, influential members of the community instead of unknown outsiders, and they are usually also familiar with disputants' personal backgrounds. Both their community stature and their personal knowledge of the disputants enhance their mediation skills. Sixth, the mediator and onlookers use informal social control mechanisms, including gossip and ridicule and possible violence, to pressure the disputants to reach a compromise.

Although all these features help make mediation effective in premodern societies, Merry (1982) says they do not characterize mediation in the United States. In this country, mediation usually takes place long after a dispute arises, allowing positions to harden, and it occurs in private. Thus, the pressure of public opinion that is so important in small societies is lacking in U.S. mediation. As well, the mediator is typically a stranger to the disputants and has no particular community stature; both these aspects reduce the mediator's effectiveness. Moreover, because the disputants usually are not part of the same network of kin and friends, they lack social incentives and pressure to reach a settlement. Instead, many mediation programs rely on the threat of court intervention to produce a settlement. To the extent this is true, mediation ironically recreates some of the problems with adjudication that led to the rise of mediation and other ADR programs. Because mediation also lacks due process, she adds, it may increase government control over individuals without legal protections.

Anthropologist Laura Nader (2002:139) argues that the growth of ADR since the 1970s reflects a larger shift in the United States "from a concern with justice to a concern with harmony and efficiency." ADR became popular, she says, in part because many judges and public officials were dismayed by the increase in litigation stemming from the expansion of civil rights, women's rights, and other legal rights during the 1960s and 1970s and by lawsuits against businesses. Because ADR lacks the adversariness that enabled litigation to advance these rights, Nader says, ADR resembles "a pacification scheme, an attempt on the part of powerful interests in law and in economics to stem litigation by the masses, disguised by the rhetoric of an imaginary litigation explosion" (p. 144). As should be evident, Nader thinks that ADR is less able than litigation to address the root causes of social problems and to achieve social justice. She is also critical of mandatory mediation, which she says is secretive and lacks legal protections and, in divorce cases, is disadvantageous for women because it ignores their lack of power in a marriage.

Other scholars extend this last concern to the use of ADR in divorce and paternity cases in which violence has been an issue (Fisher, Vidmar, and Ellis 1993; Grillo 1991; Rimelspach 2001). According to these scholars, mandatory mediation in such cases suggests that domestic violence is a relationship problem rather than a real crime and sends the wrong message to the abuser that his conduct is not criminal. Moreover, in the small, private setting in which such mediation occurs, the abuser may continue to try to control the woman, who may fear for her safety simply by being near him. In addition, successful mediation requires that both parties honestly try to compromise, yet in a battering relationship the abuser will normally be quite reluctant to do so. Moreover, the woman who is being abused may not be the best advocate for her own interests, as the fear and intimidation she feels at home is likely to continue during the mediation process. Finally, if the woman has been living apart from her abuser and in a location unknown to him, mediation may provide him the means to discover where she lives.

Proponents of mediation in these cases say that mediators are trained to handle their special circumstances and that divorce is much better treated through mediation than through the adversarial legal system, which can escalate an already bad situation (Rimelspach 2001). Thus, mediation may lessen a woman's abuse, while litigation may worsen it. As this overview illustrates, mediation in divorce cases continues to arouse much controversy, and further research is needed to understand how it may help or hurt the situation of women whose husbands are abusing them.

Summary

1. Every society has disputes that occur in many ways and for many reasons, and every society has accepted ways of processing disputes. In general, an aggrieved party must make three decisions that are key to the disputing process: (1) whether a situation is unjust, (2) whether to confront the offending party with the problem, and (3) whether to take the problem to a third party if the two disputing parties cannot resolve the problem.
2. Several methods of dispute processing exist (1) lumping it, (2) avoidance, (3) coercion, (4) negotiation, (5) mediation, (6) arbitration, and (7) adjudication. Mediation, arbitration, and adjudication all involve a third party, a significant feature to scholars of dispute processing. The involvement of a third party takes the dispute beyond the two disputing parties and, therefore, brings it to the attention of people other than the disputants themselves. In this way, the dispute becomes more public than it was before.
3. A key feature of adjudication is that it involves a win–lose outcome instead of the compromise outcome ideally achieved by mediation. As a result, adjudication does not really address the causes of the dispute and may well worsen the relationship between the parties. Many premodern societies favor negotiation and mediation over adjudication because the former are more likely than the latter to produce the reconciliation these societies favor.
4. Several sets of factors seem to explain the preferences of societies for one form of dispute processing over another. Anthropologists stress the importance of the nature of social relationships in a society for its dispute processing preferences. Small societies have relatively few people who tend to know each other very well and to value the close relationships they have. For this reason, they favor reconciliation and thus compromise outcomes in the settlement of disputes. Because large,

industrial societies thus lack an emphasis on enduring relationships, they care less about compromise outcomes and are more likely to favor arbitration or adjudication.

5. Many scholars also think that a society's legal culture matters for its choice of dispute processing procedures. Legal culture refers to a society's general views about the law and about specific aspects of the legal system and its perceptions about using the law to address disputes. The view that litigating is wrong characterizes many traditional societies and also characterized many communities earlier in American history. Japan has also been considered a society whose legal culture disapproves of litigation, but some scholars minimize the importance of Japanese culture in this regard and instead point to other factors.
6. Several individual-level factors also help explain why some individuals within a given society are more or less likely than other individuals to litigate. These include the nature of the relationship, the extent of power difference in the relationship, and personality traits such as assertiveness.
7. Early research found that the poor were less likely to use the courts, but recent research finds a more complex situation. One branch of this research says the poor and people of color use the courts less primarily because of the kinds of problems these citizens experience. Another branch of research that uses community and ethnographic studies finds low court use overall and few, if any, differences by social class and race/ethnicity.
8. Although many scholars think that until recently men litigated much more than women, the reasons for this earlier gender gap remain in dispute. Whatever the reason, the traditional gender gap in litigation seems to have narrowed and even disappeared for many types of grievances. Scholars attribute these changes to the increased education and other resources that women have gained in recent decades, their greater involvement in the business and professional worlds, and the nation's expansion of the legal rights of women.
9. Many communities earlier in U.S. history were loathe to litigate. Wishing to preserve their own traditions, fearing that litigation would disrupt their harmony by exacerbating conflict and by implying mutual mistrust, and mistrusting lawyers, these communities developed dispute processing methods that avoided litigation. They included Puritans and Quakers in colonial America, several utopian communities during the early 1800s in New England and the Midwest, and immigrant communities in the nation's large cities during the early 1900s.
10. Contemporary studies of legal consciousness have tried to understand how Americans' everyday understandings of and experiences with the law have shaped their willingness to use lawyers and courts and also how law and other aspects of society have shaped legal consciousness. These studies find that Americans are reluctant to litigate for a variety of reasons.

11. Research on rights consciousness finds that Americans have a strong sense of their rights as reflected in the Declaration of Independence and in claims of the social movements of recent decades. It also finds that Americans believe that aggrieved groups are entitled to pursue their rights through the legal and political process, even though most Americans with grievances do not do so.
12. The United States is said to have a litigation crisis, characterized by an increasing amount of tort suits and other litigation and by extremely high jury awards. Media coverage of court cases contributes to perceptions of a litigation crisis. However, an extensive body of research finds that the amount of litigation is not increasing at a rapid pace in the state courts and that extremely large jury awards are rare. This body of research thus challenges the perception of a litigation crisis.
13. Alternative dispute resolution (ADR) has become popular in the United States during the last few decades. Proponents say that it saves time and money, helps avoid the escalation of conflict that often accompanies adjudication, considers disputants' personal needs, and increases their satisfaction with the resolution of their case. Critics say ADR may not increase disputants' satisfaction, and they question whether ADR can succeed in resolving disputes as well as it does in the traditional societies studied by anthropologists.

Key Terms

Adjudication
Alternative dispute resolution
Arbitration
Avoidance
Coercion
Legal consciousness
Legal culture
Lumping it
Mediation
Negotiation

CHAPTER 5

Law and Social Control

Chapter Outline

William Golding's (1954) classic novel, *Lord of the Flies*, was a terrifying depiction of a society in which norms broke down and social order vanished. A plane crash left a group of young boys from England stranded on a remote island. Left to fend for themselves, they tried to construct a new society with what they remembered from their lives back home, but their new society obviously lacked their families, places of worship, and other institutions that had made life possible back in England. They soon degenerated into savages, as the book repeatedly called them, with fatal results.

This brief summary does not do Golding's book justice, but one of his major themes was that society can easily dissolve into chaos without adequate social controls. This was a theme of the nineteenth-century conservative intellectual movement that arose in response to the French Revolution of 1789 and the Industrial Revolution that began several decades later. The terror of the French Revolution and the drastic changes it wrought frightened many European intellectuals, who tended to come from aristocratic backgrounds. The Industrial Revolution led to the rise of large cities with neighborhoods marked by concentrated poverty, crime, and mob violence. This development, too, frightened intellectuals, who began to write of the need for strong societies that would limit individual impulses and social disorder.

Emile Durkheim, a founder of sociology discussed in Chapter 2, was one of these conservative intellectuals. He emphasized that societies are necessarily unstable if they must depend on the threat of punishment for securing people's obedience to social norms. The most stable society results, said Durkheim, when people learn to respect and obey social norms through two related processes: socialization and social bonding. Through socialization, children learn the norms of society, but they also learn that it is morally appropriate to obey these norms; to the extent this process succeeds, they obey norms not because they are afraid of punishment if they disobey, but because they feel the norms ought to be obeyed. Effective socialization is not possible, Durkheim also said, without adequate social bonding to other members of society and, in particular, to social institutions such as religion. Effective socialization and social bonding, then, create a more stable society in which social order rests on people's willingness to obey norms rather than on the threat of punishment if they disobey.

As Durkheim recognized, effective social control is essential for any society. Without such social control, norm violation, or deviance, can become quite common and, in extreme cases like *Lord of the Flies* or any number of real-life revolutions, social order can break down completely as a society descends into chaos. In the traditional societies studied by anthropologists, norm violation is usually not a great problem. Norm violation and disputes do occur, and the anthropological literature has rich descriptions of the types of deviance found in these societies (Edgerton 1976) and of their disputing processes (Nader and Todd 1978a) discussed in the preceding chapter. For the most part, however, socialization and informal social control such as anger and ridicule are so effective that deviance in traditional societies is relatively uncommon.

Socialization and informal social control certainly characterize modern societies like the United States, but their size and heterogeneity, among other factors, lead to more deviance and more disputes than are found in traditional societies. The legal system in modern societies provides a major means of handling disputes, the focus of the previous chapter, and the principal means of social control for dealing with crime, the focus of this chapter. As we shall see, consensual (or victimless) crimes—including illegal drug use, prostitution, gambling, and pornography—are of special interest in the study of the law and social control. Because most individuals involved in these behaviors are willing participants, the legal banning of the behaviors and the use of legal machinery to enforce these bans have been a source of much controversy over the years.

This chapter examines two major, related questions: (1) How effective is law in preventing criminal behavior in the general society and in preventing repeat offending by offenders after they have been arrested, prosecuted, and punished? and (2) How effective is law in dealing with consensual behaviors that are deemed criminal largely because they offend our society's sense of moral order? In addressing the latter question, we will also implicitly be asking whether our society should ban such behaviors.

LAW, DETERRENCE, AND INCAPACITATION

Perhaps the most important goal of the criminal justice system is to prevent crime. It may do so by discouraging potential offenders in the general public from committing crime or by discouraging actual offenders who are legally punished from committing repeated crime. In either case, individuals are discouraged from committing crime because they do not want to risk the many disadvantages incurred by arrest, punishment, and other legal sanctions. When they are so discouraged, **deterrence** is said to have occurred. The criminal justice system may also prevent crime by putting convicted offenders in prison or jail; while behind bars, they obviously cannot commit crime in the outside world. If the criminal justice system prevents crime in this manner, **incapacitation** is said to have occurred.

How well does the criminal justice system achieve its goals of deterrence and incapacitation? This question certainly has much practical significance, especially because U.S. criminal justice policy during the past few decades has relied heavily on a "get tough" approach that emphasizes harsher and more certain prison terms, the necessary building of more prisons and jails, the use of capital punishment, and the hiring and deployment of many additional police, all at a cost of tens of billions of dollars. To the extent these efforts have helped prevent crime, this money has perhaps been well spent; to the extent these efforts have not helped prevent crime, this money has perhaps not been well spent. For these reasons, it is important to determine how well the criminal justice system in fact prevents crime via deterrence or incapacitation.

Such determination will help to answer a key question in the study of law and social control: To what degree does law prevent criminal behavior? In studying this issue, scholars ask three related questions: (1) To what degree does law deter potential criminal behavior in the general population? (2) To what degree does law deter repeated criminal behavior by offenders who have already been arrested and punished? and (3) To what degree does law prevent criminal behavior by incarcerating convicted offenders and thereby keeping them isolated from the outside world? The first two questions speak to the issue of deterrence, while the third question speaks to the issue of incapacitation.

Although all three questions are important for practical reasons, the first two are also important for theoretical reasons. Recall from Chapter 2 the *utilitarianism* perspective of the eighteenth and nineteenth centuries that is credited to Italian economist Cesare Beccaria and English philosopher Jeremy Bentham. Both men were criminal justice reformers who believed that people act rationally and with free will and wish, above all, to maximize their pleasure and reduce their pain. Thus, they assess the potential rewards and costs of their actions and proceed accordingly. Working from this basic belief, Beccara and Bentham reasoned that because potential offenders are rational, many will be deterred from committing crime by the prospect of arrest and punishment. Accordingly, the criminal justice system needs to be punitive only to the extent that it deters potential offenders in this manner. In trying to gauge the strength of law's deterrent impact (the first two questions above), scholars are testing a fundamental assumption of utilitarianism as conceived by Beccaria and Bentham.

In the field of criminology, utilitarianism has also been called the *classical school* of criminology. Its modern equivalent is called **rational choice theory**, a general approach that is becoming more popular not only in criminology but also in sociology and other disciplines (Wallace and Wolf 2006; Wright et al. 2004). Borrowing heavily from assumptions of human behavior in the field of economics, rational choice theory, like utilitarianism, presumes that people act after weighing the potential consequences of their actions and calculating whether a given act is more likely to be beneficial or costly. When the potential benefits outweigh the potential costs, they are likely to engage in the action; when the reverse is true, they are not likely to engage in it. When applied to criminals, rational choice theory assumes that potential offenders weigh the possible consequences of their actions, including arrest and punishment, and are more likely to break the law when they perceive the risk and severity of arrest and punishment as low, and less likely to break the law when they perceive the risk and severity of arrest and punishment as high.

A logical criminological offshoot of rational choice theory is **deterrence theory** (Miller, Schreck, and Tewksbury 2006). As Beccaria and Bentham recognized, if potential offenders weigh the likely consequences of their actions and proceed accordingly, then it makes sense to think that the law can be used to deter criminal behavior by potential offenders in the general population or by actual offenders after they have been convicted and punished. This, of course, is a key assumption guiding the "get tough" approach mentioned earlier: increasing the certainty and severity of criminal punishment and the numbers of police should, all things equal, deter potential offenders from breaking the law. In assessing the deterrent impact of the law, scholars are thus testing the fundamental assumption of contemporary deterrence theory.

Deterrence: Conceptual Considerations

Before discussing the research literature on deterrence, it will be helpful to distinguish several types of deterrence and certain dimensions of criminal behavior and criminal offenders.

Types of Deterrence. A first distinction in deterrence lies between absolute deterrence and marginal deterrence. **Absolute deterrence** refers to the effect of having some law (in terms of arrest, punishment, and other legal sanctions) or of having a legal system versus the effect of no law or of no legal system. Almost all scholars would agree that the law does have a strong absolute deterrent effect. That is, criminal behavior would be much more common if we had no legal system and its attendant sanctions compared to having the legal system we now have. Absent a legal system and these sanctions, people would feel free to break the law because they realize they could do so without fear of legal punishment.

Although law undoubtedly has an absolute deterrent effect in this respect, the question of deterrence in pragmatic terms involves one of marginal deterrence. **Marginal deterrence** refers to the effect on criminal behavior of increasing criminal sanctions (e.g., the severity or certainty of punishment or the swiftness with which one is punished) versus not increasing such sanctions. As noted earlier, belief in the impact of marginal deterrence has guided much of U.S. criminal justice policy during the past few decades, and such measures as mandatory imprisonment for gun crimes and other offenses and harsher prison terms for serious crimes have been intended to reduce crime by having a marginal deterrent effect.

For these reasons, almost all research on deterrence has involved the impact of marginal deterrence. Such research in turn examines the impact of two kinds of marginal deterrence. **General deterrence** refers to the ability of law to deter offending in the general population by sending a message that potential offenders will be apprehended and punished if they break the law. **Specific deterrence** (also called **individual deterrence**) refers to the ability of law to deter criminal behavior by offenders who have already been arrested and punished because they do not want to incur the risk of additional punishment.

Types of Offenses. In assessing the deterrent effect of law, it is important to keep in mind that certain kinds of crime may be more deterrable than other kinds by the threat of arrest and punishment. In a classic article, sociologist William Chambliss (1967) distinguished between *instrumental* offenses and *expressive* offenses. Instrumental offenses are ones committed for material gain and with a relatively large amount of planning, while expressive offenses are committed for emotional reasons and are relatively spontaneous with little planning preceding them. "Street" crimes involving economic gain, such as robbery, burglary, and motor vehicle theft, tend to be instrumental in nature, while violent crimes such as homicide and assault tend to be expressive. Chambliss argued that instrumental crimes are inherently more deterrable than expressive crimes, because people committing the latter are acting so emotionally and spontaneously that they do not normally take the time to weigh the possible consequences of their actions in terms of arrest and punishment. Marginal deterrence, then, should be less effective for expressive offenses than for instrumental offenses. If, for example, a state legislature increases the penalties for aggravated assault in the hope that these harsher penalties will deter this crime, people about to commit aggravated assault will not be deterred by the harsher penalties because they are not taking the time to consider their chances of being arrested and punished.

Chambliss also distinguished between offenders with high commitment to their criminal behavior and criminal careers and those with only low commitment. For example, "cat" burglars and other so-called professional offenders are highly skilled at what they do and are very committed to their life of crime; drug addicts are also very committed, if for different reasons, to their illegal drug use. In contrast, amateur property criminals, such as teenagers who take a car for a joyride, are relatively uncommitted to a criminal career. Chambliss reasoned that the level of commitment to crime should be related to the likelihood of being deterred by possible arrest and punishment. To be more precise, offenders with low commitment to crime have less stake in their criminal behavior and thus are more likely to be deterred by the threat of legal sanctions. In contrast, those with high commitment to crime place great importance on their criminal behavior and thus are less likely to be deterred by the threat of legal sanctions.

A third distinction also helps differentiate crimes that are more or less deterrable by law. Some crimes tend to occur in public, broadly defined, while others tend to occur in private. Most "street" crimes are relatively public, while offenses such as domestic violence and some forms of illegal drug use tend to occur in the privacy of one's residence. Because public offenses are more easily detected by law enforcement agents, they should be more deterrable, all things equal, than private offenses by the threat of legal sanctions.

These basic distinctions of offenses and offenders have important implications for the degree to which law can have a marginal deterrent effect. Rational choice and deterrence theories both assume that potential offenders carefully assess the likely consequences of their actions and weigh the possible benefits (e.g., obtaining money or possessions very quickly) against the possible costs (e.g., arrest and possible imprisonment). Because potential offenders do so, they are capable of being deterred in a marginal manner; that is, they are capable of being deterred by changes (such as new mandatory sentencing or longer prison terms) that make legal sanctions more certain and/or more severe. However, Chambliss' analysis indicates that offender assessment of likely consequences is much less apt to happen for expressive crimes than for instrumental crimes because the former are committed out of strong emotions and with little or no planning. If so, expressive offenses, including most homicides and assaults, are less easily deterrable by law (in terms of marginal deterrence) than instrumental offenses, and perhaps not deterrable at all.

The other distinctions between high-commitment offenders and low-commitment offenders and between public and private offenses also have implications for the likelihood of marginal deterrence. High-commitment offenders—professional criminals, drug addicts, or those with a large stake in crime for other reasons—are much more resistant to marginal deterrence than low-commitment offenders. Similarly, private offenses are also more resistant to marginal deterrence than public offenses.

These considerations all suggest that legal sanctions may not easily deter many types of offenses and offenders. For all these offenses and offenders, marginal deterrence should be relatively low or even nonexistent. To say this a bit differently, changes in the law that make arrest and imprisonment more certain or that make punishment harsher are likely to have only a small deterrent effect, if that, on crime rates involving expressive offenses, and/or high-commitment offenders, and/or private offenses.

Other Considerations. Of course, these considerations also imply that harsher or more certain legal sanctions might well have a higher marginal deterrent effect on the remaining categories of offenses and offenders: instrumental offenses, low-commitment offenders, and public offenses. This implication seems plausible on its face, but studies of actual criminals have yielded important understandings of their decision making and behavior that call this implication into question. Two findings from this literature are particularly relevant in this regard.

The first finding comes from lengthy interviews of active burglars and robbers who are still out in the streets and of similar offenders who have been convicted and incarcerated. Because their crimes are instrumental, many of these offenders plan them to a fair extent and carry them out without undue emotion; for these reasons, they are capable of weighing the likely consequences of their actions and thus, according to rational choice and deterrence theory reasoning, are capable of being deterred by the threat of legal sanctions. Yet, the studies of the offenders just described reveal a different picture from that assumed by rational choice and deterrence theories (Tunnell 2006; Wilson and Abrahamse 1992; Wright et al. 2004). Many of these offenders either assume they will not get caught because they plan their crimes or else just give little or no thought to their chances of getting caught. Others realize they might be caught but resign themselves to this outcome because they have a fatalistic attitude, and some do not regard the prospect of prison as very unpleasant. As criminologist Samuel Walker (2006:128) summarizes the evidence from these studies, "Actual offenders do not appear to make their decision about criminal activity on the basis of a rational and carefully calculated assessment of the costs and benefits." As a result, they are relatively incapable of being deterred by the threat of legal sanctions, and thus relatively insusceptible to marginal deterrence.

The second finding comes from interviews of prison and jail inmates; the federal government has conducted out such interviews of national random samples of inmates several times during the past two decades. One set of questions asks inmates about their extent of drug and alcohol use at the time of the offense that led to their incarceration. Inmates' responses to these questions reveal a high level of drug and alcohol use while committing their crimes (Karberg and James 2005; Mumola 1999). Specifically, about one-half of convicted offenders say they were drinking or under the influence of other drugs at the time of their offense. It is fair to say that these offenders' substance abuse made it very difficult and perhaps impossible for them to carefully weigh the possible consequences of the actions they were about to commit. If so, it is fair to say that one-half of offenders are incapable of being deterred by the threat of legal sanctions because of their substance use at the time of their offense and thus not susceptible to marginal deterrence.

Here it may be argued that offenders may be more likely to be captured and arrested if they were drinking or using drugs at the time of their offense. If so, studies of convicted offenders may exaggerate the amount of substance abuse at the time of the offense, resulting in an overly pessimistic appraisal of the potential of marginal deterrence. According to the National Crime Victimization Survey (NCVS), an annual government survey of thousands of people nationwide, about one-third of victims of violent crime (typically the only ones that see their offender) perceive that their offender was using alcohol or drugs at the time they were victimized. Whether the percentage of offenders using alcohol or drugs at the time of offense is as low as one-third, as NCVS data suggest, or as high as one-half, as inmate survey data suggest, many offenders are still incapable of being deterred by legal sanctions because of their substance abuse. For these offenders, then, marginal deterrence works poorly at best and perhaps not at all.

An additional conceptual consideration again calls into question the deterrence argument. For law to have a deterrent effect, it makes sense to think that potential offenders must accurately know their chances of getting arrested and harshly punished so that they can weigh these chances against the possible rewards of their criminal behavior. If they lack such awareness, they cannot be marginally deterred by any real increases (e.g., because of new laws or new criminal justice practices) in the certainty and severity of punishment. Legislators, criminal justice professionals, and criminologists may be aware of offenders' chances of arrest and punishment, but it turns out that most offenders themselves lack such awareness. Without such awareness, then, they cannot be deterred by any increases in the certainty and severity of punishment. Sometimes, of course, a good deal of publicity accompanies a new law that imposes a mandatory minimum prison term for a specific offense or drastically increases the maximum prison term. With such publicity, it might be expected that offenders will indeed be deterred by the new law. However, research on this possibility (to be summarized below) finds that highly publicized new laws still do not have a noticeable marginal deterrent effect. If even highly publicized increases in the severity of punishment do not deter crime, then deterrence theory once again has very little support.

Three final conceptual considerations also suggest that law can have little, if any, marginal deterrent effect on the bulk of offenders (Nagin 1998a). First, as Walker (2006:111) points out, "if arrest and imprisonment are common experiences in a particular social group, they begin to lose their stigmatizing effect." If this happens, the prospect of arrest and imprisonment thus loses its deterrent effect. Because so many young African American males have been arrested during the past few decades as part of the nation's "get tough" approach on crime, many scholars feel that arrest and imprisonment have lost any deterrent effect they might have had for this group of offenders. As Walker (2006:111) puts it, "So many people in their neighborhoods are arrested and convicted that it is simply a 'normal' life experience, with little deterrent effect."

Second, many times new laws and policies are enacted to try to reduce crime, but they are not implemented in the way that legislators and policymakers would have preferred. A state may enact a law that increases the maximum prison term for a specific crime only to find that prosecutors in some cities fail to apply the law, perhaps because they disagree with it or worry that it will exacerbate prison overcrowding. A state or the federal government may provide funds for hiring more police, but some jurisdictions may deploy the police less efficiently (and thus with less deterrence) than other jurisdictions. There is much evidence, in fact, that new laws and policies are not implemented in the real world as they were intended to be implemented, a fact that again calls into question law's potential deterrence (Walker 2006).

Finally, the chances of arrest and punishment for serious crime are so low that it would be surprising if incremental (marginal) increases in these chances had a strong deterrent effect. In 2005, for example, victims of violent and property crime reported only about 41.5 percent of their crimes to the police, and police made arrests in only about 19.7 percent of these reported crimes (Catalano 2006; Federal Bureau of Investigation 2006). Putting these figures together, police made arrests in about 8 of every 100 serious street crimes. Even if new laws and policies doubled this low risk of arrest, a virtually impossible prospect, more than four-fifths of all serious crime would still not end in arrest. Moreover, while only 8 out of every 100 serious crimes end in arrest, only about 4 out of every 100 serious crimes are prosecuted, with the charges dropped for the remaining arrests, and only about 1 out of every 100 serious crimes end in incarceration (Barkan 2006b). In the United States, serious street crime is thus such a low-risk event (in terms of arrest and punishment) for offenders that it would be surprising if law did have a strong deterrent effect.

Taking into account all the conceptual aspects of crime and criminals just discussed, a realistic look suggests that marginal deterrence cannot really work for many and, perhaps, most potential offenders, despite what rational choice and deterrence theories assume. An extensive amount of empirical research, to which we now turn, supports this conclusion.

Research on Deterrence

Over the years, research on deterrence has addressed the general deterrent effect of the certainty and severity of punishment. The **certainty** of punishment refers to the likelihood of being arrested and is typically measured as the number of arrests for a given type of crime divided by the number of offenses (as measured by the federal government) for that type of crime. This resulting figure yields the percentage of all known offenses for a give type of crime that end in arrest. The **severity** of punishment refers to the harshness of prison sentences. Typical measures include the maximum sentence an offender may receive for a given crime or the actual sentences that convicted offenders receive on the average.

Early research on deterrence compared states' certainty and severity levels with their crime rates (typically measured as the number of crimes per 100,000 residents) (Gibbs 1968; Tittle 1969). These studies usually found that states with higher certainty levels had lower crime rates, and they also sometimes found that states with higher severity levels again had lower crime rates. In general, the evidence for certainty was more consistent than that for severity. Because the evidence overall was consistent with what deterrence theory would expect, this early research concluded that law (or more precisely punishment and the threat of punishment) indeed had a deterrent effect.

This conclusion was soon challenged for at least two reasons. The first reason pertained to the familiar causal order or "chicken and egg" problem that characterizes many two-variable (in this case, crime rates and the certainty or severity level) relationships (Babbie 2007). In many such relationships, it is difficult to know which variable affects the other or whether there may be a reciprocal relationship in which each variable affects the other. To take a common example from delinquency research, it is often found that adolescents with close relationships with their parents are less likely than those with more distant relationships to be delinquent. Although it is usually assumed that the nature of the adolescent–parent relationship affects the likelihood of delinquency, it is quite possible that delinquency affects the nature of the relationship. Thus, an empirical link between close relationships and less delinquency may mean that the nature of the relationship affects the amount of delinquency, but it may also mean that the amount of delinquency affects the nature of the relationship or that a reciprocal relationship exists between the two variables.

To return to deterrence research, although the early evidence linking higher certainty and severity to less crime did support deterrence theory, critics soon pointed out an alternative explanation that called this conclusion into question (Decker and Kohfeld 1985; Nagin 1978; Pontell 1984). This explanation reversed the supposed causal order of the certainty/severity–crime relationship and became known as the **system capacity** model. According to criminologist Henry Pontell (1984:6), this model assumes that "crime levels affect criminal justice practices just as much if not more than such practices affect crime." According to this way of thinking, the chances of arrest and the severity of punishment both depend on the capacity of the criminal justice system to carry out these actions. All things equal, states with higher crime rates must worry about overcrowding their prisons and jails and thus make adjustments by arresting fewer people and/or having weaker sentences (shorter prison terms or no incarceration at all) for offenders who are arrested and convicted. Similarly, police in locations with high crime rates have less time and fewer resources to investigate crimes than police in locations with low crime rates.

For these reasons, arrest rates (certainty) and the likelihood and length of imprisonment (severity) are both lower in states with higher crime rates than in those with lower crime rates, as crime rates affect both the certainty of arrest and the severity of punishment. This process yields the same inverse correlation that the early deterrence research found and calls into question its support for deterrence theory. It also has a troubling implication for the real world of crime and deterrence that was aptly stated by Pontell (1984:6) more than two decades ago: "It is in precisely those circumstances where deterrence might be most needed (conditions of high incidence of crime) that the criminal justice system is least able to effect it."

Subsequent research on general deterrence became much more methodologically sophisticated in ways that lie beyond the scope of this book but still focused on the effects of the certainty and severity of punishment. A few studies have concluded that increases in the number of police and in imprisonment rates have reduced crime rates, as deterrence theory would predict (Marvell and Moody 1994:1996). A particularly intriguing study compared the number of arrests and the number of crimes in the Orlando, FL, area during a six-month period (D'Alessio and Stolzenberg 1998). The daily number of arrests ranged from 8 to 104, with a daily average of about 54. In general, an increase in arrests on a given day was accompanied by a decrease in crime the next day, most likely because potential offenders hear of the new arrests and decide to lay low.

These studies notwithstanding, most studies suggest that general deterrence is a false hope. A review from the early 1990s concluded that the bulk of studies "provide little, if any, evidence consistent with the general deterrence perspective" (Chamlin 1991), and later reviews have reached a similar conclusion (Doob and Webster 2003; Nagin 1998b; Walker 2006). This conclusion rests on several kinds of evidence.

First, imprisonment rates have risen consistently and dramatically since the 1960s. Deterrence theory would predict that crime should decline as a result, and that this decline should be greatest when the increase in imprisonment rates is greatest. However, in reality the crime rate has not consistently decreased since the 1960s. For example, the violent crime rate rose steadily during from the mid-1980s to the early 1990s even as imprisonment was soaring. Although the crime rate declined steadily since the early 1990s before a recent leveling off, factors other than incarceration (such as an improved economy and a stabilizing of the illegal crack market) seem to account for most of this crime decline (Blumstein and Wallman 2006).

Second, laws that have increased the penalties for gun-related crimes and for certain other offenses have not usually lowered rates of those crimes. The gun-crime laws are particularly relevant in this regard. Many were accompanied by much publicity, to recall our earlier discussion of offenders' lack of awareness of certainty and severity, and thus should have had a general (marginal) deterrent effect. Yet, studies of these gun laws have generally failed to find this effect. A study that examined the effects of such firearm sentence enhancement (FSE) laws in the forty-four states that enacted them since the 1960s concluded that "on balance the FSE laws do little nationwide to reduce crime or gun use" (Marvell and Moody 1995:274). Studies of FSE laws in specific states and cities, including Massachusetts and Detroit, also fail to find a general deterrent effect (Loftin and McDowall 1981; Pierce and Bowers 1981).

Third, so-called "three strikes" laws that impose life imprisonment or at least a very long sentence for an offender committing a third felony also have not lowered the crime rate in states that have enacted such laws since the mid-1990s (Austin et al. 1999; Kovandzic, Sloan, and Vieraitis 2004). In fact, violent crime fell at a greater rate during the 1990s in states that did not enact three strikes laws than in the states that did enact them (Walker 2006). Some evidence even suggests that these laws caused *more* homicides, possibly because offenders committing their third felony and facing a life term do not want to leave any.

Fourth, the death penalty—the most severe legal punishment possible—also does not have a general deterrent effect on homicide. Many studies have examined this issue, and almost all find no support for deterrence theory (Levitt and Miles 2006; Walker 2006). States with the death penalty generally do not have lower homicide rates than states without it, and the few states that eliminated their death penalty several decades ago did not experience a homicide increase compared to the states that retained capital punishment. Some studies even find that executions increase the number of homicides in what is called the **brutalization effect** (Bailey 1998; Cochran, Chamlin, and Seth 1994). Scholars speculate that some offenders have a "death wish" and kill someone because they want to be killed by the state through execution or that some offenders are simply imitating the killing that execution involves.

In sum, law does not appear to have a general deterrent effect on criminal behavior. Absolute deterrence almost certainly does work, but marginal deterrence—the key issue for criminal justice policy—does not appear to work, in large part because potential offenders do not decide and behave in the ways that rational choice and deterrence theories assume. As Walker (2006:129) summarizes the evidence, "The commonsense notion that people will avoid unpleasant things and that we can influence their decisions by increasing the unpleasantness does not necessarily work in the real world of criminal justice."

Specific Deterrence. Even if law does not have a general deterrent effect, is it possible that it has a specific or individual deterrent effect? Recall that this type of deterrence involves offenders who have already been arrested, convicted, and sentenced and who decide not to commit additional offenses because they dread another arrest and term of imprisonment. Although it might make sense to think that law should have this type of deterrent effect, most studies, in fact, find little or no evidence for it. These studies usually compare offenders who are convicted of similar crimes but who serve prison terms of different lengths. From a methodological standpoint, the ideal study would involve randomly assigning prison inmates convicted of similar crimes to two groups: one that serves a longer term and one that serves a shorter term. If specific deterrence does occur, inmates who serve the longer term should have a lower recidivism (repeat offending) rate when finally released from prison than those who serve the shorter term.

Although it is probably impossible to win permission to conduct this type of experiment, prison overcrowding has forced states to give some inmates early release. Criminologists then have compared these inmates' recidivism with that of inmates who serve their full sentences. The released inmates generally do not have higher recidivism rates than their counterparts who serve their full terms, and sometimes they even have lower recidivism rates (Austin 1986). If inmates who stay in prison longer have higher recidivism rates, these higher rates probably derive from the greater exposure of these inmates to the criminal subculture of the prison and from the psychological effects of the "pains of imprisonment," to use Gresham Sykes' (1958) notable phrase from his classic study of the prison experience.

A similar line of research compares juvenile offenders who have their cases transferred to the (adult) criminal courts with those whose cases remain in juvenile court, where the punishment is less severe. During the past two or three decades, several studies have transferred juvenile cases involving serious violent offenses. If specific deterrence works, the recidivism rates of the juveniles transferred to criminal court should be lower than those who stay in juvenile court. However, studies of this issue find that the recidivism rates of the transferred juveniles are actually higher than those of their juvenile court counterparts, contrary to what deterrence theory would predict (Bishop 2000; Lundman 2001).

In sum, law does not appear to have a specific deterrence effect. If anything, offenders treated more harshly by the criminal justice system are more likely to commit new offenses than offenders treated less harshly. If this is true, the harsher sentencing that is a key component of get tough approach guiding U.S. crime policy during the past few decades may in the long run be increasing crime rather than reducing it.

The Issue of Incapacitation

If law does not have a general deterrent effect, perhaps it still has a strong incapacitation effect that helps prevent crime and thus achieve its goal of social control. Recall that incapacitation in criminal justice occurs when offenders are incarcerated and obviously cannot commit crimes in the outside society. While they are behind bars, then, our society is theoretically safer. Although it is certainly difficult to deny that incarcerated offenders cannot commit new crimes in the outside society, the question for criminal justice policy and the more theoretical issue of law and social control is the *strength* of this incapacitation effect. If this effect is strong—if the nation's soaring imprisonment rates over the past few decades have substantially reduced crime by incarcerating so many criminals—then the costs of such imprisonment are perhaps justified, and law can be said to exert a significant amount of social control through incapacitation. On the other hand, if this effect is weak—if soaring imprisonment rates have only reduced crime to a minimal degree—then the costs of such imprisonment are perhaps unjustified, and law cannot be said to achieve a significant amount of social control through incapacitation.

As with deterrence, incapacitation is a complex issue conceptually and one on which several types of research have yielded a large body of evidence. Here again, we must distinguish between incapacitation in the *absolute* sense and incapacitation in the *marginal* or *incremental* sense. Few people would dispute that the crime rate would rise significantly if the more than 2.2 million U.S. offenders who are incarcerated at any one time were suddenly released. As criminologist Daniel S. Nagin (1998b:365) observes, "To be sure, there would be a sizable increase in the crime rate from the wholesale release of the incarcerated population." Incarceration and, thus, law have a strong incapacitation effect in the absolute sense. However, the relevant question both for criminal justice and for law and social control is whether incarceration has a strong incapacitation effect in the incremental sense: If we put more people in prison, how much will the crime rate reduce because of incapacitation? Is this reduction relatively large, or is it relatively small?

Before turning to research on this question, a conceptual consideration suggests that any incapacitation effect of incarceration is likely to be small. Recall from our earlier discussion that only about 1 percent of serious crimes (violent and property) in a given year end in incarceration. In 2004, this amounted to about 360,000 offenders who were convicted and incarcerated for violent or property crime (Durose and Langan 2007). If somehow we were able to double this number, an almost impossible achievement, the percentage of serious crime ending in incarceration would "jump" to 2 percent. This example is admittedly somewhat simplistic because it assumes that each offender commits only one crime per year and that the offenders who were incarcerated thus account for only 1 percent of the total crime committed in that year. If each offender accounted for many more crimes on the average, then doubling the incarceration rate would more than double the number of crimes accounted for.

A small number of offenders do commit many crimes each year and are called *chronic offenders*, but most offenders commit only a handful each year. For various reasons, it is very difficult to be sure of how many violent or property offenses per year each prison or jail inmate has committed on the average, but an educated guess from research on this issue would yield an estimate in the range of 3.5 to 5 per year (Walker 2006). If so, the number of new inmates every year account for only about 5 percent at most of all serious crime in that year, resulting in an incapacitation effect that is still very low. However, it is also true that this effect accumulates as more and more people are incarcerated each year. Over time, then, the incapacitation effect from increasing incarceration might become quite sizable. The relevant questions then become: (1) what is the size of this effect? and (2) how cost-effective is the amount of incarceration needed to achieve this effect?

Two additional considerations suggest that the size of the incapacitation effect will necessarily be relatively small at best. First, much crime is committed by groups of offenders. If one of them is arrested and incarcerated, the remaining members of the group may recruit a new member or simply continue to commit crime even in the absence of the arrested member. For this reason, says Nagin (1998b:365), "it is unclear whether the incarceration of one member of a group will avert any crimes at all."

Second, and more important, it is difficult for several reasons to identify and incarcerate chronic offenders to achieve **selective incapacitation**, which should have a large effect on the crime rate. U.S. criminal justice policy thus has instead relied on **collective incapacitation** (also called **gross incapacitation**), involving the mass imprisonment of offender without regard to their history and likely future of offending. Many chronic offenders are already in prison because their high rate of offending greatly increases the chances that they will be apprehended and incarcerated. As more and more people are incarcerated, these "extra" offenders increasingly become only occasional offenders, and the incremental incapacitation effect from incarcerating them becomes smaller as the rate of incarceration increases. Because the U.S. incarceration rate is the highest in the Western world and has been so for some time, the extra offenders being sent to prison every year yield only a small incapacitation effect (Blumstein and Wallman 2006). Walker (2006:143) summarizes this problem: "As we lock up more people, we quickly skim off the really high rate offenders and begin incarcerating more of the less serious offenders. Because they average far fewer crimes per year . . . we get progressively lower returns in crime reduction . . ."

The noticeable crime decline during the 1990s occurred during a period of soaring incarceration rates. Studies of the incapacitation effect of this rising incarceration find that it accounted only for about 25 percent of the crime decline during the 1990s and that this effect was not at all cost-effective given the huge economic cost of the extra incarceration (Blumstein and Wallman 2006). The author of one of these studies called imprisonment "an incredibly inefficient means of reducing crime" (Spelman 2000:124). After concluding that the extra incarceration prevented about 100 homicides annually during the 1990s but at the cost of $1 billion per year in prison costs, another author speculated that this money would have achieved much greater reductions in crime had it been spent on such things as "preschool programs, parent training, vocational training, drug treatment, and other promising prevention programs" (Rosenfeld 2000:151). Thus, although massive imprisonment may have an incapacitation effect, this effect is relatively small and costs an extraordinarily high amount of money that, many scholars feel, would be more wisely spent on other crime control strategies.

Perhaps even worse, the massive imprisonment of the past few decades may be creating a breeding ground for even more crime as offenders are released back to their communities (Blumstein and Wallman 2006; Travis and Visher 2005). More than 650,000 inmates are released from prison every year. Once home, they face bleak job prospects and difficulty in forging relationships with law-abiding persons. As indicated earlier, many come out of prison with as many problems as when they entered prison and often with even more problems. Not surprisingly, many ex-inmates end up committing more crime, as about two-thirds of all ex-prisoners are rearrested within three years. In fact, they are between 30 and 45 times more likely, depending on the type of offense, than the general population to be arrested (Rosenfeld, Wallman, and Fornango 2005). Although incarceration may have a short-term incapacitation effect, in the long run it might well be worsening the crime problem rather than improving it. Although the social control of crime is one of the most important goals of law in American society, our current criminal justice policies and practices may ironically be contributing to crime rather than controlling it.

LAW, MORALITY, AND CONSENSUAL CRIME

Before 1965, the use of birth control was a crime in several states. Before 2003, private homosexual sex between two adults was also a crime in several states. Before 1967, marriage between people of different races was yet another crime in many states. During the colonial period 300 years earlier, it was a crime in Puritan Massachusetts to commit adultery or to curse one's parents, and the penalty (though very rarely implemented) for these horrible crimes was death.

As these examples indicate, law has often been used in the United States to enforce standards of morality, even when the behavior addressed by law involves consenting adults. Today all the behaviors just listed are legal, but the law continues to ban other behaviors involving consenting adults that are thought in many circles to be immoral and/or socially harmful: the use of certain drugs, prostitution, certain types of gambling, and certain forms of pornography. Abortion is another behavior that is the subject of many legal battles, at the heart of which is the moral question (no matter which stance you take on the abortion issue) of whether terminating a fetus is equivalent to killing a human being.

These behaviors all fall into a special category of crime that raises important issues for the understanding of law and society. This category involves crimes in which the participants engage voluntarily. As such, they do not involve unwilling victims and in this sense are different from crimes such as homicide, robbery, rape, and burglary, all of which victimize people who do not want to be victimized. When they first began receiving scholarly attention more than four decades ago, these behaviors were termed "crimes without victims" or "victimless crimes" to distinguish them from the violent and property crimes that do have (unwilling) victims. However, increasing recognition that these behaviors do have victims—that is, some of their participants do suffer in various ways and some of their participants' friends and family members also may suffer—eventually led to a preference for other terms such as "public order crimes" and "consensual crimes." To emphasize that these behaviors involve consenting participants and that their willingness to engage in these crimes has important implications for the law's attempt to ban their behaviors, we will use the term **consensual crimes** in the following discussion.

Because consensual crimes, by definition, involve participants who want to engage in these crimes, they raise two important and fascinating questions for an understanding of law and society (Packer 1968; Schur and Bedau 1974). The first of these might be called the **philosophical question**: In a free society, to what degree should the state prohibit behaviors in which people want to engage because the behaviors are perceived to be immoral and/or socially harmful? The second question might be called the **social science question**: Do laws against consensual crimes do more good than harm, or do they do more harm than good?

Philosophical Considerations

Because the field of law and society is grounded in the social sciences, it ultimately cannot answer the first question, which is a topic primarily addressed by political philosophers at least since John Stuart Mill in the nineteenth century. Mill, the famous British philosopher, espoused what today would be called a strict libertarian position, as he believed that the state should allow people (i.e., adults) to engage in any behavior they desired as long as long as the behavior did not hurt anyone else and even if that behavior might hurt its participants. Here, a famous passage from his seminal work *On Liberty* is often cited:

> [T]he only purpose for which power can be rightfully exercised over any member of a civilized community, against his will, is to prevent harm to others. His own good, either physical or moral, is not a sufficient warrant. He cannot rightfully be compelled to do or forbear because it will be better for him to do so, because it will make him happier, because, in the opinion of others, to do so would be wise, or even right. These are good reasons for remonstrating with him, or reasoning with him, or persuading him, or entreating him, but not for compelling him, or visiting him with any evil in case he do otherwise (Mill 1999 [1859]:51–52).

The only exception Mills gave to this general argument was slavery, as he thought that slavery was so evil that no people should be allowed to sell themselves into slavery even if they needed money for their families.

A full examination of Mill's treatise is beyond the scope of this book, but the general philosophical question in which his argument is grounded—to what degree should the state ban consensual behaviors that are perceived as immoral and/or socially harmful?—raises at least two subsidiary issues that merit some consideration. The first pertains to the issue of immorality. Who is to decide which behaviors are immoral (and by extension, which are moral)? If lawmakers deem a behavior immoral even though many members of the public do not, is it acceptable for lawmakers to ban the behavior? If a minority of the public engages in a behavior that a majority of the public deems immoral, is it acceptable for that behavior to be a crime only for this reason? More generally, does the state have the right to mandate morality through the law? In a free society, is that a function the state should pursue?

These are all philosophical issues that lie beyond our scope, but their resolution can have social repercussions that lie within our scope. Historically, individuals and groups with strong influence in American political, economic, and social life have used their influence to ban behaviors that they considered immoral; not coincidentally, these were behaviors in which subordinate groups—immigrants, the poor, people of color—tended to engage at greater rates. Thus, "just as beauty is in the eyes of the beholder," as the old saying goes, so can morality be in the eyes of the beholder, and dominant groups may use their influence to impose their morality through law and other means on subordinate groups.

Joseph Gusfield's (1963) classic treatise, *Symbolic Crusade: Status Politics and the American Temperance Movement*, recounted a telling illustration of this dynamic. Gusfield's book concerned the years-long effort to ban the manufacture, sale, and use of alcohol that finally led to the adoption in 1919 of the Eighteenth Amendment. According to Gusfield's account, the American temperance movement consisted mostly of white, Anglo-Saxon Protestants from small towns and rural areas, who also dominated state legislatures and the U.S. Congress. Not surprisingly, their strong religious views led them consider drinking a moral sin. Yet, their antipathy to drinking was fueled in no small part by social trends after the Civil War. In particular, as industrialization continued apace, cities grew and along with them crime and disorder. At the same time, Catholic immigrants from Ireland and Italy moved into many cities and became major components of the urban poor. These immigrants brought with them from the old country their own cultures and cultural habits, among them an affinity for drinking that horrified the small town and rural Protestants who had been living in America for many generations. In the eyes of the Protestants, moreover, the drinkers had several strikes against them: they were poor, they were urban, they were immigrants, and they were Catholic. All these prejudices combined with the drinking itself to produce even greater antipathy among the Protestants toward the use of alcohol, which they considered not only a moral sin but also a behavior that was socially harmful to families and to the very fabric of society.

The Protestants' successful attempt to ban alcohol was equivalent, wrote Gusfield, to a symbolic attack on the types of people—poor, urban, Catholic immigrants—who tended to drink alcohol. The effort to ban alcohol was finally able to succeed because the rural and small town Protestants who made up the temperance movement also controlled state legislatures and the Congress. In effect, they used their political and social dominance to enforce their moral standards on subordinate groups toward whom they held prejudicial views. As this historical example indicates, the use of law to ban behavior considered immoral raises important questions of law, power, and politics in addition to the underlying philosophical question, as framed by Mill, at stake.

The second subsidiary issue raised by this philosophical question is just as important. Most nations, including the United States, do not take Mill's strict libertarian position. They ban some consensual behaviors while allowing other consensual behaviors. This fact speaks to the second issue: In a free society, which behaviors perceived to be immoral and/or socially harmful should be prohibited, and which should be allowed?

In thinking about this question, consider the many behaviors in which people engage that pose some risk to their health or safety: driving a car or other motor vehicle; skiing, swimming, mountain climbing, or playing any number of competitive sports; eating ice cream, candy, donuts, red meat, or other food laced with saturated fats or trans fats. All these behaviors can injure you or impair your health. To take just the eating example, the average American diet is one that includes the foods just listed and is responsible for obesity and attendant health problems, including heart disease and thousands of premature deaths annually. In a free society like the United States, we are not about to ban eating these foods, and neither are we about to ban skiing, mountain climbing, or playing football, even though all of these sports cause thousands of serious injuries annually. Rather, we feel that adults in a free society should be allowed to decide for themselves whether they want to incur the risk that all these behaviors entail, even though the illness, injury, and deaths they suffer costs the nation billions of dollars in health care expenses and lost productivity and, in the worst situations, a profound emotional toll on their friends and family members. The individuals (which are most of us) engaging in these behaviors, then, are potentially hurting not only themselves, but also their friends and loved ones, and they are costing our society an untold number of dollars

Although we think people in a free society should be allowed to engage in all these behaviors despite the problems they often generate, we do not hold the same view about the behaviors that fall into the category of consensual crime. By the same token, although we allow individuals to engage in behaviors like premarital sex, adultery, and homosexuality that much of society regards as immoral, we also do not hold the same view about consensual crime behaviors. Specifically, we do not let people use certain drugs, engage in prostitution and certain types of gambling, and view certain forms of pornography. Our reasoning is that these behaviors are immoral and/or harmful to the individuals engaging in these behaviors or to their friends and loved ones. The result is that some behaviors that are perceived as immoral and/or socially harmful are prohibited by law, while other behaviors perceived as immoral and/or socially harmful are permitted. Where is the logic in the determination of which behaviors should be prohibited and which should be permitted?

The answer might be that there is little or no logic in the distinctions our society has made between these two categories of behaviors. Some legal behaviors are much more socially harmful, at least in terms of death, illness, or injury, than some illegal behaviors. Conversely, some illegal behaviors are much less socially harmful than some legal behaviors.

As an illustration, consider legal and illegal drugs. Two legal drugs, tobacco and alcohol, are responsible for an estimated 535,000 (tobacco, 435,000; alcohol, 100,000) premature deaths annually in the United States. Gastrointestinal side effects from another category of legal drug, NSAIDs (nonsteroidal anti-inflammatory drugs; examples include aspirin and ibuprofen) kill at least 16,500 people every year (Wolfe, Lichtenstein, and Gurkipal 1999). In contrast, one illegal drug, marijuana, has probably never been responsible for a single death even though about 700,000 people are arrested every year in the United States for using or possessing the drug. All other illegal drugs are responsible for an estimated 17,000 deaths annually (Mokdad et al. 2004), and many of these deaths stem from overdoses or from impurities in the illegal drugs rather than to the physically harmful effects of the drugs themselves. These 17,000 deaths are about the same number of deaths caused every year by the NSAIDs, yet there is no call to make the use of NSAIDs illegal and punishable by arrest and imprisonment.

As another illustration, consider gambling. Many forms of gambling—playing the numbers, betting on sports in most locations, and so forth—are illegal, while many other forms of gambling—state lotteries across the nation, casino gambling, and bingo and beano in churches—are legal and are also an important source of revenue. Are there any logical reasons for why some forms of gambling are illegal and other forms are legal? Are there any logical reasons for why some forms of gambling are illegal and other "easy ways" to lose your money, including risky investments in the stock market, real estate, and other ventures, are legal?

All these considerations fall far short of answering the philosophical question—in a free society, to what degree should the state prohibit behaviors in which people want to engage because the behaviors are perceived to be immoral and/or socially harmful?—with which we began the discussion, but they at least suggest some of the difficulties in applying the law to consensual behaviors in a free society.

Social Science Considerations

Now let us turn to the *social science question* listed earlier: Do laws against consensual crimes do more good than harm, or do they do more harm than good? As its name implies, this is a question on which the social sciences have much to say, even if there is good evidence to support either position implied by the question. Accurate answers to the question are important for social policy. If laws against consensual crimes do more good than harm, then it makes sense (leaving aside for now the philosophical considerations just discussed) to have the laws and perhaps even to ban other consensual behaviors. On the other hand, if laws against consensual crimes do more harm than good, even if they do some good, then it makes sense to repeal the laws or at least to take a careful look at their application.

In considering this social science question, critics of consensual crime laws often harken back to Prohibition, which lasted from the enactment of the Eighteenth Amendment in 1920 to the passage of the Twenty-first Amendment, which repealed Prohibition, in 1933 (Husak 2002; Meier and Geis 2007). After Prohibition took effect, the demand for alcohol continued, even though the best historical evidence indicates that fewer people were drinking. Seeing an opportunity to make great profits, organized crime met this demand by becoming heavily involved in the manufacture, distribution, and sale of alcohol. Its success in doing so greatly expanded organized crime's wealth, power, and influence.

The fact that alcohol was illegal led to this expansion. It also led to several related problems. First, organized crime gangs vied to control the alcohol market in specific locations. As many films have since depicted, this competition often was violent, with drive-by shootings with machine guns resulting in many deaths, including those of innocent bystanders. Second, police efforts to stop the newly illegal activity surrounding alcohol manufacture and sale were also met with violence; as many films have also depicted, the police and organized crime engaged in numerous violent confrontations, with many police dying or being wounded in the line of action. So many organized crime members, bystanders, and police were killed because of Prohibition that the number of homicides increased during this period even though fewer people were apparently drinking and thus less prone to violence. Third, official corruption increased as police and politicians took bribes or blackmail from organized crime to look the other way, and as police confiscated illegal alcohol only to sell it themselves. Fourth, although alcohol use probably decreased during Prohibition, millions of people continued to use it, even though the government spent millions of dollars to enforce Prohibition. These and other problems led our society to conclude that Prohibition was doing more harm than good, resulting in the Twenty-first Amendment and the repeal of Prohibition in 1933.

Citing the Prohibition experience, critics of consensual crime laws make several arguments to support their view that these laws do more harm than good. To simplify matters, our discussion will focus on laws against the use of drugs like cocaine, heroin, and marijuana. First, and in no particular order, the laws cost at least $40 billion every year in law enforcement, court processing, and corrections costs, with little evidence to indicate that they have appreciably lowered illegal drug use. These funds would be more effective in lowering drug use were they spent on treatment and prevention programs rather than on the legal war against drugs. Second, the legal war against drugs puts police and other law enforcement agents literally in the line of fire, with many dying and being wounded every year. Third, many of the 17,000 deaths every year from illegal drugs result from the fact that the drugs are illegal. Users often do not know exactly what they are taking and, fearing that their product will contain too little of the drug itself, take an overly large amount and die from overdose.

Fourth, many illegal drug users necessarily become involved in the world of crime to obtain their illegal substances, and some turn to crime to obtain money to pay for the inflated prices of illegal drugs (inflated because they are illegal). Fifth, as in Prohibition, much police and other law enforcement corruption results from the legal war against drugs. Sixth, if illegal drugs were made legal, they could be taxed like other products, and billions of dollars of new tax revenue would flow into the federal and state governments; much or all of this money could be used for drug treatment and prevention programs. Finally, enforcement of laws against illegal drugs and other consensual behaviors often involves unsavory practices by police and other law enforcement agents, including wiretapping and the use of informants. These practices sometimes raise significant civil liberties issues.

The issue of prostitution further illustrates some of the problems with consensual crime laws. Many people dislike the very idea of prostitution, either because they consider having sex for money immoral or because they think it is degrading for women to engage in it. Leaving aside the philosophical issue of whether we *should* have laws against prostitution for these reasons, the fact remains that prostitution is the world's "oldest profession," as the saying goes, and was common in ancient times. Given its continued popularity, with an estimated 500,000 prostitutes in the United States and an estimated 750 million sex acts annually between prostitutes and their customers (Kappeler and Potter 2005), the social science question is whether the laws against prostitution do more good than harm or more harm than good.

Streetwalking, the most common form of prostitution in response to the laws against it, subjects women to the control of and abuse by pimps and to possible abuse by their customers. It also subjects women and their customers to possible infection by sexually transmitted disease, including the AIDS virus. Although prostitution laws are often enforced to address all these problems, these problems in fact stem from the prostitution laws themselves. Suppose that prostitution were legalized and regulated with a brothel approach, similar to the situation in Nevada now outside of Las Vegas and Reno and to the many legal brothels found in American cities about a century ago. Safe sex and weekly health exams would be required. Pimps would not be involved, and the brothel's employees would be free to quit their job just like any other job. Their earnings would be taxable, adding millions of dollars annually to federal and state tax revenue. No doubt some streetwalking would still exist, but a legalized and regulated brothel approach would likely reduce streetwalking and all its problems to a considerable degree.

Repealing Consensual Crime Laws?

If laws against consensual crimes were repealed, an important question would be whether more people would engage in the behaviors that are currently banned. Consideration of this question again has a philosophical component and a social science component. The philosophical considerations discussed earlier raise an interesting issue: Why would it be so bad if more people did engage in these behaviors if they became legal? Given that people are now allowed to engage in many risky behaviors and in many behaviors that are often regarded as immoral or offensive, is it really that different for them to be engaging in the behaviors now banned by consensual crime laws? If more people did engage in the behaviors if the laws were repealed, is that not the price we pay for living in a free society? Again, a full discussion of this philosophical component lies beyond the scope of this book, but a philosophical case can be made that our society should not overly worry if more people did engage in the behaviors now banned by consensual crime laws if the laws were repealed.

Legal and political philosophers and some readers of this book might care about this intellectual argument, but the possible repeal of consensual crime laws has a social science component that is a larger concern for most people. Because many of us do not like these behaviors and consider them harmful, we naturally would be very concerned if more people engaged in them if they became legal. The social science component, then, addresses this specific question and also the related question of whether repeal of the consensual crime laws might do more good than harm even if more people engaged in the behaviors. Let us consider this question in regard to drug use, the consensual crime about which there is the most concern and the most debate.

Prohibition again provides a tantalizing clue. The best historical evidence apparently indicates that drinking did reduce during Prohibition and rose after it was repealed. This increase after Prohibition ended was a risk our nation decided to take because it decided that, overall, Prohibition was doing more harm than good and that repealing Prohibition would thus do more good than harm, even if more people started to drink. If the experience of Prohibition were to apply to the present, then it is very possible that more people would use the now illegal drugs if they became legal, more would engage in prostitution, and so forth. Yet, even if this were to happen, it might still be true that the repeal of consensual crime laws would, on balance, cause more good than harm, as was the case with the repeal of Prohibition.

At the same time, it is not entirely certain that many more people would, in fact, use the now illegal drugs if the laws against them were repealed. Three kinds of evidence are relevant here. First, during the 1970s, marijuana use in the United States became decriminalized in many states: laws against marijuana use were still in place, but the legal penalties were reduced to citations and fines. Despite the weakening of these marijuana laws, marijuana use declined in the following years. A similar phenomenon happened in The Netherlands, where marijuana use also declined after that nation decriminalized it during the 1970s (Nadelmann 1992; Reinarman, Cohen, and Hendrien 2004).

Second, although illegal drugs are widely available in many parts of the United States, use of these drugs is low. Studies of high-school seniors are illustrative. In a 2006 national survey of high- school seniors, 84.9 percent said they could obtain marijuana "easily" or "fairly easily," and 46.5 percent said the same thing about cocaine and 52.9 percent about amphetamines. Despite the evident availability of these drugs, the percentage of the seniors who had actually used them during the past year was only 31.5 percent, 5.7 percent, and 8.1 percent, respectively (Johnston et al. 2007). The disparity between the easy availability of the drugs and their actual use suggests that legalization of the drugs would not produce a large increase in their use. Third, some scholars cite the "forbidden fruit" argument. According to this way of thinking, at least some people find the use of illegal drugs and involvement in other illegal consensual behaviors attractive precisely because they are illegal. If these behaviors became legal, one of the attractions for using them would no longer exist.

Proponents of legalization make all these points, and they also argue that legalization would free up billions of dollars now spent on the legal war against drug use and other behaviors that would be more effectively spent on prevention and treatment programs. Thus, even if drug use did increase initially if drug laws were repealed, proponents say, in the long run drug use would probably decline as the nation shifted its dollars and attention away from law enforcement to prevention and treatment programs.

Taking all these arguments into account, it is difficult to predict accurately the impact of legalizing some or all of the drugs that are currently illegal. Legalization advocate Ethan A. Nadelmann (1992) concedes that legalization "is a risky policy, one that may indeed lead to an increase in the number of people who abuse drugs. . . . That risk is by no means a certainty." Because he thinks drug laws do more harm than good, Nadelmann recommends that they be repealed even if drug use would rise. Another legalization proponent, former New Mexico governor Gary Johnson, agrees with this assessment: "There are going to be new problems under legalization. But I submit to you they are going to be about half of what they are today under the prohibition model" (Kelley 1999:A7).

Because legalization is admittedly a risky policy and because the United States is not about to pursue this policy in the foreseeable future, some critics of the legal war against drugs advocate a harm reduction approach (Ritter and Cameron 2006). This approach recognizes that drugs can be harmful, but it also recognizes that the legal war against drugs is costly in many other ways. In general, it involves treating drug use more as a public health problem and less as a legal problem. In this view, drug users should not be treated as criminals; instead, they should be treated as people who need help to reduce or eliminate their use of drugs. A harm reduction approach also involves several kinds of strategies, such as the provision of sterile needles for those who inject drugs, that are designed to minimize certain harms associated with drug use. Another strategy that has become popular in recent years is the use of *drug courts*. As their name implies, drug courts are venues for the handling of cases involving drug offenders. These courts sentence offenders to treatment and counseling programs rather than to jail or prison. The evidence overall indicates that this approach costs less money than traditional incarceration and also is more likely to succeed in reducing drug use (Ritter and Cameron 2006).

We have had space here only to outline the many issues surrounding laws against consensual crimes, and fuller treatments are available elsewhere (Husak 2002; Meier and Geis 2007; Packer 1968). These laws raise fascinating questions for the interplay between law and society and promise to spark controversy and debate for many years to come.

SPECIAL TOPICS IN LAW AND SOCIAL CONTROL

As should be clear from the previous sections, the use of law to achieve social control can be fraught with controversy. This is perhaps especially true regarding how the law should address two types of offenses: murder and white-collar crime. We now turn to these two topics.

The Death Penalty Controversy

Of all the issues for law and social control in the United States today, capital punishment remains perhaps the most controversial. The United States is the only Western nation that continues to execute individuals convicted of common crime, specifically homicide. In 2006, 53 people were executed, and 3,366 inmates were on death row as of July 1, 2006 (Death Penalty Information Center 2006). These numbers represent declines since the beginning of the decade but remain high enough to ensure that the death penalty will continue to arouse controversy for years to come.

Arguments for the Death Penalty. Proponents cite several reasons for favoring capital punishment: (1) Echoing the familiar "an eye for an eye" argument, the death penalty is the appropriate penalty for someone convicted of taking a human life (the *retribution* argument); (2) The death penalty is less expensive than life imprisonment and thus saves the government a considerable sum of money (the *cost* argument); and (3) The death penalty deters homicide (the *deterrence* argument). In a 2003 Gallup Poll, respondents were asked to indicate why they favored the death penalty. Half the respondents cited a retribution reason ("an eye for an eye," "fits the crime," and "they deserve it"). The two next most common responses, each chosen by 11 percent of respondents, were the cost and deterrence arguments (Maguire and Pastore 2007).

Our earlier discussion discussed the third reason, deterrence, just listed. Almost all research on capital punishment concludes that it does not, in fact, deter homicide, and some studies even find that the death penalty may increase homicide through the brutalization effect. If capital punishment does not deter homicide, much of the explanation for this noneffect lies in the nature of homicide. Recall that homicide and much other violent crime are performed relatively spontaneously and out of strong emotions. When someone is about to kill someone else, the offender is not usually taking the time to assess the possible consequences of his (usually the offender is male) actions beyond the injury about to be inflicted. Many homicide offenders have also been drinking, again making it unlikely that they will act in the way rational choice and deterrence theories assume. Given the nature of homicide, it would be rather surprising if the capital punishment did deter this type of crime. Although many proponents favor capital punishment because they believe it deters homicide, this belief appears to be mistaken.

Recall that retribution is the most popular reason by far for public support for capital punishment. Law and society scholarship has little, if any, to say about this reason, as it normally derives from moral and/or religious beliefs, which are typically the subject of the fields of philosophy and theology and not of social science.

Two related points are worth noting here, however. First, every other Western nation that had capital punishment has long ago abolished it on the grounds that the death penalty is an inappropriate penalty for civilized nations to implement.

Second, the Bible, sometimes cited as the source of many Americans' retributive attitudes toward homicide offenders, is in fact ambiguous on the issue of executions. Some passages in the Bible justify capital punishment, while other passages argue against it. For example, two familiar passages in the Old Testament support the idea of executions: "Who so sheddeth a man's blood, by man shall his blood be shed; for in the image of God made he man" (Genesis 9:5–6), and "Life for life, eye for eye, tooth for tooth, hand for hand, foot for foot" (Deuteronomy 19:21). The Old Testament prescribes the death penalty for twenty-one capital offenses, including adultery, cursing one's parents, homosexuality, incest, premarital sex, and working on the Sabbath (Costanzo 1997). Yet, the Bible also speaks of "God to whom vengeance belongeth" (Psalms 94:1), and in ancient times various restrictions and regulations actually made it very difficult to execute anyone. Moreover, as the examples of capital offenses in the Old Testament suggest, the Bible condones capital punishment for certain behaviors that even the most devout people today would not think deserve execution. For these reasons, many theologians conclude that the Bible fails to provide a clear picture about the morality of capital punishment (Megivern 1997). Most religious bodies in the United States officially oppose the death penalty on the grounds that it emphasizes revenge over redemption and that it disrespects the sanctity of human life.

If the retribution argument in favor of capital punishment is both complex and ambiguous and largely beyond the scope of social science, the cost argument is easy to address and certainly the province of social science examination. Here the evidence is clear: the belief that capital punishment saves money compared to life imprisonment is a myth. The average life imprisonment term costs about $1 million in prison maintenance costs (about $25,000 per year × about 40 years in prison). Exceeding this cost, the average execution costs between $2 million and $3 million in legal expenses that a state has to pay since almost all capital defendants are too poor to afford their initial criminal defense and later legal appeals. Thus, each execution costs between $1 million and $2 million more than each term of life imprisonment. Applying this difference to the 53 people executed in 2006, their executions cost their states an additional $53 million to $106 million more than life imprisonment would have cost. If the 3,366 death row inmates as of mid-2006 are all executed, their executions will cost their states an additional $3.4 billion to $6.8 billion more than life imprisonment would have cost.

Arguments Against the Death Penalty. Death penalty opponents cite all the evidence just discussed, but they also raise several other arguments against capital punishment. One of their major concerns is the possibility of **wrongful executions** (Bohm 2007). It is always possible that someone will be convicted of a crime that he or she did not commit. Sometimes, a considerable amount of evidence, including eyewitness testimony, can point to someone who happens to be innocent of the crime. Even more regrettably, sometimes prosecutors or police fabricate or hide important evidence or fail to tell the truth at pretrial hearings or in a trial itself. All these problems may lead to wrongful convictions, or the conviction of a defendant who is, in fact, innocent. C. Ronald Huff (2002), a former president of the American Society of Criminology, estimates that almost 1 percent of all felony convictions are in error; this figure extrapolates to almost 10,000 of the approximate 1 million felony convictions achieved annually in the United States.

Wrongfully convicted defendants can, of course, be released from prison if new evidence exonerates them. However, if someone is wrongfully convicted of a *capital crime* (one for which the death penalty is possible) and then executed, nothing can bring this defendant back from the dead. For this reason, the possibility of wrongful executions concerns many death penalty opponents. This concern has heightened in recent years as DNA and other evidence have exonerated dozens of people on death row. As of late 2007, such evidence had led to the release of 124 death row inmates in 25 states since 1973.

Has an innocent person ever been executed in the United States? Although it is difficult to know for sure, death penalty scholars estimate that twelve innocent defendants were executed between the 1970s and end of the 1990s and that several others were executed earlier in the century (Bohm 2007; Radelet, Bedau, and Putnam 1992). The former figure represents about 3 percent of all executions during that period.

The issue of wrongful conviction and execution has received much attention during the past few decades, and this attention may have produced a decline in death penalty support (Unnever and Cullen 2005): In 2006, 47 percent of Americans favored capital punishment over life imprisonment without parole as "the better penalty for murder," compared to a much higher figure of 61 percent in 1997 (Maguire and Pastore 2007). The issue won major headlines in 2000 when the governor of Illinois, reacting to revelations that thirteen innocent people were on death row in his state, suspended all executions pending further investigation. Three years later, his concern about this problem led him to commute the sentences of all 167 death row inmates in Illinois. Explaining his decision, the governor declared, "Our capital system is haunted by the demon of error: error in determining guilt and error in determining who among the guilty deserves to die" (Wilgoren 2003:A1).

One reason wrongful convictions in capital cases occur is that the legal representation of capital defendants at their trials is often *inept*. This problem is a second argument of death penalty opponents against the death penalty. Inept or inadequate defense is a problem that characterizes the legal defense of many types of criminals (see Chapter 9), but it is an even more urgent problem in capital cases in which the defendant's life may be at stake. Because almost all capital defendants are poor, they rely for their legal representation on public defenders or on counsel assigned by the state. Most of these attorneys are untrained in capital defenses and have insufficient time or money to represent their capital clients to the fullest extent possible. Capital defenses are very complex because they involve murder charges and also because they need to focus at the trial stage on building a case for appeal if a conviction results. As a result, many, and perhaps most, capital defendants receive inadequate legal representation. Investigations of capital cases in Illinois, Kentucky, and Texas found that many capital defendants had been represented by attorneys who were later suspended or disbarred for various kinds of professional misbehavior (Johnson 2000; Perez-Pena 2000). An investigation of Texas executions during George W. Bush's term as governor found that "Texas has executed dozens of Death Row inmates whose cases were compromised by unreliable evidence, disbarred or suspended defense attorneys, meager defense efforts during sentencing and dubious psychiatric testimony" (Mills, Armstrong, and Holt 2000:A1).

A third argument against the death penalty is that its application is *racially discriminatory*. In this regard, scholars have studied decisions by prosecutors to seek the death penalty in murder cases and decisions by juries to impose the death penalty after convicting defendants in such cases. Some evidence has found that death penalty charges and sentences are more common when the defendant is African American than when the defendant is white, but overall the evidence is inconsistent on this point. However, strong and consistent evidence finds that death penalty charges and sentences are much more likely when the victim is white than when the victim is African American, suggesting that prosecutors and juries in effect place more importance on the lives of white victims than on the lives of African American victims. Studies in Georgia and South Carolina have found that prosecutors are three to five times more likely to seek the death penalty when the victim is white than when the victim is African American (Baldus, Woodworth and Pulaski 1990; Songer and Unah 2006). Reflecting this disparity, although whites comprise about 50 percent of all murder victims, 80 percent of all executions since the 1970s have been for cases involving white victims.

A final argument against the death penalty is that its use is extremely *arbitrary*. Again, this is a problem affecting criminal cases of all types, but is an even greater problem in capital cases because the defendant's life may be at stake. By saying that the death penalty is arbitrary, we mean that very similar murders (in terms of motivation, gruesomeness, and other factors) are treated very differently depending on where they occur, who the prosecutor is, and so forth. Prosecutors seek the death penalty against some accused murderers but not against others accused of similar crimes. Juries sentence some convicted murderers to death but others to life imprisonment. The whole process is arbitrary and not very logical, and critics say that "being sentenced to death is the result of a process that may be no more rational than being struck by lighting" (Paternoster 1991:183).

White-Collar Crime

The Enron scandal from earlier this decade awakened many Americans to the enormity of white-collar crime (Friedrichs 2007; Rosoff, Pontell, and Tillman 2007). Enron was an energy company whose chief executives exaggerated the company's profits and hid its losses. After this accounting fraud was disclosed, Enron went bankrupt and laid off more than 4,000 workers, many of whom lost their pensions. In less than a year, Enron's stock fell from $84 a share to less than $1 a share, and people and institutions that held its stock lost tens of billions of dollars. The company's chief executives either pleaded guilty or were found guilty after trial. One of them, former CEO Jeffrey Skilling, was sentenced to twenty-four years in a federal prison in Minnesota after being convicted of nineteen counts of conspiracy, fraud, and insider trading. Enron's founder, Kenneth Lay, was convicted of similar charges but died of a heart attack before his sentence was pronounced. He likely would have received a sentence about as long as Skilling's.

Although Skilling's sentence was a full twenty-four years, its severity was in fact an aberration in the world of white-collar crime and especially of crime committed by corporations, or *corporate crime*, where the proverbial "slap on the wrist" is more the norm than prison terms of more than two decades. Chapter 7 on law and inequality discusses this issue in further detail, but a few examples are illustrative. In 1999, several major pharmaceutical companies settled a lawsuit regarding their alleged worldwide price-fixing of vitamins over a nine-year period. As part of the settlement, they agreed to pay $1.17 billion. Although this was no small sum, it still amounted to only 20 percent of the companies' sales of the price-fixed vitamins, making their illegal activity quite profitable despite the large settlement they paid (Moore 1999). In 2004, Bank of America was fined $10 million in a case involving alleged securities trading violations. Although again this was no small sum, it still amounted to less than 1 percent of Bank of America's 2003 profit of almost $11 billion and less than .001 percent of its assets of almost $1 trillion at the time of the fine. Thus, while $10 million would have been an incredible sum of money for almost any American to pay in a fine, it was mere "chump change" for Bank of America in view of its incredible wealth.

Even when corporations engage in behavior that injures and kills people, their executives often do not go to prison. A notorious example here was the 1970s' scandal involving the Ford Pinto (to be further discussed in Chapter 7), a car susceptible to fire and explosion when hit from behind in a minor rear-end collision. At least two dozen people died in Pinto fires and explosions in such accidents, and hundreds more were burned severely. An investigation later found that Ford executives knew the car was dangerous before it went on the market but decided not to fix the problem in order to save money. Despite such behavior that led to many deaths, not a single Ford executive ever went to prison (Cullen, Maakestad, and Cavender 2006).

As these examples indicate, white-collar crime routinely receives lenient treatment from the U.S. legal system despite the long sentence that Enron executive Skilling received. Corporations can easily afford to pay the fines they receive, and few individual offenders receive lengthy prison terms, if indeed they are incarcerated at all. Such leniency is somewhat ironic, because much white-collar crime is instrumental in nature, to recall our earlier typology of instrumental and expressive crime, and thus possibly susceptible to deterrence by legal sanctions that are sufficiently certain and harsh. Most white-collar criminals, including most corporate criminals, plan their crimes in a very deliberate, rational manner and are quite capable of assessing the possible consequences of their actions and acting accordingly. Therefore, many scholars think that white-collar crime could be at least somewhat deterred if potential offenders thought they could indeed be incarcerated if their crimes were detected and prosecuted (Cullen, Maakestad, and Cavender

2006). As a writer for *Fortune* magazine, a business-oriented publication, observed, "[T]he problem will not go away until white-collar thieves face a consequence they're actually scared of: time in jail" (Leaf 2002:62).

However, for law to deter white-collar crime and especially corporate crime in this manner, three other problems must first be surmounted (Rosoff, Pontell, and Tillman 2007). The first problem is that the laws and regulations that govern corporate behavior are often weak or nonexistent, and enforcement of these laws and regulations is often inconsistent. This is, perhaps, especially true in the area of environmental pollution, where the weak laws and lax enforcement, according to one investigation, create "a system where major polluters can operate with little fear of being caught or punished" (Armstrong 1999:A1).

The second problem is that federal and state regulatory agencies and prosecutors lack the funds and personnel to do a better job of detecting and trying to control corporate misconduct. In contrast, corporations are worth billions of dollars and can easily afford expensive legal teams to resist any attempt at regulatory or legal control.

A final problem is that corporate crime often involves very complex procedures and transactions that juries and even prosecutors may not easily understand. Moreover, it is often difficult for prosecutors and regulatory agencies to determine whether a crime was in fact committed and, if so, who was responsible.

All these problems impair the ability of law to control corporate crime. Two examples illustrate the ineffectiveness of law in this regard. From 1982 to 2002, the federal Occupational Health and Safety Administration (OSHA) determined that 2,197 workers died because of unsafe or illegal workplace conditions but tried to criminally prosecute the employer in only 7 percent of the cases (Barstow 2003). Similarly, from 1992 to 2001, the Securities and Exchange Commission referred 609 cases of alleged financial crime to federal prosecutors for criminal proceedings, but only one-third of these cases resulted in criminal charges, and only one-half of these latter cases ended in someone being incarcerated (Loomis 2002).

Summary

1. Effective social control is essential for any society. In modern societies, law is a principal means of handling disputes and of dealing with crime. This chapter examines two major questions: (1) How effective is law in preventing criminal behavior in the general society and in preventing repeat offending by offenders after they have been arrested, prosecuted, and punished? and (2) How effective is law in dealing with consensual behaviors that are deemed criminal largely because they offend our society's sense of moral order?

2. Two goals of the criminal justice system are deterrence and incapacitation. In studying the degree to which these goals are realized, scholars ask three questions: (1) To what degree does law deter potential criminal behavior in the general population? (2) To what degree does law deter repeated criminal behavior by offenders who have already been arrested and punished? and (3) To what degree does law prevent criminal behavior by incarcerating convicted offenders and thereby keeping them isolated from the outside world? Answers to the first two questions are important for assessing the accuracy of deterrence theory.
3. Several types of deterrence are possible. Absolute deterrence refers to the effect of some law or of having a legal system versus the effect of having no law or no legal system. Marginal deterrence, the more relevant type for social policy, refers to the effect on criminal behavior of increasing criminal sanctions versus not increasing such sanctions. General deterrence refers to the ability of law to deter offending in the general population, while specific or individual deterrence refers to the ability of law to deter criminal behavior by offenders who have already been arrested and punished because they do not want to incur the risk of additional punishment.
4. Studies of active burglars and robbers find that they either assume they will not get caught because they plan their crimes or else just give little or no thought to their chances of getting caught. Others realize they might be caught but resign themselves to this outcome because they have a fatalistic attitude, and some do not regard the prospect of prison as very unpleasant. For these reasons, they are relatively incapable of being deterred by the threat of legal sanctions, and, thus, relatively insusceptible to marginal deterrence. Research on deterrence supports the conclusion that the marginal general deterrent effects of law on criminal behavior are small or nil. It also supports the conclusion that law does not have a specific or individual deterrent effect.
5. Law does have a small incapacitation effect that comes at a great economic and social cost. Moreover, the massive imprisonment of the past few decades may be creating a breeding ground for even more crime as offenders are released back to their communities, where they face bleak job prospects and difficulty in forging relationships with law-abiding persons.
6. Law has often been used in the United States to enforce standards of morality. Today, the law bans certain behaviors involving that are thought in many circles to be immoral and/or socially harmful: the use of certain drugs, prostitution, certain types of gambling, and certain forms of pornography. These crimes raise two important questions: (1) In a free society, to what degree should the state prohibit behaviors in which people want to engage because the behaviors are perceived to be immoral and/or socially harmful? and (2) Do laws against consensual crimes do more good than harm, or do they do more harm than good?

7. The first of these two questions is a philosophical question and raises two subsidiary issues. First, who is to decide which behaviors are immoral or moral? Second, in a free society, what behaviors perceived to be immoral and/or socially harmful should be prohibited, and which should be allowed? American citizens are allowed to engage in many behaviors that pose some risk to their health or safety. In a free society, we are not about to ban these behaviors even though they can be very risky and even deadly.
8. The second of these two questions is a social science question on which there is much evidence and also much speculation. Critics of consensual crime laws cite several harms caused by laws against consensual crime. First, and focusing on drug laws, these laws cost billions of dollars annually, with little evidence that they have appreciably lowered illegal drug use. Second, the legal war against drugs puts police and other law enforcement agents literally in the line of fire. Third, many of the 17,000 deaths every year from illegal drugs result from the fact that the drugs are illegal. Fourth, many illegal drug users become involved in the world of crime to obtain their illegal substances. Fifth, as in Prohibition, much police and other law enforcement corruption results from the legal war against drugs. Sixth, if illegal drugs were made legal, they could be taxed like other products. Finally, enforcement of laws against illegal drugs often involves unsavory practices by police and other law enforcement agents.
9. If laws against consensual crimes were repealed, an important question would be whether more people would engage in the behaviors that are currently banned. Again focusing on drugs, there is reason to believe that drug use would increase if illegal drugs were legalized, but there is also reason to believe that any increase might be relatively small. However large it might be, the harm that any increase would do might be outweighed by doing away with the harms associated with current laws against drug use.
10. Because legalization is admittedly a risky policy, some critics of the legal war against drugs advocate a harm reduction approach that treats drug use as a public health problem instead of a legal problem. This approach involves strategies like the provision of sterile needles that are designed to minimize certain harms associated with drug use. Another strategy that has become popular in recent years is the use of drug courts that sentence offenders to treatment and counseling programs rather than to jail or prison.
11. Capital punishment is perhaps the most controversial issue today in the area of law and social control. Proponents cite several reasons for favoring capital punishment: retribution, cost, and deterrence. Opponents question the use of the death penalty on moral grounds, and they also say it neither deters homicides nor costs less than life imprisonment for convicted homicide offenders. They also cite the possibility of

wrongful executions and strong evidence that the application of the death penalty is racially discriminatory and extremely arbitrary.

12. Despite recent exceptions, white-collar crime routinely receives lenient treatment from the U.S. legal system. Corporations can easily afford to pay the fines they receive, and few individual offenders receive lengthy prison terms, if indeed they are incarcerated at all. Such leniency is somewhat ironic, because much white-collar crime is instrumental in nature and thus possibly susceptible to deterrence by legal sanctions that are sufficiently certain and harsh.

Key Terms

Absolute deterrence
Brutalization effect
Certainty
Collective incapacitation
Consensual crime
Deterrence
Deterrence theory
General deterrence
Gross incapacitation
Incapacitation
Individual deterrence
Marginal deterrence
Philosophical question
Rational choice theory
Selective incapacitation
Severity
Social science question
Specific deterrence
System capacity
Wrongful executions

CHAPTER 6

Law and Social Change

Chapter Outline

One of this decade's most heated social debates has concerned same-sex marriage. States throughout the nation passed constitutional amendments that banned same-sex marriages; the state supreme court of Massachusetts ruled in 2004 that its state constitution permitted same-sex marriage; and, taking a middle ground, several states (California, Connecticut, Hawaii, Maine, New Hampshire, New Jersey, and Vermont) enacted legislation that gave civil unions or domestic partnerships between same-sex couples many or all of the legal protections enjoyed by marriages between heterosexual couples. Hundreds of gay couples married in Massachusetts after same-sex marriage became legal there, and many other couples filed for civil union or domestic partnership status in the states that granted this status new rights. The legal developments regarding same-sex marriage and civil unions were themselves a result of attention given to gays and lesbians during the last few decades, much of it resulting from their own efforts to do away with homophobia and antigay discrimination.

This brief summary does not do the same-sex marriage issue justice, but it does indicate the interplay between law and social change. Changes in society can bring about changes in law, broadly defined, and changes in law can bring about changes in society. The relationship between law and social change has been a key dimension of the study of law and society since the rise of the Historical School (see Chapter 2) during the nineteenth century. As Chapter 1 emphasized, the idea that society can affect law and that law can affect society is a basic premise of the field of law and society today. As discussed in that chapter, several key assumptions guide theory and research in the field. Two of these were that *major changes in society often bring about changes in the law*, and that *laws and legal decisions may have a potential impact on one or more aspects of society*.

As should be apparent, then, the existence of a two-way or reciprocal relationship between law and social change is a defining component of the law and society canon. Friedman (2004a) likens this reciprocal relationship to the process and aftermath of building a bridge. Suppose there is a community, he says, on the banks of a wide river that is serviced only by a slow ferry. The residents put pressure on their government to build a bridge, and the bridge eventually gets built. Now that traffic easily goes across the bridge everyday, the community begins to change. Some people begin to live on the other side of the river, and more people, whichever side they live on, begin to commute to jobs on the other side. The ferry stops operating, and the bridge becomes so dominant a feature of the residents' existence that it "affects their behavior, their way of thinking, their expectations, their way of life" (p. 16). It becomes difficult for them to even imagine life before the bridge was built. The bridge, Freidman observes, is a metaphor for law and the legal system. The bridge was built because of social forces and social change—in this case citizen pressure on government—and, once built, "it began to exert an influence on behavior and attitudes" (p. 16). Similarly, law may change because of changes and pressures in the larger society, and, once it does change, it then begins to influence behavior and attitudes. Thus, although this chapter discusses the influence of social change on law and that of law on social change separately, the reciprocal relationship suggested by the bridge metaphor should be kept in mind.

Before moving on, it will be useful to introduce some social science jargon with which many readers may already be familiar. In a potentially causal relationship between two variables (e.g., race affects income), the **independent variable** is something that affects or influences a dependent variable, and the **dependent variable** is something that is affected or influenced by an independent variable. In the race and income relationship, race is the independent variable and income is the dependent variable. These two terms are often used in discussing the relationship between law and social change. Thus, when we say that changes in society may bring about changes in law, we are treating social change as the *independent variable* and legal change as the *dependent variable*. On the other hand, when we say that changes in law may bring about changes in society, we are treating legal change as the *independent variable* and social change as the *dependent variable*.

This chapter reviews the large variety of work on law and social change. We start by looking at the impact of changes in society on changes in law and then examine the impact of changes in law on changes in society.

The final section discusses law and social movements, an emerging subfield within the larger law and social change rubric that is attracting increasing attention from scholars of law and society and of social movements.

THE IMPACT OF SOCIAL CHANGE ON LAW: LAW AS DEPENDENT VARIABLE

Literature on the impact of social change on legal change falls into two broad types based on the scope of the changes involved. The first type is in the tradition of "grand theory" and examines how and why broad social changes produce far-reaching changes in the nature of a legal system, legal reasoning, and other fundamental dimensions of law. The second type has a somewhat more narrow focus, as it examines how and why certain social changes produce new legislation, new court rulings, new legal procedures, or other rather specific aspects of law. We will discuss both types of impact in this section.

Social Change and Fundamental Legal Change

The model for the first type, on the impact of broad social change on fundamental legal change, comes from the Historical School (see Chapter 2). Although the different theorists who made up this school took very different approaches, all were concerned with a basic social science question: how and why did law change as society became more modern (Cotterrell 2004)?

One of the Historical School's figures, English professor Sir Henry Maine (1822–1888), answered this question in his influential book *Ancient Law* (Maine 1864), in which he explored the evolution of law from ancient times to modern (nineteenth century) times. Recall from Chapter 2 that Maine is famous for his view that law changed *from status to contract*. In older societies, relationships were governed by power (or status) based on the relative social standing of the individuals in the relationship. The most extreme of relationships in terms of power differences was slavery. Over time, these traditional power-based relationships were replaced by agreements that were more voluntary and increasingly based on verbal and then written contracts. For Maine, then, a key dimension of modernization involved the rise of contractual relationships based on the voluntary agreement of the individuals in the relationship.

Other members of the Historical School included the three key founders of sociology whose work was briefly introduced in Chapter 2: Durkheim, Weber, and Marx (with collaborator Engels). Because their work on the impact of social change on law is so historically important and is still influential more than a century later, we examine it here in some detail.

Emile Durkheim: The Rise of Restitutive Law. French scholar Emile Durkheim (1858–1917) developed several themes that continue to resonate in social science theory and research today. Our discussion here is limited to the aspects of his work that are relevant for law and social change.

Like several other scholars of his time, Durkheim sought to understand how modern (i.e., nineteenth-century) societies differed from the traditional societies characteristic of ancient times and still found today in many parts of the world studied by anthropologists. He wrote extensively about the social bonds that characterize both types of societies and spelled out his argument in his influential 1893 book, *The Division of Labor in Society* (Durkheim 1933 [1893]). In small, traditional societies, he wrote, people are very similar to each other in thought and deed, and these societies are said to be *homogeneous*. Social order in these societies arises from this similarity and homogeneity. Durkheim called this type of social order **mechanical solidarity**. Larger, modern societies obviously differ in many ways from their traditional counterparts. For Durkheim, a key difference was that modern societies are more *heterogeneous*, as people are more different from each other in their beliefs, values, and behaviors. Social order is, thus, more difficult to attain and maintain, said Durkheim, but is still possible because people have to depend on each other for the society to work. Durkheim called this type of social order **organic solidarity** and said it arises from the interdependence of roles that is a hallmark of modern society. This interdependence, he said, creates a solidarity that retains much of the bonding and sense of community found in traditional societies.

According to Durkheim, these two types of solidarity in turn affect the type of law a society has. In traditional societies, a deviant or criminal act offends the *collective conscience*, Durkheim's term for a society's belief and value system. Because the collective conscience in traditional societies is so strong (since people have similar thoughts and values), the response to deviance in such societies is especially punitive, as people react emotionally to an act that offends them. The type of law found in these societies, said Durkheim, was repressive, and he coined the term **repressive law** to characterize law in traditional societies. In modern societies, the collective conscience is weaker because people have different beliefs and values. Their response to deviant and criminal acts is thus less punitive and in fact takes the form of restitution, as it involves compensating an injured or aggrieved party for the harm done to them. Durkheim used the term **restitutive law** to characterize law in modern societies.

Durkheim's connection of the type of law found in a society to the nature and extent of social bonds it exhibits, and thus to its level of modernization, was a key insight for early law and society thinking and continues to have historical importance for the study of law and social change. Unfortunately, later scholarship (remember that Durkheim presented his argument in 1893, long before the advent of modern anthropological research) indicated that Durkheim may have misinterpreted the relationship between type of society and type of law. An oft-cited study by Richard Schwartz and James C. Miller (1964) used data on fifty-one societies that had been studied by anthropologists. This data set allowed Schwartz and Miller to determine whether each society had one or more of the following features of relatively modern legal systems: counsel, mediation, and police. Of these, police corresponds most closely to what Durkheim meant by repressive law, and mediation corresponds most closely to what he meant by restitutive law.

After determining the pattern of features that characterized each society, Schwartz and Miller noted how modern the societies were in other respects, for example, whether they had a division of labor (role specialization) and/or money. They then determined that police, as a proxy for repressive law, tended to be found only in the most modern societies, the opposite of what Durkheim argued, while mediation, as a proxy for restitutive law, was found in many traditional, premodern societies, again the opposite of what Durkheim argued. This finding led the authors to conclude, "[T]hese findings seem directly contradictory to Durkheim's major thesis. . . . Thus Durkheim's hypothesis seems the reverse of the empirical situation in the range of societies studied here" (p. 166).

Although Schwartz and Miller's study suggested that Durkheim may have reversed the relationship between modernity and type of law, a decade later another scholar challenged their refutation of Durkheim's hypothesis. Upendra Baxi (1974) argued that Schwartz and Miller erred in using the presence of police as a measure of repressive law because the presence of police by definition requires the existence of role specialization. Measured this way, repressive law could *only* be found in modern societies. For this reason, Baxi argued, the earlier two authors unwittingly "virtually insure that their counter-thesis is correct" (p. 647), because in simple societies that have no role specialization, police could hardly be expected to be found. Although Baxi did not mention it, his argument suggests that the use of a different measure of repressive law, such as whipping or other physical punishment, would have constituted a more appropriate test of Durkheim's hypothesis.

Although Baxi made a valid point, most scholars do think that Durkheim misinterpreted the modernity and law relationship. Although the old Tarzan movies and any number of other films and books have depicted traditional societies as savage in the way Durkheim envisioned, many traditional societies studied by anthropologists in fact rely on restitution to settle disputes and deal with their members who violate social norms (see Chapter 4). By the same token, many modern societies are very punitive in their approach to law. Although physical punishment is no longer used to bring criminals to justice, their incarceration is obviously much more punitive than restitutive, and the death penalty is still used in the United States. Although Durkheim may have misconstrued the law-modernization relationship, his thesis remains valuable for stimulating scholarship on law on social change and more generally on the idea that law reflects a society's beliefs, values, and social structure.

Max Weber: The Rise of Rational Law. Like Durkheim, German scholar Max Weber (1864–1920) developed many themes that continue to influence sociological theory and research. Our discussion focuses on the aspects of his work relevant for an understanding of law and social change.

Weber (1978 [1921]) recognized that as societies become more modern and complex, their procedures for accomplishing tasks rely less on traditional customs and beliefs and more on rational (which is to say rule-guided, logical, and impersonal) methods of decision making in which the means have a reasonable connection to the ends and vice versa. The development of rational thinking, he said, allowed complex societies to accomplish their tasks in the most efficient way possible. For Weber, then, the key hallmark of modern society is the development of **rationality**.

Weber used the concept of rationality to understand how societies changed legally as they became more modern and complex. One major change involves the type of power characteristic of a society. In many traditional societies, he wrote, the major type of power is **traditional authority**. As its name implies, traditional authority is power that is rooted in traditional, or long-standing, beliefs and practices of a society. It exists and is assigned to particular individuals because of that society's customs and traditions. Individuals enjoy traditional authority for at least one of two reasons. The first is inheritance, as certain individuals are granted traditional authority because they are the children or other relatives of people who already exercise traditional authority. The second reason individuals enjoy traditional authority is more religious: their societies believe they are anointed by God or the gods, depending on the society's religious beliefs, to lead their society. Traditional authority is most common in preindustrial societies, in which tradition and custom are so important, but it continues to exist in some modern monarchies where a king or queen enjoys power because he or she comes from a royal family.

An important aspect of traditional authority is that it is granted to individuals regardless of their qualifications. They do not need any special skills to receive and wield their authority, as their claim to it is based solely on their bloodline or divine designation. An individual granted traditional authority can be intelligent or dull, fair or arbitrary, and exciting or boring, but the individual receives the authority just the same because of custom and tradition. As not all individuals granted traditional authority are particularly well qualified to use it, societies governed by traditional authority sometimes find that individuals bestowed it are not always the best leaders.

If traditional authority derives from custom and tradition, **rational-legal authority**, a second type of power, derives from law and is based on a belief in the legitimacy of a society's laws and rules and in the right of leaders to acting under these rules to make decisions and set policy. It is a hallmark of modern democracies, in which power is given to people elected by voters and the rules for wielding that power are usually set forth in a constitution, charter, or other written document. Whereas traditional authority resides in an individual because of inheritance or divine designation, rational-legal authority resides in the office that an individual fills, not in the individual per se. Thus, the authority of the president of the United States resides in the office of the Presidency, not in the individual who happens to be president. When that individual leaves office, authority transfers to the next president. This transfer is usually smooth and stable, and one of the marvels of democracy is that officeholders are replaced in elections without revolutions having to be necessary. Even if we did not voted for the person who wins the Presidency, we accept that person's authority as our president after the electoral transition occurs.

Rational-legal authority even helps ensure an orderly transfer of power in a time of crisis. When John F. Kennedy was assassinated in 1963, Vice President Lyndon Johnson was immediately sworn in as the next president. When Richard Nixon resigned his office in disgrace in 1974 because of his involvement in the Watergate scandal, Vice President Gerald Ford (who himself had become Vice President after Spiro Agnew resigned because of financial corruption) became president. Because the U.S. Constitution provided for the transfer of power when the Presidency was vacant and because U.S. leaders and members of the public accept the authority of the Constitution on such matters, the transfer of power in 1963 and 1974 was relatively smooth and orderly.

A third type of power discussed by Weber is **charismatic authority**, which stems from an individual's extraordinary personal qualities and from that individual's hold over followers because of these qualities. Such charismatic individuals may exercise authority over a whole society or only over a specific group within a larger society. They can exercise authority for good and for bad, as this brief list of charismatic leaders indicates: Joan of Arc, Adolf Hitler, Mahatma Gandhi, Martin Luther King, Jr., Jesus Christ, Prophet Muhammad, and Buddha. Each of these individuals had extraordinary personal qualities that led their followers to admire them and to follow their orders or requests for action. Weber emphasized that charismatic authority is often less stable than either traditional authority or rational-legal authority. The reason for this is simple: Once a charismatic leader dies, the leader's authority dies as well. Although the leader's example may continue to inspire people, it is difficult for another leader to come along and command people's devotion as intensely. After the deaths of Joan of Arc and the other charismatic leaders just named, no one came close to replacing them in the hearts and minds of their followers.

Charismatic authority can reside in a person who came to a position of leadership because of traditional or rational-legal authority. Over the centuries, several kings and queens of England and other European nations were charismatic individuals as well (while some were far from charismatic). A few U.S. Presidents—Washington, Lincoln, both Roosevelts, Kennedy, Reagan, and, despite his affair with an intern, Bill Clinton—also were charismatic, and much of their popularity stemmed from various personal qualities that attracted the public and sometimes even the press. Ronald Reagan, for example, was often called the "Teflon President," because he was so loved by much of the public that accusations of ineptitude or malfeasance simply rubbed off him (Lanoue 1988).

Weber's emphasis on the rise of rationality and more specifically on the development of rational-legal authority in modern society was a major contribution to the understanding of law and society. Weber, in fact, used the development of law to illustrate the development of rationality. He wrote that legal procedures can be either *rational* or *irrational*, with the former characterizing modern societies and the latter characterizing traditional societies. Rational legal procedures involve the use of logic and reason to reach legal decisions and achieve other goals, while irrational legal procedures are based on magic or faith in the supernatural, including religion. In another distinction, legal procedures can also be *formal* or *substantive*. Formal in this context means that legal decisions are based on established rules, regardless of whether the outcome of a decision is fair or unfair. Substantive means that a legal decision takes account of the circumstances of individual cases in order to help ensure a fair outcome.

Weber said that law has become both more rational and more formal over time. Thus, most modern legal systems are characterized by formal rationality: Legal decisions are based on logic and do not consider whether their outcomes are fair or unfair, only whether the outcome makes sense in view of the facts and other circumstances of a case. This type of law, of course, creates the tension (see Chapter 1) that sometimes occurs between logical decision-making and justice and fairness. As the case in Chapter 1 involving misdiagnosis of brain cancer made clear, some decisions make absolute sense in view of the law governing a case and the facts involved in a case but at the same time yield outcomes that many of us would consider unjust, unfair, or at least unfortunate.

Karl Marx and Friedrich Engels: Law as Domination. Karl Marx (1818–1883) was a founder of sociology, but he was also a towering figure in the history of social and political thought. He and his frequent collaborator Friedrich Engels (1820–1895) left a body of written work that has influenced the course of history and also the development of sociology, political science, economics, and other disciplines.

In understanding the modernization of society, Marx and Engels' chief concern was the rise of capitalism as societies' economies evolved from ones that were based on agriculture to ones that were based on industry. According to Marx and Engels, every capitalist society is divided into two classes based on the ownership of the **means of production**—tools, factories, and the like. In a capitalist society, the **bourgeoisie** or ruling class owns the means of production, while the **proletariat** or working class, the other class, does not own the means of production and instead is oppressed and exploited by the bourgeoisie. This difference creates an automatic conflict of interests between the two groups. Simply put, the bourgeoisie is interested in maintaining its position at the top of society, while the proletariat's interest lies in rising up from the bottom and overthrowing the bourgeoisie to create an egalitarian society. As many readers may know, Marx and Engels thought that the effort of the bourgeoisie to maintain its top position involved the oppression of the working class.

How did law fit into Marx and Engels' thinking? Although law was not one of their chief concerns, they thought it aided the ruling class's oppression of the working class in two ways (Cain and Hunt 1979; Collins 1982). First, law helps preserve private property. Although this might mean that law benefits anyone, including the working class, who owns private property, in reality the ruling class owns almost all private property, and, in this way, the law benefits this class much more than the working class. Second, law provides legal rights for all and thus creates a façade of justice that obscures working-class oppression and helps the working-class to feel good about their society when in fact they should be angry about their oppression. In this way, law contributes to *false consciousness* that prevents the working class from realizing its revolutionary potential. Marx and Engels predicted that the working class would eventually revolt and create a true communist society in which everyone was equal and where the state, including the law, would eventually not be needed.

Related to Marx and Engels' views about law were their views about crime, which their various books and articles viewed in at least three ways. Sometimes they depicted crime as a natural and logical reaction of the working class to the squalid conditions in which they lived under capitalism. Thus Engels (1993 [1845]:48) wrote, "The worker is poor; life has nothing to offer him; he is deprived of virtually all pleasures. . . . What reason has the worker for not stealing? . . . Distress due to poverty gives the worker only the choice of starving slowly, killing himself quickly, or taking what he needs where he finds it—in plain English—stealing." In other writings, however, Marx and Engels depicted crime as political rebellion by the working class. In this regard, Engels (1993 [1845]:49) once wrote, "Acts of violence committed by the working classes against the bourgeoisie and their henchmen are merely frank and undisguised retaliation for the thefts and treacheries perpetrated by the middle classes against the workers." Although Engels likened crime to rebellion, he thought that it was unlikely to succeed as rebellion because it is only an individual act and because it leads to arrest and incarceration.

The third way in which Marx and Engels viewed crime by the poor was much more negative, as they sometimes called criminals the *lumpenproletariat*, by which they meant "the social scum, the positively rotting mass" of street criminals (Marx and Engels 1962 [1848]:44). Elsewhere Engels (1926:23) also called the lumpenproletariat "an absolutely venal, and absolutely brazen crew." It should be obvious from their language that Marx and Engels viewed the lumpenproletariat very negatively, partly because they thought it lacked the class consciousness that was needed for a working-class revolution.

As noted above, Marx and Engels devoted much less attention to law than to other aspects of capitalism that help oppress the working class.

However, as Chapter 2 indicated, contemporary scholars have drawn on Marx and Engels' views about law and capitalism to develop a general Marxian perspective on law and society.

Roberto Mangabeira Unger: The Development of Legal Order. Long after the work of the theorists identified with the Historical School, other scholars have studied law and social change at the broad level in the grand theory tradition. A notable effort was Roberto Mangabeira Unger's (1976) *Law in Modern Society*. Unger first distinguished several types of law—customary law, bureaucratic law, and legal order—and then sought to explain the historical conditions that helped lead to the latter two types. *Customary law* consists of shared patterns of interaction and "mutual expectations that ought to be satisfied" (p. 49) and is the type of law most often found in small, traditional societies studied by anthropologists (see Chapter 2). *Bureaucratic law* consists of "explicit rules established and enforced by an identifiable government" (p. 50). A precondition for such law, then, is the existence of a state, or a society ruled by a central person of power and his or her staff; as such, states and bureaucratic law (e.g., ancient Rome) represent the next stage of political and legal development after traditional societies and customary law. Unger notes that both customary law and sacred law served historically to limit the scope and influence of bureaucratic law.

Legal order, the third and most modern type, is law that is autonomous from the state and general (or applicable to everyone) regardless of any individual's power, wealth, or personal connections. Autonomy and generality are, of course, the ideal of the rule of law (see Chapter 3) characteristic of Western democracies. Unger notes that legal order emerged in postfeudal Europe and that two preconditions made this emergence possible. One was *group pluralism*, or the existence of many social and political groups that compete with each other for power, with no one group consistently dominant. These groups need legal order, as defined by Unger, to ensure that no one group becomes dominant and that all groups play by the same rules, so to speak. In postfeudal Europe, the rise of the merchant class led to a struggle among three groups for power: the monarchy, the aristocracy (wealthy landowners), and the merchants. The latter two groups wanted a legal order to limit the monarchy's power, and the legal order that resulted gave all three groups some protection while also limiting their potential to achieve dominance. In effect, the emergence of a legal order was a compromise to help group pluralism continue without any one group succeeding.

The second precondition for the emergence of legal order was a *reliance on a higher divine or universal law* "as a standard by which to justify and to criticize the positive law of the state" (p. 66). This meant that a precondition for the emergence of legal order was a belief in natural law (see Chapter 3). This belief, and the conclusion among postfeudal Europeans that their emerging legal order was in accordance with the precepts of natural law, helped to legitimize and thus strengthen their new legal order. Unger adds that group pluralism and a belief in higher law were both necessary preconditions for the new legal order that emerged; either precondition by itself would not have been sufficient to produce this new, historically monumental belief in the rule of law.

Unger adds that law continued to change as Western democracies became *postliberal societies* and as their governments became so-called welfare states by intervening more than they did previously to redistribute economic resources and to regulate private transactions to ensure fairness and to prevent various social harms. Thus, the state has become more proactive in using the law to help disadvantaged groups in society, as law in Western democracies has become more concerned with substantive justice (see Chapter 3) and with policy considerations. These trends, he argues, had an important consequences for law: "They repeatedly undermine the relative generality and autonomy that distinguish the legal order from other types of law, and in the course of so doing they help discredit the political ideals represented by the rule of law" (p. 197). Ironically, then, according to Unger, the use of law by Western democracies to help disadvantaged groups has helped to undermine the rule of law for which these democracies are so renowned.

Social Change and Specific Legal Developments

The field of legal history is very fertile, and many studies exist of the social forces and developments that led to specific changes in legislation and in other various aspects of law and the legal system (the second type of impact of social change on law distinguished earlier) in the United States, Great Britain, and elsewhere. We obviously do not have room here to discuss all these studies, but a closer look at a few of them will help to indicate how social change has affected law in various eras. We begin with two examples from British history and then turn to some examples from U.S. history.

The Law of Vagrancy. An interesting legal issue today concerns the treatment of the homeless or other people who are vagrants or beggars. Some jurisdictions have been fairly proactive in arresting homeless people for vagrancy or other similar crimes, while others have adopted a "live and let live" policy that allows the homeless to live in the streets while offering them various social services. Critics of arrest policies say they violate the civil liberties of the homeless if they are breaking no other laws and tie up police and other legal resources unnecessarily. On the other hand, proponents say these policies make the streets safer and help downtown communities and business to thrive. New York City is one of the municipalities that has been following a proactive arrest policy, much to the dismay of homeless advocates. In 2002, one of the city's police officers refused to arrest a homeless man sleeping in a garage. The officer was suspended without pay for a month and was eventually put on probation for one year (The New York Times 2004). At the time of his suspension, a homeless man near the garage defended the officer: "He's a good guy—he's got a heart. He knows it's not a crime to be homeless, and the NYPD should be ashamed of itself" (Getlin 2002:A18).

In fact, it was not a crime to be vagrant early in the common law, but a law of vagrancy eventually was enacted in response to certain social changes. As recounted by sociologist William J. Chambliss (1964), the key decade in the history of the law of vagrancy was the 1340s. Before this time, people in England could beg or loiter without fear of arrest and legal punishment. In 1348, however, the bubonic plague devastated England, with about half the population eventually dying. The aristocracy (rich landowners) of England suddenly had a scarcity of labor, with far too workers remaining to work on their land. This meant that the landowners would have to compete for the relatively few workers who remained and, because of simple supply and demand considerations, they would have to raise the wages they offered.

To avoid doing so, they decided to increase the supply of labor by forcing people to work who previously did not work. They did this by inducing Parliament to enact the nation's first vagrancy law in 1349. This law made it a crime to beg, and it also made it a crime to move from one place to another to find better employment. The first provision meant that more people would have to work, and the second meant that they could not easily look for a job with better wages. Both provisions served to keep wages lower than they would have been otherwise. Chambliss (1964:68) writes that the new vagrancy law was "designed for one express purpose: to force laborers . . . to accept employment at a low wage in order to insure the landowner an adequate supply of labor at a price he could afford to pay." According to Chambliss's analysis, then, a terrible social change, the plague, led to the law of vagrancy that, though greatly changed since, still arouses controversy almost seven centuries later.

The Law of Theft. Suppose you bought a plasma TV from a store and arranged to have it delivered. If for some reason the driver of the delivery truck decided to keep the TV, you would certainly not be surprised if the driver were arrested for theft, and you might find it difficult to imagine that this act would not be illegal. Yet, in the early history of common law, this type of theft—the taking and possessing of an item by someone delivering it to the person who had purchased it—was *not* considered a crime punishable by law. As recounted by Jerome Hall (1952), the origins of the body of law that made this action a crime reflected the needs of an emerging mercantile (trade) economy, the forerunner of capitalism.

By the middle of fifteenth century, writes Hall, mercantilism was growing rapidly in England and much of the rest of Europe, whose economies before this time, of course, had been feudal and agricultural. (Recall Unger's discussion, discussed above, of the growth of mercantilism for the development of legal order.) At the heart of mercantilism is a transaction involving the selling of goods, often made by a third party, to a purchaser. Often purchasers will transport the goods home themselves (as you might with a plasma TV you had bought), but obviously sometimes the seller will have the goods delivered to the purchaser. During the early fifteenth century, this was routinely done by a *carrier*, the name given to the person, usually someone who was poor, who would drive a horse-drawn cart that carried the goods. During the time the carrier was transporting the goods, legal doctrine held they were technically in the carrier's possession. As such, if the carrier decided to keep the goods, no crime was committed, as it was thought that the carrier owned the goods while they were under the carrier's control. Fearing the loss of their job or violent reprisals by the seller or the purchaser, most carriers did not keep the goods for themselves even if doing so would not have been illegal. However, enough carriers did abscond with the goods that this type of behavior had become a significant problem for the growing mercantile class by the middle of the fifteenth century. However, it was not a problem for most poor people, who could not afford to buy and sell goods and who in some cases were the carriers absconding with the goods.

Perhaps not surprisingly, the issue reached the courts. In 1473, the landmark *Carrier's Case* established new and very consequential legal doctrine by declaring that it was now a crime for anyone transporting goods to keep the goods. According to Hall, this ruling protected the mercantile class's interests and reflected their growing power in English society and the lack of power among the poor. In effect, Hall wrote, the establishment of this new legal doctrine reflected the needs and influence of incipient capitalism. More generally, in this manner economic changes in society produced an important change in legal doctrine.

The Rise of Workers' Compensation. Workers who suffer an injury in their workplace or become ill because of workplace conditions are entitled to workers' compensation. This is a system financed by business contributions in which workers' medical expenses are ideally taken care of without the workers having to threaten to go to court. Workers are entitled to such compensation even if the injury they suffer stems from their own carelessness. For example, a construction worker who accidentally breaks a finger while using a hammer is entitled to workers' compensation even if the worker caused her or his own injury. Most readers probably favor workers' compensation in theory, and several have probably used it themselves or at least know friends or relatives who have used it. Workers' compensation is certainly so familiar that it almost seems as American as apple pie. Yet, workers' compensation has existed for only about a century, and before the late 1800s workers could not expect to have employers pay their medical expenses for injuries and illnesses suffered in the workplace. In fact, many Americans back then might have thought it illogical for employers to have to do so. The change in attitudes about this issue and the actual development of workers' compensation represent an interesting example of the impact of social change, in this case the advent of the Industrial Revolution, on a specific legal development.

As recounted by law and society scholars Lawrence M. Friedman and Jack Ladinsky (1967) in an oft-cited article, "Social Change and the Law of Industrial Accidents," existing tort law at the beginning of the nineteenth century (and just before the dawn of the Industrial Revolution) allowed individuals to sue other individuals who injured them through such behaviors as assault, trespass, or slander. In addition, according to a legal doctrine called the *law of agency*, an individual was also allowed to sue an employer if the individual was harmed by the behavior of one of the employer's agents (employees): An innkeeper's employee might rob a guest; a pub's employee might allow meat to spoil and make diners ill. Even though the employer in such cases had not injured the individual in question, the employer was still held legally responsible for the malfeasance or negligence of the employee. As a practical matter, for the injured individual to recover any costs, it made much more sense for the claimant to be allowed to sue the employer than the employee, who typically would be fairly poor and in no position to pay medical expenses or other damages to the claimant.

The beginning of the Industrial Revolution some years later resulted in many more workplace injuries because industrial work (e.g., in factories and mines and on railroads) was much more hazardous than agricultural work. Existing tort law and the law of agency could have been logically adapted to apply to these injuries such that workers would have been allowed to sue their employers, but this did not happen initially. A key development here was the establishment of the **fellow-servant rule** in an 1837 English case, *Priestly v. Fowler*. In that case, Fowler, a butcher, had instructed one of his employees, Priestly, to deliver some goods that another employee (or servant) had loaded onto a cart. Because the cart was in fact overloaded, it collapsed on the way, and the accident broke Priestly's leg. Undoubtedly realizing that Fowler was in a much better position to pay damages than the servant who had overloaded the cart, Priestly sued Fowler rather than the servant. Rejecting his claim, the court's ruling in the case established the fellow-servant rule, which held that an employer was not responsible for injuries to an employee that were caused by the actions of another employee. An employee could sue the employer only if employer had personally caused the injuries. Yet, because the employer was hardly ever physically present in factories and mines and on railroads, the employer could never be held legally responsible for any injuries that did occur. Friedman and Ladinsky (1967:53) note the implications of this new legal doctrine:

> In work accidents, then, legal fault would be ascribed to fellow employees, if anyone. But fellow employees were men without wealth or insurance. The fellow-servant rule was an instrument capable of relieving employers from almost all the legal consequences of industrial injuries. Moreover, the doctrine left an injured worker without any effective recourse but an empty action against his co-worker.

Five years after *Priestly v. Fowler*, the Massachusetts state supreme court rendered another influential decision in *Farwell v. Boston & Worcester Railroad Corporation* in 1842. In this case, a switchman's negligence caused a train engineered by Farwell to run aground; Farwell lost a hand because of the accident. Probably again recognizing it made more financial sense for him to sue the railroad company rather than the switchman, Farwell filed a suit for damages. The court ruled against him on the following grounds. Because some occupations are especially dangerous, employers have to promise more pay to induce potential employees to take on these occupations. In effect, then, the free market has already compensated these employees for doing this dangerous work, and the employer thus owes them no further compensation if, in fact, they are injured on the job. Workers know they are taking on the risks of their jobs for greater pay and do not deserve to be compensated if their jobs do in fact become risky.

The legal reasoning of both *Priestly v. Fowler* and *Farwell v. Boston & Worcester Railroad Corporation* was adopted throughout the United States, and, by the middle of the nineteenth century, American legal doctrine did not allow workers to sue employers for injuries caused by the negligence of other workers. This situation eventually changed for several reasons. One reason was the sheer number of industrial accidents that began to occur after the Civil War as industrialization continued apace and workplaces became more dangerous for more and more workers. By 1900, about 35,000 workers were dying every year from workplace injuries and another 2 million were suffering workplace injuries. A second reason was the start of the *contingency fee system*, in which attorneys take on lawsuits and do not receive any payment unless their plaintiff wins the case. This allowed injured workers or their families to sue without incurring legal expenses, and many did win their lawsuits despite existing legal doctrine. Their victories encouraged other injured workers to sue as well. As Friedman and Ladinsky (p. 61) observe, "Whether for reasons of sympathy with individual plaintiffs, or with the working class in general, courts and juries often circumvented the formal dictates of the doctrines of the common law." A third reason was labor unrest. Beginning in the 1870s, workers protested, went on strikes, and engaged in other agitation, and labor history is filled with many accounts of labor violence and strife during this period (Lens 1973; Taft and Ross 1990). One source of workers' dissatisfaction was their dangerous workplaces and the lack of compensation for the injuries they incurred.

All these factors eventually weakened the fellow-servant rule, and new legislation began to give injured workers greater rights to sue their employers. Some states, for example, established the *vice-principal doctrine* by giving workers the right to sue an employer when the workers were injured by the negligence of someone (the vice-principal) acting in a supervisory capacity for the employer, since this supervisor was no longer a mere fellow servant. Some states also gave workers the right to sue employers for injuries incurred because of unsafe workplaces and unsafe tools.

These legal developments, along with the rising number of injury lawsuits and growing labor unrest, led the business community to begin to favor a system of relatively automatic compensation that would not involve litigation and in which negligence would not have to be proven. Such a system, businesses thought, would save money by avoiding expensive litigation and the threat of large damages by juries, and it would also help quell labor unrest. In 1910, the president of the National Association of Manufacturers appointed a committee to study this type of system. Wisconsin passed the first workers' compensation law in 1911, and most states had a similar law by 1920. In 1948, Mississippi became the last continental state to adopt a workers' compensation system.

Changes in Family Law. The family is one of our most important institutions and has undergone great change throughout history. Changes in the family in turn have produced changes in law regarding the family, or *family law*. A brief sketch of some of these changes during the nineteenth century and during the past few decades will illustrate this important aspect of law and social change in the United States.

Nineteenth-Century Changes. Several important changes in family law occurred during the nineteenth century because of the growing numbers of the middle class and their ownership of land and other property with some value. As Friedman (2004b:20) observes,

> The United States was, in a sense, the first middle-class country. . . . The United States was the first country in which ordinary people owned some capital: a farm, a plot of land, a house. Questions of title, inheritance, and mortgage do not enter the lives of people who have nothing and own nothing—serfs, tenant farmers, and the like. Once people have property, once they own something, they become consumers of the products of the legal system. Now family law becomes significant for them. A man is dead; he owned an eighty-acre farm. Is this woman his widow? Are these children legitimate heirs?"

As Friedman indicates, the fact that so many Americans, compared to people in other nations, owned property meant that it was important for them to have "clear legal lines of ownership" (Friedman 2004b:32). This in turn created pressures to develop three new (as of the nineteenth century) concepts in family law: common-law marriage, legal adoption, and judicial recognition of divorce. We examine each of these briefly.

Common-law marriage. Before the eighteenth century, a betrothed couple in England or the American colonies who wished full legal recognition of their marriage was required to have notice of the marriage announced publicly by church officials (a practice called the *banns*) and then to have the marriage performed in a church ceremony with witnesses to the marriage. Despite these requirements, many couples could not afford the cost of the wedding ceremony, were not members of a church, lived in rural areas where no clergy were present, or else were estranged from their parents and ran off together. All such couples then lived together as husband and wife (the modern term would be cohabitation) and were widely regarded by their communities as a married couple. These *informal* marriages (also called *clandestine* or *irregular marriages*) received some legal recognition but did not enjoy full inheritance and property rights (Grossberg 1985).

England tightened the law in 1753 with the enactment of the Marriage Act, which prohibited informal marriage by requiring parental consent, banns announced in a church for three consecutive Sundays, and religious ceremonies; Quakers and Jews were exempted from these requirements (Friedman 2004b). Although the Marriage Act did not apply to the American colonies, the colonies generally adopted its standards but also allowed licenses from magistrates to substitute for banns. Despite these requirements, informal marriages continued to occur during the colonial period and into the nineteenth century for the same reasons they had occurred before England's Marriage Act. The legal status of these marriages in the new nation was unclear, as they enjoyed only some of the legal rights as "official" marriages and, in particular, did not enjoy full inheritance and property rights. Given the decentralized government of the colonies and then the new nation, the legal status of informal marriage also varied from one jurisdiction to another.

As the nineteenth century progressed, however, informal marriage in the United States began to achieve full legal recognition, thanks to court rulings and new legislation, as *common-law marriage* (Friedman 2004b). Here the leading case was *Fenton v. Reed* (4 Johns. 52) in 1809, in which the New York Supreme Court granted full legal recognition to the informal marriage of the defendant, Elizabeth Reed, who had sought such recognition so that she could become the beneficiary of her husband's Revolutionary War pension. The court declared that no formal wedding ceremony was required for the couple to be considered as having legally married. Although a significant court ruling in Massachusetts a year later rejected the idea of common-law marriage, the concept eventually was adopted in most of the states.

Part of the reason for the new legal recognition of common-law marriages was the traditional American affinity for individual freedom and privacy and distaste for state authority (Grossberg 1985). Another major reason was the need to establish clear lines of ownership of property in a growing middle-class society: "[T]he common law marriage was hardly a historical accident. . . . Money, land, and inheritance: these were the points at issue. The common law marriage was a device for settling claims to property. It protected 'wives' of informal unions and their children when the marriage ended with the death of one party, usually the 'husband.' That was its major function" (Friedman 2004b:20). Other functions included protecting the reputations of common-law wives, since sex outside marriage was considered extremely immoral for women, and preventing children from being labeled as bastards.

Judicial recognition of divorce. The need to establish clear property rights also led to the judicial recognition of divorce. During the colonial period, legal divorce was rare; in effect, a couple could get a divorce only if their colony's governing assembly passed legislation that granted the divorce. A similar situation characterized England during this period, as divorce could be obtained only by an act of Parliament, a fact that effectively limited divorce to the aristocracy and other wealthy individuals. As the nineteenth century began, American couples in many states similarly could get divorced only if their state legislature passed an act to grant the divorce. As the century progressed, however, Americans began to demand faster and less expensive divorces because more of them had begun to own a farm, a house, or other property. If a marriage dissolved, they needed legal recognition of this dissolution to make clear the lines of property ownership.

Responding to this pressure, many states (following the examples of Massachusetts and Pennsylvania in the mid-1780s) began to pass laws that permitted judicial divorces, in which a spouse would sue for divorce on one or more of several grounds recognized by the new laws. Common grounds included adultery, cruelty, and desertion. Officially, this meant that divorce would still be difficult to obtain in the majority of marriages that had dissolved for none of these reasons, but, in practice, couples wishing divorce often pretended that one of these grounds existed in order to win a divorce. In some cases, a husband would register at a hotel and remove most of his clothing. A woman (not his wife) would then arrive and remove most of her clothing. Then a photographer would arrive and take a photo of them sitting on the bed that would later be used in court to prove adultery. The woman would then collect a small payment from the husband for participating in the charade, get dressed, and depart (Friedman 2004a).

By the mid-twentieth century, several states had liberalized their divorced laws by allowing couples to obtain a divorce on more lenient grounds such as incompatibility or by granting divorce if couples were no longer living together. No-fault divorce was a significant development that arose during the 1970s in response to growing pressure for faster and less expensive divorce and dissatisfaction by many judges and attorneys with the subterfuge that had been used to obtain divorce (Friedman 2004b; Jacob 1988). To obtain a no-fault divorce, one of the spouses merely had to attest that she or he wanted to end the marriage. California passed the nation's first no-fault law in 1970, with most other states following suit.

Legal adoption of children. Legal adoption of children was a practice found in ancient Rome, where it was used to ensure that a family without its own children would be able to continue its name, and it was also a practice found in civil law nations centuries later. Before the nineteenth century, however, legal adoption was a concept unknown in the English common law, which gave blood relationships primacy in property rights (Grossberg 1985). This was true in Great Britain as well as in the American colonies and then the new states. Many American children, of course, were raised by adults who were not their natural parents. Some children became apprentices and for all intents and purposes were raised by adults who were not their natural parents. More often, children would lose their parents because of death or abandonment. Death of a parent was a frequent event, as many women died during childbirth during the nineteenth century, and many parents died from disease or accidents. For all these reasons, orphans abounded and would then be cared for by relatives or other individuals (or by orphanages). However, none of the children raised by new parents had legal rights to inherit property from them.

The beginnings of legal adoption are found in laws enacted by state legislatures during the first half of the nineteenth century that named a specific child as the heir of a specific parent. Some of these children were relatives, and some had been born to the parent out of wedlock. A Kentucky law in 1845 allowed one Nancy Lowry to "adopt . . . her step son, Robert W. Lowry, Jr. . . . as her own child, who, in all respects, shall stand in the same legal relation to her as if she were his mother in fact." The law further stipulated that Robert was her heir "as if he were her own personal issue, born in lawful wedlock" (quoted in Friedman 2004b:99). Other states enacted legislation that authorized local courts to give legal status to children taken in by adults who were not their birth parents. Massachusetts enacted the first actual adoption law in 1851 that stipulated a process to be followed. This model soon spread to most other states, and by the end of the nineteenth century, adoption law was widespread (Grossberg 1985). Once again the ownership of property by increasing numbers of the middle class was a key social force, as it created pressures to give children of non-birth parents legal status to allow them to inherit.

Recent Developments. During the last few decades, other changes in family law have occurred because of several types of social and technological change, including developments in reproductive technology, the increase in cohabitation (persons living together without being married), and, as indicated earlier, because of the focus of gay and lesbian advocate groups on same-sex relationships. As these changes all have occurred, the law has had to respond to complex issues.

For example, as scientists developed techniques such as *in vitro* fertilization, in which a sperm and an egg are combined in a laboratory and the resulting embryo implanted in a woman's uterus, family law has struggled to adapt. Before this technology was developed, a large dimension of family law quite naturally concerned custody of children after divorce. Almost overnight, the rights of the embryos and their ownership in case of divorce became significant legal and moral issues. Couples often create several embryos in the hope that one or more will successfully implant and yield a birth. However, sometimes a divorce occurs while one or more of the embryos remain frozen. Should conventional custody provisions for (already born) children apply to frozen embryos? In the area of custody law, should frozen embryos be treated exactly like children? If a woman gives birth to an implanted embryo conceived with an anonymous donor's sperm, does that donor have any parental rights? Courts have had to deal with these issues for several years now, and the body of law in this area is still evolving. The use of birth surrogates and anonymous egg or sperm donors who later decide that they have an interest in their child complicates the legal and ethical questions already at stake.

The title of a report on changes in family law from reproductive technology aptly summarizes this development: "Modern Life Stretching Family Law: US Courts Grapple with Nontraditional Custody Issues" (Paulson 2004). A law professor cited in the report explained how courts are responding to these issues: "Courts are more willing now to try and think creatively about the issues of parenthood and family because [nontraditional arrangements] are a much more common thing. Family law cases, which for a long time were regarded as not very interesting from a theoretical point of view, are becoming much more complex" (quoted in Paulson 2004:1).

The report described several cases involving embryo/birth custody. In one Pennsylvania case, after four triplets from an implanted embryo were born, four individuals claimed custody as the babies' parents: the egg donor, the surrogate mother who carried the embryo, and the couple (which included the sperm donor) who paid the surrogate to do so. In a preliminary ruling, a judge awarded primary custody to the surrogate mother, but an appellate court later awarded custody to the biological father (Ayad 2006). In a California case, a woman was supposed to be implanted with an embryo conceived with anonymously donated sperm, but instead was implanted with an embryo conceived with a client's sperm and meant for the client's wife. The error was discovered after the woman's son was born, and the sperm donor and his wife claimed custody. A judge decided that she was the legal mother and that the sperm donor was the father. The woman won a suit against the fertility clinic for $1 million.

There have also been several cases involving the disposition of frozen embryos after a couple gets divorced. In a Tennessee case that is often cited, a couple had produced seven embryos that were frozen for later possible implantation. The couple had no legal document to outline what would happen with the embryos if the couple divorced. During their divorce, both spouses claimed custody of the embryos; the mother wanted to save them for later possible use, and the father claimed ownership so that he would not have to become an unwilling biological father. The trial court said that the embryos were "human beings existing as embryos" and awarded custody to the mother so that one or more of them could yield a birth. An appellate court reversed this ruling and gave custody to the father, who, the court said, had a constitutional right not to become a father against his will. The Tennessee Supreme Court affirmed this decision (*Davis v. Davis*, 842 S.W.2d 588 [Tenn. 1992]). In a Massachusetts case, a couple created four embryos in 1991 but then divorced in 1995. Each spouse claimed custody of the embryos. Citing the Tennessee decision, a trial court ruled for the father. In 2000, the Massachusetts Supreme Judicial Court affirmed that ruling in a unanimous verdict, stating that no individual should be forced to become a parent against her or his wishes (Goldberg 2000).

Cases involving same-sex couples or cohabiting heterosexuals have led courts in several states to adopt the concept of *de facto parenthood*, which goes beyond biological parenthood to consider a couple's original intent regarding birth and parenting, and the concept of *psychological parenthood*, which takes into account the bonds adults have to their child even if they are not the child's biological parent. A series of California Supreme Court cases decided in August 2005 illustrates the issues that the law has needed to address in the absence of legal recognition of same-sex marriage (Egelko 2005). All the cases involved lesbian partners with one or more children who later ended their relationships. In one case, a woman donated an egg to her lesbian partner, who later had twin girls. They raised their daughters together for six years but then separated. When the birth mother moved with the twins to Massachusetts, her former partner had no claim to custody or visitation because she had signed a standard hospital form for anonymous donors that waived any parental rights. She filed suit in California to regain these rights. The California Supreme Court ruled in favor of the plaintiff by declaring that both women were the twins' legal parents and thus had the same rights and obligations enjoyed by and required of heterosexual parents.

The court's decision was issued the same day as two other decisions involving reproductive technology in which the court similarly granted legal parenthood status to same-sex partners. In one of these two cases, a lesbian couple had separated, and the plaintiff had asked the court to order her former partner to pay child support for the plaintiff's biological children. In declaring that same-sex parents had the same legal obligations as heterosexual parents, the court's decision ordered the defendant to pay the requested child support. In the other case, two lesbian partners had signed a pre-birth agreement affirming that both would be parents of the child that one of them was bearing. After the partners separated when the child was almost two years old, the birth mother refused to let her former partner visit the child or share custody. The court's ruling granted the former partner these rights.

Although courts in other states had also granted visitation and other parental rights to same-sex couples with children who later ended their relationships, the California Supreme Court's rulings went further in granting same-sex parents all the rights and responsibilities of heterosexual parents, including custody, inheritance, and insurance coverage. A spokesperson for the California State Attorney General's office applauded the rulings: "These rulings recognize that these children have the same rights as the children of opposite-sex couples in maintaining ties to the people who helped raise them and presumably love them. The rulings also properly recognize the diverse nature of family relationships in today's world" (Egelko 2005:A1). Echoing a similar point, the legal director of the National Center for Lesbian Rights observed, "Same-sex couples are now able to procreate and have children, and the law has to catch up with that reality" (Paulson and Wood 2005:1).

Technological Changes and the Law. Reproductive technology, of course, is not the only technology that has changed during the past few decades. Personal computers, cell phones, DVD players, and other devices were unknown three decades ago, just as the motor vehicle was unknown just over a century ago. All these innovations and discoveries have led to various changes in the law because technological changes create new situations and problems that demand legal attention.

The invention of the automobile illustrates this dynamic. As automobiles became more common at the beginning of the twentieth century, many new traffic laws were necessary to govern the many aspects involved in their use, including, of course, obeying speed limits, using turn signals, and stopping at stop signs and traffic lights. New regulations were necessary to govern the manufacture of this new type of machine, the issuing of drivers' licenses, and the width of roads. Many laws have since been passed to enhance motor vehicle safety, including the requirements that all vehicles must contain seat belts, air bags, and a rear brake light separate from the taillights. The advent of the automobile also led perforce to automobile accidents, personal injuries, and deaths; this development led in turn to an increased number of personal injury lawsuits and played no small role in the development of the automobile insurance industry with its own set of regulations.

Before the automobile was invented, the invention of the railroad had a similar impact on the law. In particular, railroad accidents significantly influenced the development of tort law, most of which involves personal injury, during that period. As Friedman (1984:144) explains, "Indeed, the law of torts was insignificant before the railroad age of the nineteenth century—and no wonder. This branch of law deals above all with the wrenching, grinding effects of machines on human bodies. It belongs to the world of factories, railroads, and mines—in other words, the world of the Industrial Revolution. Basically, then, the railroad created the law of torts." In this regard, recall *Farwell v. Boston & Worcester Railroad Corporation*, the 1842 case discussed earlier that helped establish the fellow-servant rule in the United States that effectively denied injured workers financial compensation.

The railroad thus had an enormous impact on tort law, but it also (like the automobile later) necessarily led states and municipalities to pass new legislation and regulations to guide the bourgeoning railroad industry. Trains were dangerous instruments, and the railroad industry was rife with financial corruption. Various state railroad commissions were implemented to regulate the railroad industry, but these ultimately proved ineffective when confronted with the great wealth and power of the major railroad corporations. As a result, the federal government was forced to intervene with the establishment of the Interstate Commerce Commission in 1887. The legislation that created the commission imposed certain controls on the railroad industry, but did not go as far as railroad critics wished (Friedman 1984).

The advent of the personal computer during the 1980s revolutionized modern society in many ways, but, for our purposes, it also led to many legal changes. As just one example, before the computer era and the Internet, traditional copyright law could not imagine someone being able to make a copy of a copyrighted song or album and then making it instantly available to tens or hundreds of thousands of people worldwide for their free use. Copyright laws in the United States and other nations needed to be changed to encompass and ban such behaviors. For this reason, 160 nations met in 1996 to discuss revisions to their copyright laws, which at that time, according to a news report, were "under technological siege" (Lewis 1996:A1). Since then, the U.S. Supreme Court and other appellate courts have considered many cases dealing with file sharing and other possible copyright infringement, and the music industry succeeded in winning legislation that prohibits the uploading of copyrighted music to shared Web sites for free downloading. As another example, before the computer era, *hacking*, or breaking into Web sites or private computer files, was obviously a behavior that did not yet exist. When it began, existing criminal law did not cover this new type of crime, and new laws had to be written. Another new behavior was *cyberstalking*, or the use of computers (through e-mail and hacking) to stalk someone, often out of sexual interest. State governments and finally the federal government had to pass anti-cyberstalking laws to ban and punish this new type of crime in the computer era. Perhaps not surprisingly, "stalkers still roam on the Internet," according to the headline of a recent news report (Zeller 2006), and cyberstalking continues to be a problem.

THE IMPACT OF LAW ON SOCIAL CHANGE: LAW AS INDEPENDENT VARIABLE

About a generation ago, law was hailed as an effective vehicle for social change. The U.S. Supreme Court's 1954 decision in *Brown v. Board of Education* declared school racial segregation unconstitutional and helped usher in the Southern civil rights movement that changed the nation. In 1964 and 1965, the U.S. Congress passed the Civil Rights Act and Voting Rights Act, respectively. Among other provisions, the Civil Rights Act banned discrimination in employment and public facilities on the basis of race, gender, religion, and national origin, while the Voting Rights Act abolished literacy tests that parts of the South had used to prevent African Americans from voting. Again and again, the U.S. Supreme Court ruled in favor of the civil rights movement in cases that arose from arrests of civil rights demonstrators and from other events in the South. These and other legal actions on behalf of civil rights and civil liberties spurred many idealistic young people to apply to law school and indicated that law could be used to ameliorate many long-standing social problems and to benefit disadvantaged social groups. Reflecting this view, new social movements since that time, including the women's movement, the environmental movement, and the gay and lesbian movement, that arose during the 1970s all tried with considerable success to mobilize the law to help achieve their goals. In all these ways, law was seen, and is still seen by many, as a potent tool to improve society.

This view of law as a vehicle for social change may be overly optimistic, however. Although *Brown v. Board of Education* aroused a nation, school segregation remained entrenched in many parts of the South for more than a decade. Racial discrimination continued for years after passage of the 1964 Civil Rights Act. Most states do not yet have laws that fully protect gays and lesbians from discrimination, and the environment continues to be endangered despite noble and persistent efforts, many of them involving litigation, by environmental advocacy groups.

This brief overview raises two important related questions: (1) How effective is law (or to be more precise, a change in the law) as a vehicle for social change? and (2) Under what conditions is law (or legal change) more or less effective in achieving social change? There is no easy answer to the first question: although much evidence suggests that law can be a very effective vehicle for social change, other evidence indicates that law is a rather ineffective tool for social change. A fair if indecisive conclusion is that law is a somewhat effective vehicle for social change whose effectiveness varies according to certain conditions. A major goal of this section is to elucidate these conditions and, more generally, the factors that promote or impede the impact of law on social change.

Aspects of the Law → Social Change Relationship

Before proceeding further, we need to distinguish certain aspects of the legal change → social change relationship.

Sources of Legal Change. Legal change can involve new legislation, a new court ruling, a new executive order, new administrative law, or changes in legal procedure. Whatever form it has, legal change emanates from several sources. Two major initiators of legal change may be distinguished. First, some legal changes are initiated primarily by a social reform group or larger social movement, as when the NAACP Legal Defense and Education Fund initiated the case that ended with the *1954 Brown v. Board of Education decision* (Klarman 2007). This type of legal change is the subject of recent scholarly attention, and we examine it in a separate section below. Second, some legal changes are initiated primarily by legislators or other government officials themselves, although in practice they are usually responding at least in part to the concerns of certain interest groups including political lobbyists.

The social change desired varies from one situation to another, but in general involves changes in people's behavior or attitudes or changes in social policy or the structure and functioning of social institutions such as the family.

Direct and Indirect Impact of Law. In another distinction, the impact of law on social change, and more specifically on behavior to simplify the discussion, may be either direct or indirect (Dror 1969; Handler 1978; Scheingold 1974). *Direct* social change occurs when legal change (e.g., new legislation or a court ruling) itself affects behavior. Changes in behavior happen for some of the reasons underlying people's obedience to law more generally (Tyler 2006) (see Chapter 2), with the particular reason depending on the law and behavior in question: (1) fear of legal sanctions that vary from law/behavior to the next but may include a fine, arrest, or loss of business among other sanctions; (2) a felt obligation to obey the new law simply because it is law (recall here Weber's concept of legitimate authority); (3) peer pressure or other informal legal sanctions. When scholars try to determine how effectively law changes behavior, they are often trying to determine the direct impact of law for any of these reasons.

Law may also have an *indirect* impact on social change. This can occur in either of two ways. First, a change in law may first affect a social institution, and the resulting changes in this social institution may then bring about changes in behavior or attitudes (Dror 1969). Suppose, for example, same-sex marriage one day becomes a legal option in the majority of the states thanks to new legislation, state constitutional amendments, or court rulings. If, as would almost certainly happen (and as has happened in Massachusetts, the only state to permit same-sex marriage), more same-sex couples then get legally married, the increase in such marriages would be a direct effect of this legal change. If the growing familiarity of same-sex marriage then increases public acceptance of such marriage, this latter change would be an indirect effect of the initial legal change.

Attitude change of this sort occurs for certain social psychological reasons. One reason involves the important concept of *cognitive dissonance* (Festinger 1957; Hyman and Sheatsley 1964). When people hold beliefs incompatible with their situation, they experience discomfort, or dissonance, that prompts them to change their beliefs in order to reduce their dissonance. Legal change may thus lead to attitude change through cognitive dissonance. To take an example from William K. Muir, Jr.'s (1967) study of attitude change after the U.S. Supreme Court banned prayer in the public schools in its 1963 *School District of Abington Township v.Schempp* decision, some educators who favored school prayer experienced cognitive dissonance after *Schempp*, especially because they considered themselves patriotic and respected the Supreme Court. To reduce their dissonance, these educators changed their beliefs and agreed with the Court that prayer did not belong in the public schools.

In addition to its effects through changes in a social institution, law may have an indirect impact in a second way: a legal change may give a disadvantaged group a sense of legal entitlement by suggesting that its claims and grievances are entitled to legal redress and thus a new hope that social change is possible. Both these perceptions in turn may spur the group into political action that aims to achieve beneficial social changes. We return to this process below in our discussion of law and social movements.

With the distinction between the direct and indirect effects of law on social change in mind, we now turn to certain problems and issues that may limit the impact of law on social change. These limitations pertain more to law's direct impact on social change than to its indirect impact.

The Limits of Law as a Social Change Vehicle

Attempts to use the law as an instrument of social change often encounter legal, political, and social obstacles; the particular obstacles depend to some degree on the nature of the social change desired. Although we focus here on obstacles facing government-initiated attempts to use the law for social change, we will examine the obstacles facing social reform group-initiated attempts in the section below on law and social movements.

A major problem that limits the impact of law arises from the fact that many of the legal changes initiated by government aim to change people's behavior: to reduce drunk driving, to reduce illegal file sharing on the Internet, to eliminate polygamy (to use an example discussed just below), and so forth. Some of these efforts involve behavior to which people are strongly committed for cultural, moral, financial, or other strongly held reasons. As the previous chapter discussed for certain types of crime, such behavior is often resistant to change by new legislation or court rulings. This problem is exacerbated when the behavior in question tends to occur in private rather than in public, as it can then be very difficult to detect. Of course, many practices are both private and strongly held, and these are precisely the kinds of behaviors that may be resistant to legal change.

An oft-cited example here is polygamy among the Mormons in Utah in the nineteenth century (Gordon 2002). According to Mormon religious beliefs, it was appropriate and even expected for men to have several wives. Most of the rest of the nation considered polygamy immoral, and the resulting conflict took on important legal dimensions in 1862, when Congress enacted the Morrill Act that made polygamy a federal crime. Polygamy continued, however, as many Mormons simply disobeyed the new law. Several years later, a Mormon who was arrested and convicted for violating the Morrill Act took his case to the U.S. Supreme Court (*Reynolds v. United States*, 98 U.S. 145 [1878]) and argued that his conviction violated his First Amendment rights to freedom of religion. The Court upheld his conviction, the Morrill Act remained the law of the land, and Congress passed a series of measures designed to strengthen its anti-polygamy effort. Yet even after these developments, polygamy persisted until the Mormon church officially abandoned it in 1890. Even after it did so, however, some Mormons continued to practice polygamy for many more years.

Another problem in using the law to achieve social change involves the consistency of the response by legal and political authorities to a legal change. For a new law to have an impact on behavior, police, judges, and other legal and political authorities need to apply the law consistently. If police fail to make arrests under the new law or judges fail to sentence people as the new law dictates, the impact of the law will be diluted. A primary reason for the failure of criminal justice authorities to apply a new criminal law stems from the system capacity considerations discussed in the previous chapter. They realize that a greater number of arrests and/or longer prison terms will overload the criminal justice system to its detriment and, in the long run, endanger public safety. In less common reasons, some legal authorities may fail to apply a new law because they simply disagree with its substance or because they think compliance will divert their time and attention from more important matters. Whatever the reasons, the problem of inconsistent application is yet another obstacle that limits the impact of certain legal changes.

Examples of this problem abound. One is drunk driving, which, as Walker (2006) observes, involves a behavior committed by people who, as representative of the general public, should fear all the consequences (including publicity and family members' reactions) of arrest and harsh sentences and thus change their behavior if new legislation cracks down on drunk driving through mandatory sentencing and other measures. However, new legislation of this type often does not accomplish its intended impact for a variety of reasons (Ross 1992; Walker 2006). Sometimes short-term reductions in traffic accidents do result from such new legislation, but over time these reductions dissipate as accidents return to their previous levels. Drivers may drink less or otherwise not drink and drive in the wake of new legislation, and police may initially increase their stops and arrests of drunk drivers. Within a relatively short time, however, things get back to normal. People rightly recognize that their chances of being stopped for any one incident of drunk driving are very low, and begin to drink and drive the way they did before the new legislation. In another problem, police do not like traffic stops for many reasons, including the dangers they pose to police and the time they take away from crime-control activity, and thus eventually shy away from aggressive patrolling of drunk driving after publicity surrounding the new drunk driving crackdown has faded. Even if drunk drivers end up in court, plea bargaining and other decisions by prosecutors and judges in the interests of system capacity help drunk drivers avoid the harsh mandatory sentencing that is often part of new legislation aimed at drunk driving. Thus, although many reasons may explain the relatively limited impact of drunk driving legislation, the inconsistent application of such legislation by police, prosecutors, and judges is one important reason.

Another example of inconsistent response by legal authorities involves the "three strikes" laws passed in the 1990s that mandated life imprisonment or very long prison terms for someone convicted of a second or third felony.

Although these laws have not reduced serious crime for the reasons outlined in the previous chapter's discussion of the problems with deterrence, the lack of consistent application is another reason that minimizes any conceivable effect they could have had. Although many states enacted "three strikes" laws, Walker (2006:60) observes that outside of California "prosecutors in most states were simply not using the law." Instead, they used the possibility of a three-strikes prosecution to persuade defendants and their attorneys to agree to a guilty plea. Moreover, in California, prosecutors in Los Angeles used the law quite often, but their counterparts in San Francisco used it only rarely. As this example illustrates, legal officials may adapt to legal changes in ways legislators and other parties responsible for the legal changes did not intend. Legislation that dramatically increases the potential penalty for a crime may force attorneys and judges to devise ways of getting around the law to avoid various system capacity and related problems. This situation reflects an important dynamic: "An increase in the severity of the penalty will result in less frequent application of that penalty" (Walker 2006:62). This dynamic helps explain why the impact of certain legal changes will not achieve their intended impact.

Problems in Assessing Legal Impact

It is often difficult to accurately assess the actual impact that legal change may have. Several reasons account for this difficulty.

First, legal impact may occur in the short term but not persist beyond an initial phase. The public, organizations, or other targets of legal change may initially comply and behave in the way a new law intends, but over time they may revert to their previous behavior. Legal authorities may initially actively enforce a new law but later slacken their enforcement. Drunk driving legislation provides a telling illustration of these problems.

Second, even if legal change achieves its intended impact, in the long run it may have negative unintended effects, as the example of 1960s Congressional legislation and the South's voting tendencies indicates. Although, as noted earlier, the 1965 Voting Rights Act increased voting levels among Southern African Americans, it had an unintended indirect effect that no doubt dismayed the Democratic-controlled Congress, which passed the voting legislation, and President Lyndon Johnson, who signed it. The Voting Rights Act was so unpopular in the South, a strong Democratic region, that many white Southerners began to switch their allegiance to the Republican Party (Abramsky, 2007). This process accelerated under Presidents Nixon and Reagan until most Southern states finally became "red" states, or Republican strongholds at the federal level.

Third, if behavior or attitudes change after some legal change, it is sometimes difficult to know why these changes occurred and, more specifically, whether these changes stemmed from the legal change. For example, say a state legislature passes a law that raises the financial and other legal penalties for speeding. Six months later, a study is published that finds that traffic tickets for speeding are down and that (through camera surveillance) fewer cars are in fact speeding. Although this study does suggest that the law worked as intended, it is possible that other factors were at work. Suppose gasoline had become more expensive around the time of the new law; if so, drivers might have decided to drive more slowly in order to save gas and money. If the six-month study period included a harsh winter, it is possible that car speeds lowered because road conditions were hazardous. Studies of legal impact also often lack adequate control groups that would permit them methodologically to rule out these kinds of nonlegal factors. Thus, in our traffic example, if speeding apparently reduced in the state that passed the new law, it would be helpful to study any possible speeding changes in adjoining states. If speeding in these states also declined in the absence of any new speeding legislation, this would suggest that some other factor prompted the speeding reductions in the state that did have the new legislation.

The example of zero-tolerance policing in the 1990s provides a real-life example of the need for control groups in studying legal impact. Such policing involves frequent arrests for minor offenses such as loitering and disorderly conduct; the expectation is that some of the people arrested will have committed (and be likely to commit) more serious offenses, and this very visible policy of arrest will deter other persons from committing crime. After New York City initiated zero-tolerance policing in the 1990s, its crime rate plummeted, and the New York mayor and police commissioner received national credit for their city's crime rate decline. However, crime declined in several other cities, including Boston, Dallas, Los Angeles, San Antonio, and San Diego, that did not institute zero-tolerance policing, and New York's crime rate had actually begun to decline *before* it began zero-tolerance policing. These developments all suggest that New York City's policy might not deserve the credit it received (Karmen 2001).

A fourth and related problem regarding accurate assessment of legal impact concerns the actual reasons for any behavior changes that do occur. Recall that people have many reasons for obeying the law: fear of punishment, belief in the legitimacy of law, habit, peer pressure, and self-interest. Thus, although a legal change may affect behavior, it is sometimes difficult to determine exactly why the behavior did alter because of the legal change. An example of a stop sign illustrates this problem (Kidder 1983). Suppose a stop sign is installed at an intersection that earlier had only a blinking yellow light that most drivers largely ignored. The stop sign does seem to work: Drivers at least slow down in the typical "rolling stop" before proceeding through the intersection, and some drivers stop completely. What we cannot easily determine is exactly why they are slowing down or stopping because of this "legal change." Do they fear that a hidden police car will suddenly appear and result in a traffic citation? Do they stop (or slow down) because they feel obliged to obey the law? Do they stop simply because they are accustomed to stopping at stop signs? Is there someone in the car with them who would react negatively if the driver did not stop, as research suggests (Feest 1968)? Are they stopping primarily because they are concerned about a possible collision if they do not stop? As all these possibilities suggest, people may obey a new law for a variety of reasons, and even if the law does have its intended impact, we may not be able to determine the exact reasons for the impact.

A final problem concerns the intent of the legal change whose impact is being assessed. In order to know whether a legal change has its intended impact, it is obviously necessary to first know what specific impact was intended. However, the intended impact that might be inferred by the public or by researchers may not, in fact, be the impact that the initiators of the legal change actually intended. Kidder (1983) observes, for example, that lawmakers who put up stop signs seemingly want to reduce traffic accidents. However, it is possible that their real intent, or at least part of their intent, is to increase municipal revenue from the many new traffic fines that people who run the new stop signs will now have to pay. In a less cynical possibility, Kidder (1983:140) also observes that many new laws or court rulings are compromises stemming from political struggles and thus have not just one intent but "many intents." Thus, a researcher who wants to study the impact of a new law or court ruling may not know which intended impact to study.

Conditions That Maximize the Potential Impact of Legal Change

The many problems, obstacles, and issues discussed in the preceding sections indicate why the impact of law varies from one situation to another and also why this impact may be lower than the initiators of a legal change may prefer. In a classic article, William M. Evan (1965) emphasized that the impact of law is likely to be highest under certain conditions that are not always possible to achieve. We present each of these conditions with some explanation.

1. ***The source of the law should be perceived as authoritative and prestigious.*** Evan distinguishes four sources of lawmaking: administrative, executive, judicial, and legislative. Because the average citizen probably perceives the legislature to be the proper source of lawmaking and thus the most authoritative and prestigious of all four sources, laws from legislatures are apt to have more impact than laws from the other three sources. The impact of executive orders would be the next highest, followed in turn by administrative decisions and finally by judicial decisions. All things equal, Evan added, legislative laws are especially desirable when the intended social change is likely to encounter great resistance. For this reason, he hypothesized that Southern school desegregation might have occurred more quickly if it had been ordered by Congress (which, of course, was not about to do so) than by the Supreme Court in the *Brown* decision.

2. ***The rationale for the new law should emphasize its continuity and compatibility with existing institutionalized values.*** Emphasizing that a new law conforms to the values embodied in, say, the Declaration of Independence or the Constitution should help ensure obedience to it.
3. ***Publicity surrounding a new law should emphasize that similar laws have proven helpful elsewhere with few or no adverse effects.*** This again helps overcome potential resistance. To take a contemporary example, advocacy groups urging legalization of same-sex marriage might point to the experience of Massachusetts after its highest court granted legal status to same-sex marriage in 2004: no evidence has shown that this ruling weakened the institution of heterosexual marriage or otherwise had any negative practical impact.
4. ***A new law should mandate that any changes it requires should occur quickly rather than slowly.*** Evan feels that the requirement of fairly swift compliance would help ensure compliance with a new law by minimizing the opportunity for organized resistance to the law to build. In this regard, one possible reason for the successful resistance to *Brown v. Board of Education* is that the decision allowed Southern communities to proceed "with all deliberate speed," a phrase that gave these communities time to plan and implement their successful resistance to the decision.
5. ***Legal authorities must apply and enforce a new law consistently and without hypocrisy or corruption.*** We have already discussed this problem earlier. In discussing this condition, Evan noted the corruption of police during the Prohibition era that helped ensure that Prohibition would fail.
6. ***Positive as well as negative sanctions should be used to help ensure compliance with a new law.*** Negative sanctions—punishment—may prove effective, but they can also cause resentment and other problems that undermine compliance with a new law. A large body of research from the field of social psychology finds that positive sanctions—rewards—are often more effective than negative sanctions in affecting behavior. Most new laws threaten legal sanctions and do not promise any reward for compliance. Evan speculated that school desegregation might have occurred sooner in the South if the federal government had granted funds for teachers' salaries and school construction and even tax rebates to communities that desegregated their schools.
7. ***Effective protection and resources should be provided for the rights of individuals who would suffer if other people violate or evade the new law.*** Sometimes, as happened in the South after *Brown*, people have to go to court to win rulings designed to force compliance with a new law. Evans believed that these efforts are more likely both to occur and to succeed if government agencies or private organizations provide the necessary funds and legal resources needed for these efforts.

Taken together, then, these seven conditions theoretically help ensure that a new law will achieve its maximum impact. In practice, it is often difficult for all of these conditions to exist for any new law. As noted earlier, the Congress was not about to order school desegregation in the South in 1954 when the Supreme Court rendered its *Brown* verdict. The fact that the order for school desegregation came from the Supreme Court rather than from the Congress, a more authoritative source of lawmaking, helped maximize Southern resistance to school integration.

LAW AND SOCIAL MOVEMENTS

Social movements are a familiar part of the landscape in the United States and other nations. Although many definitions of social movements exist, they may be regarded as sustained, collective efforts by individuals and groups lacking political power and influence to achieve social, economic, political, and/or cultural change (Meyer 2007). To achieve such change, social movements use a variety of strategies and tactics including civil disobedience, marches, rallies, and other types of protest. Although social movements are, perhaps, most known for protest, they often also use the law to help advance their aims. More specifically, they file lawsuits to change or end a government or business practice or policy, to alter the structure or functioning of a social institution like the family and so forth. Their use of the law for this purpose is called *social movement litigation* or, more popularly, **legal mobilization** (Zemans 1983). Earlier parts of this chapter briefly discussed examples of legal mobilization by the Southern civil rights movement and other movements of the last several decades.

At the same time, law is also used *against* social movements in what may be called the *legal control of movements* or, more popularly, **legal repression**. Law enforcement agents may infiltrate or conduct other kinds of surveillance on movement groups and individuals, and police may arrest, and district attorneys may prosecute activists who engage in illegal acts of protest. Sometimes, activists are arrested or otherwise legally harassed even if they have not broken the law. As we shall see, the Southern civil rights movement was again a focus for such legal repression.

In all these ways, law often plays a vital role in the struggle between social movements and their opponents, and the interaction between law and social movements forms a key dimension of the law and social change nexus (Marshall 2005a). Accordingly, this section reviews important issues and findings from the law and social movement literature. Reflecting common usage, we will use the terms *social movements* and *social reform groups* interchangeably.

Use of Law by Social Movements

A basic issue in the study of legal mobilization is whether it provides social movements an effective tool for achieving their objectives. Scholars disagree regarding the impact of legal mobilization in this regard. Some think that law is a potentially very effective tool for helping social movements (Eskirdge 2001), while some think that law is ineffective and even counterproductive for movements (Brown-Nagin 2005), with one scholar famously asserting that the courts offer social movements only a "hollow hope" for achieving change (Rosenberg 1991). Still other scholars take a middle ground—one that is becoming more popular—in concluding that legal mobilization is sometimes effective and sometimes ineffective, depending on the particular change sought and other conditions. As a leading scholar in this area, Michael McCann (2006:35) concluded,

> Legal mobilization tactics do not inherently empower or disempower citizens. Legal institutions and norms tend to be Janus-faced, at once securing the status quo of hierarchical power while sometimes providing limited opportunities for episodic challenges to and transformations in the reigning order. . . . How law matters depends on the complex, often changing dynamics of the context in which struggles occur.

Legal mobilization has proven to be a fairly effective strategy for several social movements and social change efforts during the past few decades. The *Brown v. Board of Education* decision was a significant legal victory for the early civil rights movement, even if many Southern communities resisted school desegregation for more than a decade. The environmental movement has also achieved significant legal victories (Coglianese 2001). The equal employment opportunity movement succeeded in winning federal legislation and court rulings that banned employment discrimination based on gender, national origin, race, and religion. This general prohibition, along with subsequent victories in lawsuits brought later against businesses allegedly violating EEO laws, is often cited as illustrating the potency of legal mobilization (Burstein 1994). Litigation has also been an effective tactic for the animal rights movement, whose lawsuits have helped win it publicity, influenced public opinion, and pressured alleged animal abusers to change their practices to avoid the economic cost of dealing with the lawsuits and the practical cost of possibly losing them (Silverstein 1996).

The Limits of Legal Mobilization. Despite victories such as these, many scholars, as noted above, question the effectiveness of legal mobilization for social movements. Several concerns underlie their skepticism.

First, the courts and other law-generating branches of government are often unsympathetic to the claims of social reform groups and the larger social movements to which many such groups belong. This lack of sympathy makes it difficult for social reform groups to win legal victories. In the 1970s, for example, groups opposed to nuclear power plants brought several lawsuits to halt construction of various plants, but judges routinely refused to grant the legal relief sought by the antinuclear groups (Gyorgy 1979). And although the U.S. Supreme Court eventually came to the aid of the Southern civil rights movement, most state courts and lower federal courts were hostile to the legal claims of the movement (Barkan 1985).

Second, social reform groups typically lack the financial and legal resources and acumen typically enjoyed by the corporations, government agencies, and other established interests that are often the targets of legal mobilization. Although law theoretically offers them a level playing field to present their grievances, this disparity again makes it difficult for social reform groups to win legal victories. This problem is compounded by the fact that litigation is both time-consuming and very expensive. These twin problems aggravate the resource disparity just mentioned and again limit the potential of legal mobilization.

Third, certain rules of judicial procedure and legal doctrine may also pose various obstacles (Handler 1978). Social reform groups need legal *standing* (see Chapter 2) to initiate a lawsuit. The early environmental movement often encountered standing issues in its attempts to win litigation to reduce pollution, save endangered species, or achieve other environmental victories (Large 1972). Courts must also have proper *jurisdiction* to hear an issue; not every issue is justiciable at all, and some issues are justiciable only by certain courts (for example, state versus federal). During the Vietnam War, various groups brought lawsuits to have the war ruled unconstitutional, but courts routinely either refused to hear these suits or else ruled against the plaintiffs on the grounds that the war was a *political question* and thus not one that courts should consider (Ely 1993).

Fourth, even if movements win court rulings and other legal changes, the targets of these changes often vigorously resist these changes. Thus, legal victories by a social movement may be a victory in name only and not actually prompt much change in social policy or institutional behavior. This problem is especially acute when the movement's legal victory involves a court ruling, as courts lack effective enforcement powers. An oft-cited example of this difficulty again comes from the civil rights era, when Southern governments effectively resisted *Brown v. Board of Education* and other federal court rulings through a variety of legal stratagems. As one scholar observed,

> It is ironic that the white South was extremely successful in minimizing the impact of the desegregation decisions of the federal courts without arousing the indignation of the rest of the nation. . . . [African Americans came to realize] that although they had won a new statement of principle they had not won the power to cause this principle to be implemented (Killian 1968:70).

Fifth, litigation strategies by social movements may ironically impede their goals by channeling activists' energy into legalistic pursuits that seek only minor changes and turning their attention away from various types of protest that may prove more effective in the long run (Marshall 2005a). The contemporary environmental movement has been criticized in this regard for pursuing an overly legalistic strategy with narrow goals at the expense of a more transformative strategy relying more heavily on the grassroots organizing and protest tactics that proved effective a generation ago (Coglianese 2001). The histories of the labor and gay and lesbian movements also provide examples of legalistic strategies pursued at the possible expense of protest and other political action that may have proven more effective (Anderson 2005; Forbath 1991).

A final problem involving legal mobilization recalls the structuralist Marxist argument outlined in Chapter 2. Even if social movements win court rulings or new legislation, the impact of these victories may be primarily symbolic if, as often happens, they yield only minor improvements to serious, intractable problems (Sarat and Grossman 1975). These minor improvements may nonetheless appease social reform groups and their constituents and help legitimate the power structure and the *status quo*. In this manner, legal victories of social reform groups may be Pyrrhic victories that in the long run help preserve social inequities in power, wealth, and other resources.

An influential study reflecting many scholars' skepticism about the effectiveness of legal mobilization was Joel Handler's (1978) examination of litigation by the civil rights, consumer, environmental, and welfare movements of the 1960s and 1970s. Finding that their legal mobilization produced only limited success at best, Handler (1978:209) concluded,

> In sum, social-reform groups find it difficult to obtain tangible results directly from law-reform activity. It can be accomplished, and numerous cases have been discussed where such results have been obtained, but, on the whole, special circumstances are needed. . . . Social-reform groups seek out the courts because they are weak and have lost in the political process, and there is only so much that courts can do by way of direct, tangible benefits.

Indirect Benefits of Legal Mobilization. Although scholars thus disagree on the extent to which legal mobilization can achieve significant legal victories with far-reaching tangible results, they generally do agree that legal mobilization may provide social movements and social reform groups with significant *indirect* benefits (McCann 2006; Scheingold 1974). Perhaps the most important indirect benefit was noted earlier: even if legal mobilization does not produce significant tangible results in and of itself, it may still give aggrieved groups a sense of legal entitlement by suggesting that their claims and grievances are in fact their legal rights. This sense may in turn give them new hope for social and political change and spur members of these groups to work for such change. As Friedman (1975:234) puts it: "Vindication of rights feeds on its own success. Success in one claim encourages and reinforces more claims and gives others a model and hope of success."

A widely cited example is once again the 1954 *Brown v. Board of Education* decision. As already mentioned, *Brown* did not quickly end Southern school segregation: Southern schools so resisted the decision through legal and political machinations that only 7 percent of African American children in the South were attending desegregated schools by 1961. At the same time, however, *Brown* galvanized African Americans in the South, among whom its effect was said to be "electric" (Lomax 1962:74). It gave them new hope for change by suggesting the federal government would now pay attention to the evils of segregation and is widely credited with helping to spur the massive civil rights protests that captured national attention in the ensuing decade (Morris 1984). Other civil rights litigation efforts had a similar mobilizing effect on movement activism in the South (Polletta 2000).

A more recent example of litigation acting as a catalyst for political action involves the movement for gender pay equity. In a study of this movement, Michael McCann (1994) found that the pay increases won by the movement were relatively modest. More important for the women workers he interviewed was the increased sense of political empowerment the women gained from the legal victories that won them their modest pay increases. They gained a new sense of their rights as workers and a new sense of their discrimination as women workers. Their increased rights consciousness in turn spurred them to become more active in their labor unions and to seek reform in areas beyond pay equity, including better working conditions and maternity leave.

In sum, the indirect, catalytic effects of legal mobilization may in the long run prove more consequential for social movements than the direct, tangible effects of legal victories themselves. Legal mobilization may be especially effective in this regard when it can be used as "a course of institutional and symbolic leverage against opponents" (McCann 2006:29). The targets of legal mobilization litigation may fear both the legal expenses of defending themselves against movement lawsuits and the negative publicity that often accompanies such litigation. In this way, legal strategies can complement protest and other political strategies to most effectively advance movement goals.

If legal change inspires a disadvantaged group in this fashion, it is also true that it may inspire entrenched groups to resist any social change that would weaken these groups' power and influence. Thus, legal change may indirectly both promote social change by inspiring aggrieved groups into political action and impede social change by spurring established interests into resisting such efforts. The Southern civil rights movement again illustrates this latter dynamic, as *Brown* not only inspired civil rights protest but also galvanized Southern white resistance to desegregation and helped set the stage for the white power structure's intransigence over the next decade and beyond (Klarman 2007). In another example, although the Supreme Court legalized most abortions in *Roe v. Wade* in 1973, this decision created a backlash that led to a sustained effort to overturn *Roe* or otherwise limit abortion rights that continues to this day (Maxwell 2002). The possibility of this indirect effect that helps spur a countermovement is yet another reason underlying the skepticism of some scholars of the value of legal mobilization for social movements.

Use of Law Against Social Movements

Law may sometimes be an effective tool for social movements for the direct and especially indirect reasons just discussed, but it can also be an effective weapon used against social movements by the government and other targets of movement activity. The legal repression or control of social dissent and social movements takes several forms.

First, corporations and other targets have sued social reform groups and other citizen efforts for slander or libel or for otherwise making statements or engaging in actions that may threaten profits or other goals pursued by the plaintiff. These lawsuits have been named *strategic lawsuits against public participation* (SLAPP) (Pring and Canan 1996). Real estate developers have sued groups attempting to stop developments that the groups feel may cause environmental or other problems. After Oprah Winfrey publicly questioned the safety of U.S. beef in regard to "mad cow" disease in the mid-1990s, the cattle industry sued her. Although she won the case, most defendants in SLAPP suits are hardly as wealthy or well-known. SLAPP suits themselves, and even the mere threat of being sued, may thus intimidate citizens into silence and are said by critics to undermine the First Amendment right to of freedom of speech.

Second, police, FBI, and other law enforcement agents may conduct surveillance on members of protest groups or on people associated with other kinds of social change efforts (Marx 1988). For example, police may infiltrate protest groups and pretend to be legitimate members of a group, and they may wiretap the groups' phone calls or conduct other kinds of monitoring. One of the most notorious examples of such surveillance was COINTELPRO, an FBI counterintelligence program that began in the 1940s and lasted until the 1970s. As part of this program, the FBI monitored tens of thousands of U.S. citizens who were exercising their First Amendment rights as part of the civil rights, Vietnam antiwar, and other social movements of that era (Cunningham 2004). Targets of the FBI spying program included Dr. Martin Luther King, Jr., and Beatle John Lennon, but almost all the people monitored were "regular" citizens. The CIA and many other federal, state, and local law enforcement agencies also engaged in similar spying. Although Congress passed legislation in the 1970s that limited surveillance of this type, a decade later the FBI spied on about 2,400 individuals and many organizations opposed to U.S. policy in Central America (Gelbspan 1991). The following decade it was revealed to be monitoring the activities of gay rights groups (Hamilton 1995). After 9/11, the National Security Agency not only began monitoring phone calls received within the

United States but also phone calls outside the United States. This monitoring was conducted without judicial warrants. Although the White House said the monitoring was a necessary part of the effort to combat terrorism, critics said it violated federal law designed to protect individual privacy (Risen and Lichtblau 2005).

A third form of legal control of social dissent involves the arrest and prosecution of activists (Barkan 2006a). Some protesters are arrested for committing acts of civil disobedience, while others arrested on trumped-up charges as targets of legal harassment. Whatever the reason for their arrest, protesters may be forced to spend large amounts of time, money, and energy defending themselves against criminal charges, and the threat of arrest and prosecution may also help deter other people from joining in protest efforts. Arrests of protesters may also stigmatize them and the movement to which they belong in the eyes of the news media and of the public.

During the Southern civil rights movement, thousands of activists were arrested, many of them for acts of civil disobedience but many also for such innocuous activities as merely walking down the street. In jail, they had to face the taunts and possible violence of white racist guards and inmates. Guilty verdicts were a foregone conclusion as the South experienced "a wholesale perversion of justice, from bottom to top, from police force to supreme court" (A. Lewis 1966:289). Although violence by police and townspeople received national headlines and TV coverage and greatly helped the civil rights movement achieve its goals, cities that relied strictly on mass arrest and incarceration were able to defeat civil rights forces without the negative publicity that accompanied the use of violence against civil rights activists elsewhere in the South (Barkan 1984).

In some movements, however, protesters who are arrested and prosecuted have been able to take advantage of their trials to win publicity for their cause. During the Vietnam War, antiwar defendants in several trials were able to discuss their views about the war and occasionally win acquittals and sympathetic news media coverage (Barkan 1985). During the abolitionist movement that preceded the Civil War, several abolitionists were arrested and prosecuted for helping fugitive slaves escape from the South. Juries in Massachusetts and New York sometimes acquitted these defendants, and the newspaper publicity these acquittals received is thought to have strengthened the abolitionist cause (Mabee 1969). Susan B. Anthony's trial a quarter-century later also won headlines. Anthony, the famous women's rights leader, was arrested in 1872 in New York for voting because, as a woman, she was not permitted to vote. At the end of her trial in Canandaigua, NY, a year later, she delivered a statement before sentencing after the judge ordered the jury to find her guilty: "I shall earnestly and persistently continue to urge all women to the practical recognition of the old revolutionary maxim, 'Resistance to tyranny is obedience to God.'" Her prosecution and eloquent summation received much attention and are thought to have helped the women's suffrage movement (Friedman 1971).

Summary

1. The relationship between law and social change is a key dimension of the study of law and society. Social change can affect law, and legal change can affect some aspect of society. When scholars study how that changes in society may bring about changes in law, they treat social change as the independent variable and legal change as the dependent variable; when they study how changes in law may bring about changes in society, they treat legal change as the independent variable and social change as the dependent variable.
2. Literature on the impact of social change on legal change falls into two broad types based on the scope of the changes involved. The first examines how and why broad social changes produce far-reaching changes in the nature of a legal system, legal reasoning, and other fundamentally underlying dimensions of law. The second type has a more narrow focus and examines how and why certain social changes produce new legislation, new court rulings, new legal procedures, or other rather specific aspects of law.
3. Emile Durkheim thought that mechanical solidarity and repressive law characterize traditional societies while organic solidarity and restitutive law characterize modern societies. Although Durkheim's views helped reinforce the idea that the type of law a society has depends on certain features of the society itself, later scholarship indicated that Durkheim may have misinterpreted the relationship between type of society and type of law and in particular reversed the relationship that actually exists.
4. Max Weber emphasized rationality as a key feature of modern society and used this concept to understand how societies changed legally as they became more modern and complex. Whereas traditional authority characterizes small, traditional societies, rational-legal authority, based on a belief in the legitimacy of a society's laws in the right of leaders acting under these rules to make decisions, characterizes modern society. Legal decision-making in general becomes more rational, involving the use of logic and reason, as societies become more modern.
5. Karl Marx and Friedrich Engels emphasized that law helps the ruling class preserve its dominance in several ways. First, law helps preserve private property. Second, law provides legal rights for all and thus creates a façade of justice that obscures working-class oppression and helps the working class to feel good about their society. In this way, law contributes to false consciousness that prevents the working class from realizing its revolutionary potential.
6. Roberto Mangabeira Unger, a contemporary scholar, attributed the development of legal order to certain preconditions that characterized postfeudal Europe. One of these features was conflict among the

bourgeoning merchant class, the aristocracy, and the monarchy. The second precondition was a reliance on a higher divine or universal law as a standard to justify and legitimate state law.

7. Specific legal changes also result from various social changes. As one example, in fourteenth-century England, the massive loss of life from the bubonic plague led to England's first vagrancy law in 1349 that criminalized both begging and moving from one place to another to find better employment. As a second example, in fifteenth-century England, the growth of the merchant economy led persons transporting goods that had been purchased to keep the goods for themselves. This problem in turn led to the Carrier's Case of 1473 that made this type of behavior a crime.
8. Technological changes other than reproductive technology also affect the law. When automobiles became more common about a century ago, new traffic laws and new regulations governing the manufacture of automobiles were necessary. Many laws have since been passed to enhance motor vehicle safety. The invention of the railroad several decades before the automobile also led to many new laws and regulations. The invention and widespread use of the personal computer and the Internet have again led to many legal changes; copyright law especially has had to adapt.
9. The actual impact of law on social change is open to debate. Sometimes, law has been an effective vehicle for social change, but sometimes it has been ineffective. The impact of law may be both direct and indirect. Direct social change occurs when legal change itself affects behavior. Indirect social change can occur in either of two ways. First, a change in law may first affect a social institution, and the resulting changes in this social institution may then bring about changes in behavior or attitudes. Second, a legal change may give a disadvantaged group new hope that social change is possible and spur it into political action.
10. Attempts by the government to use the law to effect social change face several obstacles. These obstacles include (a) public resistance to court orders or other legal changes and (b) inconsistent application and enforcement of a new law by legal authorities.
11. Accurate assessment of the actual impact of legal change is difficult for several reasons. First, legal impact may occur in the short term but not persist beyond an initial phase. Second, legal change may achieve its intended impact but in the long run prove a Pyrhhic victory because it helps shore up existing inequalities. Third, it is sometimes difficult to know why certain social changes occur and, more specifically, whether these changes stemmed from the legal change in question. Fourth, although a legal change may affect behavior, it is sometimes difficult to determine exactly why the behavior did change because of the legal change. Fifth, it is not always clear what specific impact of a legal change was actually intended by the initiators of the legal change.

12. William M. Evan presented several conditions that maximize the potential impact of legal change. These include (a) The source of the law should be perceived as authoritative and prestigious; (b) The rationale for the new law should emphasize its continuity and compatibility with existing institutionalized values; (c) Publicity surrounding a new law should emphasize that similar laws have proven helpful elsewhere with few or no adverse effects; (d) A new law should mandate that any changes it requires should occur quickly rather than slowly; (e) Legal authorities must apply and enforce a new law consistently and without hypocrisy or corruption; (f) Positive as well as negative sanctions should be used to help ensure compliance with a new law; and (g) Effective protection and resources should be provided for the rights of individuals who would suffer if other people violate or evade the new law.
13. The use of law by social movements is called legal mobilization. Some scholars think that law is an effective tool for helping social movements, while some think that law is ineffective and even counterproductive for social movements.
14. Legal mobilization may provide social movements with significant indirect benefits by giving aggrieved groups a new sense of legal entitlement and a new hope for social change. For these reasons, legal mobilization may spur members of these groups to work for such change.
15. Law has been used as an effective weapon against social movements. Targets of citizen groups have sometimes sued them for slander, libel, and other allegations. Police, FBI, and other law enforcement bodies have conducted surveillance on members of protest groups and other social reform groups. Finally, the arrest and prosecution of activists may tie up their time, money, and energy in legal defense, stigmatize their movement, and deter other people from participating in the movement.

Key Terms

Bourgeoisie
Charismatic authority
Dependent variable
Fellow-servant rule
Independent variable
Legal mobilization
Legal order
Legal repression
Means of production
Mechanical solidarity
Organic solidarity
Proletariat
Rationality
Rational-legal authority
Repressive law
Restitutive law
Traditional authority

CHAPTER 7

Law and Inequality

Chapter Outline

- Law and Inequality in the American Past
  - *–Social Class*
  - *–Race and Ethnicity*
  - *–Gender*
- Contemporary Evidence
  - *–Social Class*
  - *–Race and Ethnicity*
  - *–Gender*
- Summary
- Key Terms

The last three chapters discussed several important functions that law ideally serves for society: dispute processing, social control, and social change. As we have seen, law does not always fulfill these functions as optimally as it might and even causes many problems as it tries to fulfill them. Still, the picture that emerges from studying these functions is that law aims to benefit society as a whole and often, though not as often as we would like, succeeds in doing so.

However, there is another side to law that is much less favorable. Although law often benefits society, it can also reflect and reinforce the social inequality found in the larger society. This general view is the basis for the conflict perspective on law and society that goes back at least to the work of Marx and Engels. The previous chapter discussed two classic studies that are often cited to support the conflict perspective: William Chambliss' (1964) analysis of the development of vagrancy laws in fourteenth-century Britain and Jerome Hall's (1952) analysis of the development of theft law in fifteenth-century Britain. Chambliss showed how vagrancy laws were used to depress wages and bolster the aristocracy's wealth when a severe labor shortage arose after the massive, terrible deaths from the bubonic plague in 1348, while Hall showed how a growing practice of the poor, the appropriation of goods by carriers, was criminalized in 1473 to support the economic needs of the bourgeoning middle class.

Although both studies remain historically and theoretically important, these laws arose in a nation and era that were certainly not democratic, leaving unclear the relevance of these studies for democratic nations. Evidence of law and inequality in the United States would thus provide more compelling evidence for the conflict perspective and for more specific contemporary perspectives on law, society, and inequality (critical legal studies, critical race theory, and feminist legal theory; see Chapter 2) and also have important implications for social and legal policy.

Focusing on the United States, then, this chapter examines two related issues in the study of law and inequality: (1) To what degree does inequality characterize the operation of civil and criminal law and their corresponding justice systems? and (2) To what degree does the legal system reinforce or contribute to inequality in the larger society? Our examination of both issues, which for simplicity we call the *law and inequality thesis*, will center on social class, race and ethnicity, and gender, which together comprise the key dimensions of social inequality and are discussed in many texts and readers (e.g., Hurst 2007; Kivisto and Hartung 2007). We begin with an historical overview of law and inequality in the United States before turning to the contemporary era.

LAW AND INEQUALITY IN THE AMERICAN PAST

American history is replete with examples of the two related aspects of law and inequality just outlined: (1) the use of law to foster social inequality and (2) the impact of inequality on the origins of law and the operation of the legal system. This section sketches this historical evidence for social class, race and ethnicity, and gender, respectively. Scholars increasingly emphasize the intersection among these dimensions of inequality (Kivisto and Hartung 2007), and some of the historical examples will illustrate this intersection.

Social Class

The United States has long been regarded as a land of equal opportunity where people who work hard can pull themselves up by their bootstraps and achieve the American dream of economic success. Despite this lofty ideal and any number of actual success stories over the decades of people moving out of poverty into economic comfort, the United States since the colonial period has also been characterized by persistent inequality based on social class and exemplified by differences in wealth, power, and prestige. Sociologists call such inequality **social stratification**. Although debate continues on the number of social classes in the United States, for our purposes it is sufficient to distinguish just two classes: (1) the poor or working class and (2) the upper class, that very small fraction of the population that owns the bulk of all wealth and enjoys considerable influence over political, social, and economic policy. (Other terms used for these classes are the *powerless* and the *powerful* or, more simply, the *poor* and the *rich*.) With this distinction in mind, to what degree did the law in the American past help the rich keep the poor in their place and more generally preserve a socially stratified society? Put another way, to what degree has law in the past "reflected the interests of the ruling class," to cite one scholar's view (Shelden 2001:62)? To try to answer this question, we focus on the nineteenth century, the subject of a considerable amount of work on law and inequality.

The Labor Movement. Much evidence for the law and inequality thesis comes from the history of the U.S. labor movement after the Civil War. Through years of hard struggle, the labor movement eventually won significant legal victories (via Congressional legislation and favorable rulings by the U.S. Supreme Court and other appellate courts) in the twentieth century; these victories included the establishment in 1916 of the eight-hour workday and the prohibition in 1938 of most child labor. Yet, in the decades after the Civil War, when labor-management conflict reached a peak during the 1870s and 1880s thanks to growing unemployment, inferior wages, and dangerous working conditions, management was repeatedly able to use the law as an effective tool against labor.

For example, it was a crime in most locations in that era for workers to go on strike. Although many workers thus refused to go on strike, many others went on strike anyway and were arrested. Most were then convicted and jailed, but more than a few were acquitted by sympathetic juries. In response, management decided it needed a new legal strategy that would allow strikers to be imprisoned without benefit of a jury trial. This strategy was the labor injunction: management began obtaining injunctions, with judges quite willing to grant them, that barred workers from going on strike. When many workers did so anyway, they were arrested not for striking but for violating the injunction and were then immediately imprisoned without a right to a jury trial (Frankfurter and Greene 1930). As a result, the labor injunction quickly became "an especially deadly threat against labor" (Friedman 1985:487), leading the Democratic Party to condemn in its 1896 platform "government by injunction as a new and highly dangers form of oppression" and to criticize the judges issuing the injunctions for acting "simultaneously as legislators, judges and executioners" (Friedman 1985:488).

Law was also used against the labor movement in other ways. Sidney L. Harring (1977) recounts the use of **tramp acts** to suppress unemployed workers in the post–Civil War period. During the 1870s, hundreds of thousands of workers began traveling to find gainful employment. Their ranks grew rapidly for several reasons. First, a series of economic depressions beginning in the early 1870s plunged millions of workers into unemployment and forced thousands of them to travel from one city to another in search of work. Second, as industry moved westward, so did jobs like railroad building, crop harvesting, and lumberjacking, and people seeking work naturally had to travel to the locations in which these jobs were available. Third, after immigrants came to the United States, they often had to travel to find work. Fourth, obeying the famous dictum "go west, young man, go west," many people moved out West to find work and/or adventure. Fifth, some individuals disliked working in the confines and poor working conditions of factories and preferred life on the road and the occasional job.

All these developments meant that hundreds of thousands of *tramps*, as those traveling to look for work were called, "took to the rails" or otherwise traveled to look for work. According to Harring (1977:880), companies feared the potential for labor unrest from these tramp "armies" and, in particular, worried that tramps were "political troublemakers who incited otherwise well-disciplined local workers to class violence." Tramps were soon merged in the public's mind with common criminals and seen as posing a great danger to the safety of common citizens.

In response, by the end of the nineteenth century almost every state passed *tramp acts* that banned traveling without visible means of support and subjected tramps to arrest and imprisonment at hard labor. New Jersey passed the first tramp act in 1876 and was quickly followed by other northeastern states. In addition to making traveling for work a crime, the tramp acts also typically reclassified misdemeanors committed by tramps as felonies subject to much harsher legal sanctions. The penalty stipulated by the New York tramp act for these reclassified offenses was three years at hard labor.

Partly for this reason, during the 1890s the New York act helped suppress tramps in Buffalo, then the nation's eleventh largest city and a major hub of industry and commerce. As economic downtowns increased the ranks of the city's unemployed, companies became increasingly worried about tramps. More than 2,000 people were arrested in 1891 for being tramps and almost 5,000, about 40 percent of all arrests, were arrested in 1894. Harring attributes this increase to a major railroad strike in Buffalo in 1892, as police sought to break the strike by arresting workers for tramping offenses, and to an economic depression in 1893 and 1894. According to Harring, the Buffalo experience, replicated elsewhere during the 1890s, "demonstrated the repressive potential of the Tramp Acts to impose severe sentences on large numbers of workers for trivial offenses" (p. 906). This potential was increased, he says, by the legal experience of tramps after arrest. Guilt was a foregone conclusion: tramps were not allowed jury trials, and judges quickly found them guilty after accepting the arresting officer's testimony as more compelling than the tramp's testimony.

Policing the Dangerous Classes. The tramp acts reflected two related social class dynamics in the development of law during the nineteenth century. The first dynamic was a common perception of the poor as an inferior class of people who were especially prone to criminal behavior and thus a major threat to the wealthy and more generally to the social order. A popular label applied to the poor, the **dangerous classes**, reflects these perceptions. Perhaps the first author to use this phrase was French social scientist Honoré-Antoine Frégier, who wrote in 1840 that the poor "always have been and always will be the most productive seedbed for all sorts of malfeasants; it is they whom we refer to as the dangerous classes" (quoted in Toth 2003:151). The label was soon imported to the United States and ingrained into the national consciousness as American cities and concomitant poverty and crime grew rapidly amid rapid industrialization. Fear of the poor's criminality was widespread and played a fundamental role in the development of law and criminal justice during much of the nineteenth century (Adler 1994a; Shelden 2001).

The second dynamic was the rise of the modern police force in the 1830s and 1840s. Before this time, American cities were typically patrolled by constables and night watchmen, most of them working part-time and many working as unpaid volunteers. As cities grew rapidly during the 1830s and 1840s, mob violence by the poor became a common occurrence, leading many cities to establish organized police forces (Walker 1998). The extent to which these new police then served to suppress the poor remains in vigorous debate among criminal justice historians.

Many leftist scholars (e.g., Harring 1993; Shelden 2001) believe that the police developed in response to an inordinate fear of the dangerous classes and then were used "primarily as an agent of class control, suppressing working-class militancy and enforcing middle-class notions of public morality," according to Samuel Walker's (1998:66) summary of this viewpoint. Walker disagrees with this assessment for several reasons. First, the police came from and represented working-class, immigrant neighborhoods, whose priorities they represented while on the job. Second, the police often failed to arrest or otherwise act against striking workers because of their own working-class backgrounds. Third, the police often failed to enforce laws against drinking, gambling, prostitution, and other crimes against public morality. Thus, says Walker, although the wealthy elite may have intended the police to act as agents of class control, the police in reality did not routinely do so. Even so, Walker concedes that the police did perform this role to a considerable extent: they "were brutal and often disrupted strikes and labor-union organizing efforts" (p. 67), and police from one ethnic background (e.g., English) would harass and use their nightsticks against working-class citizens from other backgrounds (e.g., Irish or Italian). Middle-class and upper-class citizens did not suffer such police behavior.

Interesting evidence of the police role in class control comes from a study of arrest rates in 1900 for public drunkenness. Using data from the 50 largest cities, M. Craig Brown and Barbara D. Warner (1995) found that these arrest rates were higher in cities with higher proportions of immigrants even after they took into account city differences in alcohol use. The authors concluded that "social control is not entirely driven by levels of crime, but is in part a response to potentially threatening groups" (p. 94).

Other evidence underscores the degree to which police were indeed used to break up strikes and otherwise control labor unrest despite the preference of some companies to use their own guards for this purpose. Harring (1993) observes that the last two decades of the nineteenth century were an era filled with labor strikes. New York City had more than 5,000 strikes involving almost 1 million workers during this period, while Chicago had more than 1,700 strikes involving almost 600,000 workers. These strikes, argues Harring, led to a doubling or tripling of the size of the police forces in these and other cities with great numbers of strikes. Municipal police arrested and/or beat striking workers, guarded company property, and sometimes arrested union activists on false charges. In these and other ways, says Harring, "the local police were most often the major antistrike institution, did effective antistrike work, and almost always took an aggressive stand against the workers and in favor of the corporations" (p. 558).

In sum, it would be too extreme to say that the police in the nineteenth century functioned solely to suppress the poor and working class on behalf of a wealthy power elite. In many ways, the police did surprisingly little in the way of law enforcement or much else, as many officers routinely shirked their duties because they knew (in the absence of communication technology that would permit their superiors to monitor their activities and whereabouts) that they could do so with impunity (Walker 1998). When they did enforce criminal laws, the actual and potential victims of crime they were trying to protect were often the poor and working class, who then, as today, are disproportionately the victims of violent and property crime. Yet, it would also be too extreme to say that the police never engaged in the suppression of the poor and working class. Again and again, they were used to suppress strikes and other forms of labor activism and to control of the movement and behavior of the unemployed workers who were better known as tramps. The use of police in this manner provides important evidence for the law and inequality thesis as applied to social class.

Race and Ethnicity

Because race and ethnicity are deeply intertwined with social class, it is sometimes difficult to determine whether the negative life experiences of people of color stem more from their race/ethnicity or from their disproportionately low socioeconomic status. This issue notwithstanding, there is ample evidence from the American past of legal inequality based on racial and ethnic prejudice and discrimination. As the late W. Haywood Burns, former dean of the law school at the City University of New York and president of the National Conference of Black Lawyers, summarized this evidence, "Indeed, the histories of the African, Asian, Latin, and Native American people in the United Sates are replete with examples of the law and the legal process as the means by which the generalized racism in the society was made particular and converted into standards and policies of social control"

(Burns 1998:279). Most of the discussion below focuses on African Americans and Native Americans because of the long history of legal and other oppression they have experienced since the colonial era.

African Americans. The history of slavery and of its aftermath provides much evidence for Burns' observation. It is commonly known that the U.S. Constitution endorsed slavery by stipulating in Article I that slaves would count as three-fifths of a person in determining a state's representation in the House of Representatives. However, it is less well known that a body of law (called *slave law*) supporting the practice of slavery arose during the colonial period and continued until slavery ended with the Civil War (Friedman 2004a). Because England did not have slaves, slavery was not a part of the English law that the colonists adapted in the new land. However, as slavery became more entrenched after slaves were brought to the colonies before the mid-1600s, new colonial laws were written to govern the status, selling, and buying of slaves. Slaves were treated like any other commercial product, as "human beings who were owned as if they were dogs or cattle, and could be bought and sold and rented" (Friedman 2004a:29–30). Among other provisions, slave law specified that the children of slave mothers were also slaves themselves, and it said that slaves could not marry or own property. It also specified the status of freed slaves (and also blacks who had not been slaves), and generally made clear that they lacked the rights of full citizens. Because white Southerners feared slave revolts, much of the Southern criminal justice system was also geared to slavery. As Walker (1998:24) observes, "Controlling slaves was one of the major purposes of the criminal justice system, and court proceedings were designed to reinforce the idea that masters had full authority over slaves and servants." Some Southern cities established slave patrols, a forerunner of the organized police forces that, as noted earlier, developed in Northern cities during the 1830s and 1840s.

Additional types of laws subjugated or discriminated against African Americans in other ways well into the nineteenth century (Burns 1998). In 1855, the Kansas state legislature passed a law that mandated the legal penalties for a man convicted of raping a white woman. A white man could be sentenced to a prison term of no more than five years, while an African American would be castrated and also have to pay for the cost of this procedure. Northern states during this time had many laws restricting the educational, employment, and housing opportunities for free blacks (Burns 1998).

In 1857 the U.S. Supreme Court issued its infamous decision in *Dred Scott v. Sanford* that endorsed slavery. Scott was a slave who sued for his freedom after his master took him to a free territory. The Court ruled 7-2 against Scott and declared that African Americans, whether enslaved or free, could never be U.S. citizens under the Constitution. The majority opinion by Chief Justice Roger B. Taney declared in part, "[The black race] has for more than a century before been regarded as beings of an inferior order, and altogether unfit to associate with the white race, either in social or political relations; and so far inferior, that they had no rights which the white man was bound to respect . . ." (quoted in Burns 1998:282).

Although the Civil War ended slavery, the South responded in 1865 with the notorious Black Codes, a series of laws that subjugated African Americans so much that they mirrored the antebellum slave codes. Among other codes, vagrancy laws were instituted to keep freed slaves from moving from the land on which they had toiled under slavery, and other laws made it a crime for someone to offer a freed slave a better job and for freed slaves to grow their own crops or to enter a town without permission. In effect, the Black Codes sought to keep freed blacks informally enslaved even though official slavery had ended. Outraged by the Black Codes, the federal government sent troops into the South in 1866, and the period called Reconstruction began. Although Southern blacks then enjoyed relative economic and political opportunity, the end of Reconstruction in 1877 led to the rise of Jim Crow laws that restored Southern segregation and legalized racial inequality. This legal development included the denial of voting rights for African Americans, who "were no longer slaves; but most of them were little better than serfs. . . . The American south had become a caste system, in which all whites outranked all blacks; a system in which a black man or woman had virtually no chance of fulfilling what for whites would be the normal American ambitions and dreams" (Friedman 2004a:71). The Jim Crow laws culminated in the U.S. Supreme Court's infamous 1896 *Plessy v. Ferguson* decision, in which it ruled 7-1 that "separate but equal" racially segregated public facilities were Constitutional. *Plessy* cemented the legal segregation that had characterized the South since 1877 and that was to continue into the 1960s until it finally ended thanks to the civil rights movement.

Native Americans. Law also played a key role in the historical subjugation of Native Americans. As political scientist Suzanne Samuels (2006:267) observes, "[T]hroughout our nation's history, we have used the law to undermine the rights of Native American peoples, particularly with respect to their tribal lands." Beginning in the early colonial period and continuing well after the Civil War, white settlers and eventually U.S. troops waged a virtual war on Native Americans, with deaths from combat and European-introduced disease reducing the Native American population from about 1 million in the early 1600s to less than 250,000 by 1900 (Venables 2004). The goal was to seize the vast amount of land on which Native Americans lived. Native Americans everywhere were persuaded or forced to sign *treaties*, legal documents that ceded control of much of their land to whites and confined them to reservations; almost 400 treaties were signed from 1778 to 1870 (Samuels 2006).

Despite these problems, treaties still preserved wide tracts of land for Native Americans and gave them other rights, but over the years, various legal efforts undermined these benefits. For example, the Congressional Removal Act of 1830, signed by President Andrew Jackson, eventually led to the relocation of Cherokees and other Native Americans in the Southeast to across the

Mississippi River. They were forced to march the whole way in what became known as the *Trail of Tears*, with thousands dying en route from disease and malnutrition. In 1903, the U.S. Supreme Court's ruling in *Lone Wolf v. Hitchcock* (187 U.S. 553) gave Congress the power to invalidate the land treaties and to take Native American land without proper compensation (Clark 1999). The *Lone Wolf* decision was so disadvantageous for Native Americans that it "has often been called the Dred Scott of Indian law" (Wildenthal 2002:113).

The criminal justice system also discriminated against Native Americans in early America. During the colonial period, for example, Native Americans living in white towns had to obey curfews and other limits on their behavior, and whites who killed Native Americans generally received more lenient sentences than Native Americans who committed the same type of killing (Walker 1998). In effect, judges concluded that murders of Native Americans were less reprehensible when they were committed by whites than when they were committed by other Native Americans.

Chinese Americans. The Chinese, too, were victims of legal discrimination in the American past. Chinese immigrants began to move into California and other western areas in large numbers to try to strike it rich after the 1849 California Gold Rush and to work on the burgeoning railroad construction industry. Because whites feared job competition from the Chinese, anti-Chinese sentiment grew rapidly. This prejudice led in 1882 to the federal Chinese Exclusion Act, which banned Chinese immigration for ten years and then evolved into a permanent ban in 1902 (Lee 2003). Many Western states also passed laws banning marriage between whites and Chinese (Friedman 2004a).

Drug Laws. Further historical support for the law and inequality thesis comes from the origins of several U.S. drug laws. In this regard, Chapter 5 recounted the symbolic attack of the temperance movement on Catholic immigrants, and the history of Prohibition cannot be properly understood without appreciating the role that racial and ethnic prejudice thus played (Gusfield 1963; Lerner 2007). This basic lesson of the history of Prohibition is reflected in the history of laws that banned the use of three other drugs: opium, cocaine, and marijuana (Goode 2008; Musto 1999).

Opium was a quite popular drug during the later nineteenth century. It was found in a large number of patent medicines that were advertised to reduce headaches, menstrual cramps, toothaches, depression, and other problems. Historians estimate that about 500,000 Americans were addicted to opium by the beginning of the twentieth century. There was concern about opium just as there is concern now about alcohol and nicotine, but it was a drug that was tolerated and widely used. However, the anti-Chinese prejudice just described heightened public concern about opium owing to the Chinese cultural practice of smoking the drug in opium dens, their equivalent of a bar, pub, or tavern. With the aid of newspaper stories and politicians' pronouncements, rumors spread that the Chinese were enticing white boys and girls into the dens to be sexually molested and turned into opium fiends. The mayor of

San Francisco warned that the opium dens were increasingly "frequented by white males and females of various ages" and criticized them as "against good morals and contrary to public order" (Gieringer 2007:E1). In response to this anti-Chinese hysteria, San Francisco passed the Opium Den Ordinance in 1875. This act, the nation's first drug law, prohibited opium dens but did not criminalize the private use or sale of the drug. Several other California cities followed suit, and the state legislature enacted a statewide ban in 1881. Anti-Chinese prejudice continued to fuel concern about opium and helped lead to the 1914 Harrison Narcotics Act, a Congressional bill that imposed severe limits on the legal manufacture, sale, and use of opium and cocaine.

As its inclusion in the Harrison Act might imply, cocaine was another popular drug at the turn of the century that was used in many over-the-counter products and even as an anesthesia during surgery. Coca-Cola, first sold in 1886, included cocaine and quickly became popular because people presumably felt so good after drinking the new product. However, concern about cocaine began to grow, and Coca-Cola stopped including cocaine in its formulation in 1903. Some of this concern was fueled by prejudice against African Americans, who, it was felt, would become especially violent, acquire super strength, and become invulnerable to bullets if they used cocaine. Although there was obviously no evidence to support such nonsense, the concern was real and helped lead not only to the 1903 decision to exclude cocaine from Coca-Cola but also to the Harrison Act eleven years later.

Marijuana was yet another drug where racial prejudice fueled attempts to criminalize the drug. During the late nineteenth century, marijuana, like opium and cocaine, was a popular ingredient in a variety of patent medicines and was touted for the relief it gave to migraine headaches, menstrual cramps, and toothaches. Marijuana use (smoking) increased when Mexicans came into the United States after the Mexican Revolution of 1910 and brought with them a cultural practice of marijuana use. These new immigrants tended to settle in southwestern and western states where, like the Chinese before them, they competed with whites for jobs. White prejudice against Mexican immigrants increased during the Great Depression years of the 1930s and took many forms, including a focus on their marijuana use. Newspapers ran many stories warning their readers that marijuana use would turn Mexican Americans violent and cause them to murder and rape white people. The concern about marijuana, fueled by anti-Mexican prejudice, helped lead to bans on marijuana use in 29 states by 1931 and the federal Marijuana Tax Act of 1937 that made marijuana use a crime.

Gender

The law and inequality thesis finds ample historical support in the legal treatment of women. Simply put, for much of the nation's history women were second-class citizens in the eyes of the law, and the law served in many ways to reinforce their political and social subordination. As two law professors,

Nadine Taub and Elizabeth M. Schneider (1998:328) observe, "The Anglo-American legal tradition purports to value equality, by which it means, at a minimum, equal application of the law to all persons. Nevertheless, throughout this country's history, women have been denied the most basic rights of citizenship, allowed only limited participation in the market place, and otherwise denied access to power, dignity, and respect."

Women could not vote before 1920, when the Nineteenth Amendment gave them the right to vote, and they could not serve on juries in many locations before 1975, when the U.S. Supreme Court ruled in *Taylor v. Louisiana* (419 U.S. 522) that excluding women from juries was unconstitutional. At the beginning of the nineteenth century, married women could not own property, sign contracts, or make wills, although by the end of the century they could do all these things thanks in part to efforts of the women's rights movement of that era. Yet, employment discrimination against women was lawful until it was prohibited by the 1964 Civil Rights Act, and other kinds of gender discrimination were also lawful until a series of Supreme Court decisions in the 1970s found such discrimination unconstitutional.

In short, then, gender discrimination in a variety of realms was legal for most of U.S. history. Underlying the law's support for this discrimination was the widespread view that women are naturally different from and subordinate to men, and perhaps especially, biologically destined to be mothers and homemakers. The oft-cited opinion in an 1872 Supreme Court case, *Bradwell v. Illinois* (16 Wall 130), illustrates this reasoning. As Chapter 2 recounted, Myra Bradwell had sought admission to the bar in Illinois and was fully qualified to practice law. The Illinois Supreme Court rejected her petition because it feared her husband might be sued if someone took issue with something Bradwell did as a lawyer. She appealed to the U.S. Supreme Court, which upheld the Illinois court's decision. The concurring opinion by Justice Joseph P. Bradley bears repeating here for its sexism:

> . . . The civil law, as well as nature herself, has always recognized a wide difference in the respective spheres and destinies of man and woman. Man is, or should be, woman's protector and defender. The natural and proper timidity and delicacy which belongs to the female sex evidently unfits it for many of the occupations of civil life. The constitution of the family organization, which is founded in the divine ordinance, as well as in the nature of things, indicates the domestic sphere as that which properly belongs to the domain and functions of womanhood. The harmony, not to say the identity, of interests and views which belong, or should belong, to the family institution is repugnant to the idea of a woman adopting a distinct and independent career from that of her husband. . . . The paramount destiny and mission of woman are to fulfill the noble and benign offices of wife and mother. This is the law of the Creator. . . . (*Bradwell v. Illinois* [1872] 83 U.S. 130).

As this passage implies, women were thought to be little more than wives and mothers and, reflecting the common law concept of **coverture**, the property of their husbands. For much of U.S. history, this notion gave husbands the right to beat or rape their wives with impunity. Throughout most of the nineteenth century, men were allowed to hit their wives if the purpose was to correct or punish (the *chastisement* defense) supposed misbehavior by their wives and if any injuries the husband inflicted were not very serious (Siegel 1996). Men began to lose this right in 1871, when two states, Alabama and Massachusetts, made it illegal for a man to beat his wife (Lemon 1996), and by the end of the century most states no longer recognized chastisement as a valid defense to charges of wife battery. By the end of the century, appellate courts had also begun to recognize that men were also not allowed to rape their wives, but the so-called *spousal exemption* for rape continued in many states until well into the twentieth century and allowed men to rape their wives without fear of legal punishment. Thus, for much of this nation's history, the law supported the right of husbands to beat and rape their wives and, in this manner, contributed to women's social subordination and criminal victimization.

According to the known historical record, it was not until 1978 that any American man was criminally prosecuted for raping his wife (Heinemann 1996). By 1996, only 11 states had entirely eliminated their spousal rape exemptions, and by 2005, this number had increased to 20. In the remaining 30 states, husbands either remain exempt from marital rape prosecutions under some circumstances, for example, when a wife is asleep or mentally impaired, or receive smaller penalties for marital rape than for other types of rape (Bergen 2006). According to sociologist Raquel Kennedy Bergen,

> The existence of some spousal exemptions in the majority of states indicates that rape in marriage is still treated as a lesser crime than other forms of rape and is evidence of societal patriarchy. This perpetuates marital rape by conveying the message that such acts of aggression are somehow less reprehensible than other types of rape. Importantly, the existence of any spousal exemption indicates an acceptance of the archaic understanding that wives are the property of their husbands and that the marriage contract is still an entitlement to sex (Bergen 2006:2).

In these and other ways, then, the historical record provides compelling evidence for the law and inequality thesis in regard to gender. For much of this nation's history, the law both reinforced and contributed to women's subjugation by reflecting a **separate sphere** ideology that women belonged at home and men belonged in the working world outside the home (Grana 2002). This belief is still with us: in the 2006 General Social Survey, a national poll of adult Americans, 35.1 percent of respondents agreed that "it is much better for everyone involved if the man is the achiever outside the home and the woman takes care of the home and family" (author's analysis). As we shall see below, however, the law in many respects no longer reflects this belief, or at least reflects it much less than it used to.

CONTEMPORARY EVIDENCE

Although the evidence from the American past for the law and inequality thesis seems undeniable, what about the present? Does law still reflect and reinforce the many dimensions of inequality, or is the American ideal of equal justice under law, signified by the familiar blindfolded statue of Lady Justice, finally a reality? A realistic assessment indicates that equal justice under law is still far more of an ideal than a reality despite important changes in the law since the mid-1900s. We now turn to this assessment.

Social Class

The Lady Justice statue symbolizes that the administration of justice should be neutral and blind to everything but the facts of a case and the law governing a case; the personal backgrounds of litigants, including their financial wealth or poverty, should not matter. In theory they do not matter, but in practice they do make a difference because the wealthy can and do take advantage of the legal system in ways that the poor can only dream of. This is true of both civil justice and criminal justice. Simply put, the wealthy have many more advantages than the poor have in both systems of justice.

Civil Justice. To begin with the civil law, recall from Chapter 4 that adjudication can be very expensive in part because it requires the hiring of skilled attorneys and, depending on the case, other sorts of experts and consultants. For any given case, then, adjudication often favors the litigant with greater financial resources (e.g., the wealthy). Thus, even though law is supposed to be neutral and impartial, wealth can indeed make a considerable difference.

In one of the most cited articles in the law and society literature, Marc Galanter (1974) discussed the advantages that explain "why the 'haves' come out ahead," to borrow his article's title. He wrote that litigants can be divided into two basic categories: repeat players and one-shotters. **Repeat players**, as their name implies, use the law and courts on a recurring basis, while **one-shotters** use the law and courts very rarely or, at best, only occasionally. Although repeat players and one-shotters can be either individuals or organizations, in practice repeat players tend to be organizations and one-shotters tend to be individuals. The most common configuration, Galanter wrote, involves repeat players bringing legal action against one-shotters: a finance company versus a debtor, a landlord versus a tenant, the IRS versus a taxpayer. One-shotters may also bring legal action against one another: a divorce or child custody hearing, someone suing a neighbor over a property dispute. Occasionally, a one-shotter may bring legal action against a repeat player (e.g., an injured victim versus an insurance company or manufacturer of the product causing an injury, a tenant versus a landlord, a libeled individual versus a publisher). In the final configuration, a repeat player brings legal action against another repeat player (e.g., a labor union versus a company, a regulatory agency versus a corporation).

With some exceptions, Galanter wrote, repeat players tend to be "haves" in regard to wealth, power, and prestige, and one-shotters tend to be "have-nots." Thus, he said, repeat players are usually "larger, richer and more powerful" than one-shotters (p. 103). As such, repeat players have important advantages (e.g., wealth and influence) over one-shotters even before they enter the legal arena. Importantly, their status as repeat players gives them further advantages (discussed just below) over one-shotters. Repeat players thus have a double advantage in terms of both their wealth and influence and their status as repeat players. For this reason, Galanter wrote, "a legal system formally neutral as between 'halves' and 'have-nots' may perpetuate and augment the advantages of the former" (p. 104).

What are some of the advantages repeat players enjoy from using the law and courts so often? First, because they know how the legal system works and what documents and other evidence they need to protect their interests, they have "advance intelligence" (Galanter 1974:98) that allows them to shape transactions to their benefit by, say, having the other party sign a contract or submit a security deposit. Second, they employ or can afford to hire very skilled attorneys. Third, they can also afford to pay for protracted litigation, a significant advantage since most litigation involves many delays. Fourth, they enjoy "facilitative informal relations" (p. 99) with various members of the legal and political systems that may help them in many ways. Fifth, the stakes in any one case are relatively smaller for repeat players because of their wealth and other resources and larger for one-shotters because they lack such resources. For this reason, repeat players will be more willing than one-shotters to take risks to win a case. Finally, the legal and political institutions that handle legal claims are passive: they await action by a litigant to initiate a claim. This fact, said Galanter, gives a significant "advantage to the claimant with information, ability to surmount cost barriers, and skill to navigate restrictive procedural requirements" (p. 119). Once a legal case begins, a judge is again passive, as it is the duty of the litigants to gather and present evidence. This fact again works to the advantage of repeat players: "Parties are treated as if they were equally endowed with economic resources, investigative opportunities and legal skills. Where, as is usually the case, they are not, . . . the greater the advantage conferred on the wealthier, more experienced and better organized party" (pp. 120–21).

For all these reasons, then, the "haves"—the wealthy and powerful organizations and individuals who tend to be the repeat players—should come out ahead, and the "have-nots"—the poor and working class—should come out behind. Research on this hypothesis is surprisingly lacking, as social class has been somewhat neglected in studies of litigation outcomes

(Munger 2004). Still, the evidence from the studies that have been performed does support the law and inequality thesis (Carlin, Howard, and Messenger 1966; Farole 1999; Songer, Sheehan, and Haire 1999; Wald 1965; Wanner 1975). To try to help the "have-nots," most of whom cannot afford to hire an attorney, groups like the Legal Aid Society, which has offices throughout the United States, and poverty attorneys funded by the federal Legal Services Program begun in the mid-1960s, handle civil cases for the poor without charge. Since the 1960s, these groups and attorneys have won important legal and political victories for the poor, and these victories have helped the "have-nots" come out a little less behind than otherwise would have been true (Harris 1999). These victories notwithstanding, considerable evidence from the civil justice system supports the law and inequality thesis in regard to social class.

Criminal Justice. Political scientist Herbert Jacob (1978:185) once observed that criminal courts are "fundamentally courts against the poor." This was so, he wrote elsewhere, because "the behaviors most severely punished by governmental power are those in which persons on the fringes of American society most readily engage. . . . Crimes (especially white-collar crimes) committed by other segments of the population attract less public attention, less scrutiny from the police, and less vigorous prosecution" (Eisenstein and Jacob 1977:289).

To assess the law and inequality thesis in regard to social class with evidence from the criminal justice system, two issues must be addressed. First, once the poor and working class enter the criminal justice system as suspects and defendants, how do they fare compared to wealthier suspects and defendants accused of similar offenses? Second, as Jacob's observations remind us, to what extent is there disparate treatment of crimes by the poor and of the white-collar crimes typically committed by the wealthy?

To answer the first question, several studies assess whether income is related to the chances of being convicted, of being imprisoned upon conviction, and of receiving a longer prison term if imprisoned (Chiricos and Waldo 1975; D'Alessio and Stolzenberg 1993; Myers 2000). This line of research, in which the typical study analyzes quantitative data on hundreds of suspects or defendants, does not find a relationship between income and these criminal justice outcomes once factors such as prior criminal record and offense seriousness are taken into account. To some observers, this lack of a relationship demonstrates that the criminal justice system treats defendants equally regardless of their social class (Chiricos and Waldo 1975). To the extent this is true, the criminal justice experience does not provide evidence for the law and inequality thesis in regard to social class.

However, other observers say this conclusion is misleading because almost all criminal suspects and defendants are poor or near-poor; wealthy people rarely commit robberies or burglaries or steal motor vehicles. For this reason, quantitative studies of suspects and defendants cannot yield valid tests of social class differences in legal treatment (Shelden 1982). Better evidence of social class differences arises from the very few cases in which wealthy people are arrested and prosecuted of homicide and other "street" crimes. Once arrested, they immediately enjoy many advantages over the much more typical defendant who is poor or near-poor. As Jacob (1978:185–186) pointed out,

> Those few defendants who are not poor can often escape the worst consequences of their [criminal] involvement. . . . [T]hey can afford bail and thus avoid pretrial detention. They can obtain a private attorney who specializes in criminal work. They can usually obtain delays that help weaken the prosecution case. . . . They can enroll in diversion programs by seeking private psychiatric treatment or other medical assistance. They can keep their jobs and maintain their family relationships and, therefore, qualify as good probation risks. They can appeal their conviction (if, indeed, they are convicted) and delay serving their sentence.

The cases of two celebrity athletes, O.J. Simpson and Kobe Bryant, illustrate many of these advantages. Simpson, a former football star and TV and movie celebrity, was prosecuted during 1994 and 1995 for the alleged brutal murders of his ex-wife and her friend. Any typically poor defendant accused of similar murders would have almost certainly pleaded guilty or, at best, would have been represented by a public defender or assigned counsel, after which he would have had a very short trial ending almost certainly with a conviction. Simpson was wealthy enough to avoid this fate. His so-called "dream team" legal defense cost an estimated $10 million and included the hiring of private detectives and forensic experts. It is fair to say that Simpson's wealth enabled him to achieve his controversial jury acquittal (Barkan 1996). Bryant, a professional basketball player with the Los Angeles Lakers, was accused of rape and prosecuted during 2003 and 2004 before all charges were dropped after the alleged victim refused to testify. Bryant's legal defense probably cost several million dollars (Saporito 2004). Although it is often difficult to win convictions in rape cases regardless of the defendant's poverty or wealth (Frohmann 1991), Bryant's wealth no doubt gave him a considerable advantage over a poor defendant in a similar circumstance.

If wealthy defendants enjoy many advantages, the average poor defendant often cannot even obtain adequate legal representation. Sociological studies and journalistic accounts since the 1960s provide clear evidence that the legal representation of poor defendants is substandard (Blumberg 1967a; Downie 1972; Fritsch and Rohde 2001; Strick 1978). Because they cannot afford their own attorneys, indigent defendants rely, depending on the jurisdiction, on public defenders or assigned counsel. These attorneys typically have far too many cases to be able to spend much time on any one case. "Factory" or "assembly line" justice is said to prevail in the urban criminal courts, with attorneys first seeing their clients only moments before appearing before a judge. A study at the beginning of this decade found that many defendants in Georgia languish in jail for months without ever seeing an attorney, and then spend only a few minutes with an attorney before their trial (Firestone 2001). About the same time, a newspaper investigation of New York City defendants similarly concluded, "Thirty-eight years after the United States Supreme Court ruled in Gideon v. Wainwright that indigent defendants have a right to legal counsel, New York City offers representation to the poor that routinely falls short of even the minimum standards recommended by legal experts" (Fritsch and Rohde 2001:A1). Five years later, a judicial commission found that a similar situation existed in so many jurisdictions in New York state that defendants were deprived of their constitutional right to legal representation throughout the state (Hakim 2006). The substandard legal representation of the poor combines with the many advantages enjoyed by the wealthy to provide considerable evidence for the law and inequality thesis in regard to social class.

This evidence is reinforced when white-collar crime is considered, to turn to the second question with which this section began. By any standard, white-collar crime, defined as "illegal or unethical acts . . . committed by an individual or organization, usually during the course of legitimate occupational activity, by persons of high or respectable social status for personal or organizational gain" (Coleman 2006:7), takes a terrible toll in terms of life, health, safety, and money and property (Reiman 2007; Rosoff, Pontell, and Tillman 2007). For example, although the United States has had about 16,000–17,000 homicides annually during this decade, preventable deaths from such sources as unsafe workplaces, unsafe products, and unnecessary pollution and surgery probably number in the tens of thousands every year. White-collar crime thus accounts for many more deaths per year than does homicide. Similarly, although the economic loss from all property crime (robbery, burglary, motor vehicle theft, and larceny) is just under $20 billion annually, economic loss from property crime runs into the hundreds of billions of dollars annually. In terms of death, injury, and economic loss, then, the toll of white-collar crime thus far exceeds the toll of street crime (Barkan 2009).

Despite this fact, white-collar criminals generally do not receive legal sanctions commensurate with their crimes and, more to the point, generally receive much more lenient treatment than street criminals convicted of far less harmful offenses. The title of a popular white-collar crime text illustrates this point: "the rich get richer and the poor get prison" (Reiman 2007). Here, the example of the Ford Pinto is often cited (Cullen, Maakestad, and Cavender 2006). First sold in 1971, the Pinto proved to be a dangerous car because it was vulnerable to explosion after being hit from behind in a minor rear-end collision. Evidence later came to light that Ford knew the Pinto was dangerous even before the car went on the market and knew the car needed a $11 heat shield to protect the gas tank from explosion. Ford did a cost-benefit analysis and determined it would cost more to repair the tens of thousands of

Pintos it planned to manufacture and sell than to deal with lawsuits that would arise from the inevitable deaths and severe burns that would occur if it chose not to repair the car. Based on this determination, Ford decided not to repair the car, and this decision eventually meant that up to 500 people died when their Pintos were hit from behind by motor vehicles traveling at relatively low speeds. Although Ford executives were thus responsible for all these deaths, not a single executive ever went to prison.

An interesting study by Robert Tillman and Henry N. Pontell (1992) compared the sentences received in California by grand theft defendants and by physicians and other health care professionals prosecuted for Medicaid fraud. Although the median economic loss from the Medicaid fraud was 10 times greater than that from the grand thefts, only 38 percent of the Medicaid defendants went to prison, compared to 79 percent of the grand theft defendants. Thus, Medicaid defendants were only half as likely as their grand theft counterparts to be incarcerated even though the economic loss from their crimes was 10 times greater than that from the grand thefts.

In sum, whether we examine social class differences in the legal treatment of suspects and defendants accused of common street crimes, or differences in the legal treatment of street criminals versus white-collar criminals, considerable evidence supports the law and inequality thesis. Although justice is ideally blind to social class and other personal characteristics, the rich enjoy many advantages in the criminal justice system, and the poor experience many disadvantages, in part because the legal system fails to provide them with adequate counsel and other resources for their defense.

Race and Ethnicity

Evidence for the law and inequality thesis also exists in regard to race and ethnicity although, as we shall see, there is much more research on the criminal justice dimension of this thesis than on the civil justice dimension.

Civil Justice. If, as noted in the previous section, there is relatively little research on social class and litigation outcomes, there is even less research on race and ethnicity and litigation outcomes. Studies do exist of racial and ethnic differences in the use of the courts, and Chapter 4 indicated that research on this issue is inconclusive. To recall, some studies find that people of color are less likely to use the courts because of their lack of legal knowledge and sense of hopelessness stemming from poverty (Merry 1979; Moulton 1969), other studies attribute their low use of the law to the kinds of problems they experience (Silberman 1985; Silbey 2005), and other studies find low court use overall regardless of race and ethnicity and few differences by race and ethnicity (Bumiller 1988; Greenhouse 1986; Merry and Silbey 1984).

However, the focus of these studies has been on racial and ethnic differences in the use of litigation rather than on racial and ethnic differences in the outcome of litigation. Still, because people of color are disproportionately poor, they should be at a disadvantage in the civil justice system compared to their white counterparts. Additional research is needed to determine whether they also experience disadvantage because of discrimination and prejudice based on their racial and ethnic backgrounds.

Criminal Justice. Although there is a lack of research on race and ethnicity and civil justice outcomes, there is much research on race and ethnicity and criminal justice outcomes, especially in regard to African Americans. Despite or perhaps because of the amount of this research, scholars continue to debate its conclusions. Some think the belief that the criminal justice system is racially discriminatory is a "myth" (Wilbanks 1987). In their view, while some decisions by criminal justice professionals may be racially discriminatory, that does not necessarily mean or prove that the criminal justice system as a whole practices systematic discrimination. Other scholars think that racial and ethnic discrimination in the criminal justice system is very common and systematic and not just a myth (Mann 1993). Where does the truth lie? A very cautious conclusion takes a middle ground between the two viewpoints just outlined: racial and ethnic discrimination in criminal justice is too widespread to be just a myth, but it is also not common enough to be considered systematic. Instead, its manifestation and extent vary from one context to another and depend on a variety of circumstances (Walker, Spohn, and DeLone 2007). We have space here to briefly review some of the evidence on these issues, looking first at policing and then at sentencing.

Two relevant issues in policing concern racial profiling in traffic and street stops and racial discrimination in arrest. In many traffic stop studies, the proportion of African Americans and Latinos among drivers who are stopped, searched, ticketed, and/or arrested is compared with the proportion of African Americans and Latinos in the general population (residents of a town or city or drivers on a particular stretch of highway). Often these data indicate that people of color are disproportionately stopped, searched, and so forth, with one review of these studies concluding, "Nearly all of the publicly available reports and studies that we are aware of reveal disparities in the percentages of minority citizens who are stopped, cited, searched, or arrested as compared to selected benchmarks" (Engel and Calnon 2004:55). In related work, a national government survey asked a large random sample of Americans to report on their own traffic stops by police. Race was not related to the probability of being stopped, but it was related to the probability of being searched once stopped: among young males, African American and Latino drivers who were stopped were about twice as likely as their Anglo white counterparts to be searched (Smith and Durose 2006). However, findings from either kind of traffic stop research do not necessarily demonstrate racial profiling because it is possible, at least for the sake of argument, that African Americans and Latinos are, in fact, more likely than whites to commit traffic offenses, leading to greater stops, or to give police legitimate suspicion (as with evidence of drugs) to search them once stopped (Smith and Durose 2006).

In view of these methodological problems, two studies provide more compelling evidence of racial profiling in traffic stops (Harris 2002). A study in Maryland found that African Americans comprised only 17.5 percent of drivers on major highways but more than 72 percent of drivers stopped and searched by state troopers; a similar study in New Jersey found that African American and Latino drivers comprised only 13.5 percent of drivers on major highways but about 73 percent of drivers stopped and searched by state troopers. These disparities are so huge that it is very difficult to believe that they arose entirely from any bad driving tendencies among African Americans or Latinos.

The police stop pedestrians in addition to stopping drivers. Once they stop a pedestrian, they may frisk the individual for weapons or drugs. While the police have the right to make such stops under certain circumstances, if they are more likely to stop people of color simply because of the latter's race/ethnicity, that is racial profiling. Evidence on the extent of such profiling is mixed overall: some studies find racial profiling in street stops, but other studies do not. According to a recent review of this evidence (Walker, Spohn, and DeLone 2007:134), a study of New York City street stops (Fagan and Davies 2001) is the "most sophisticated study" of this issue. This study was initiated by the New York state attorney general and concluded that the police disproportionately stopped African Americans and Latinos even after taking into account neighborhood crime rates and racial demographics.

The evidence on race/ethnicity and arrest is also complex. In 2006, African Americans accounted for about 32 percent of all arrests for serious street crime (violent and property offenses) and for 39 percent of arrests for just violent crime. Because African Americans comprise only about 13 percent of the U.S. population, they are disproportionately arrested. Does this fact occur because they commit higher rates of crime or because the police are biased against them, or might it occur from both these possibilities?

Criminologists disagree in their conclusions regarding these possibilities, but a rough consensus is that higher African American crime rates do, in fact, account for the bulk of the disproportion in their arrest rates and that racial discrimination by police in arrest plays only a small role (Tonry 1994; Walker, Spohn, and DeLone 2007). Several kinds of evidence support this conclusion. First, important observational studies of police behavior have not found racial discrimination in arrest (Black 1980; Riksheim and Chermak 1993). Second, national victimization surveys, in which aggravated assault victims are, among other things, asked to indicate the race of the person who assaulted them, yield proportions of African American offenders that are roughly equal to those appearing in arrest statistics. Third, self-report studies, in which respondents are asked to indicate what crimes they have committed, indicate that African Americans do, in fact, commit relatively higher rates of serious street crime (Farrington, Loeber, and Stouthamer-Loeber 2003).

Despite this body of evidence, other evidence does point to at least some racial discrimination in arrest (Walker, Spohn, and DeLone 2007). Some observational studies have found such discrimination even if other such studies have not (Smith, Visher, and Davidson 1984). Some studies find that the police are more likely to make an arrest when the victim is white than when the victim is African American, indicating that the police consider crimes with white victims more important than those with black victims (Smith 1986). Some studies also find that the police tend to arrest African Americans and Latinos even if the evidence against them is relatively weak, whereas they do not arrest whites unless the evidence against them is fairly strong (Petersilia 1985). Summarizing the arrest evidence, Walker et al. (2007:125) conclude,

> In sum, patterns of arrest by race are extremely complex. Racial minorities, especially African Americans, are arrested far more frequently than are whites. Much of this disparity, however, can be attributed to the greater involvement of minorities in serious crime. . . . Even after all the relevant variables are controlled, however, some evidence of arrest discrimination against African Americans persists.

Race and sentencing research is also complex. Several influential studies conclude that a defendant's race does not affect the sentence he or she receives once the defendant's prior record and the seriousness of the alleged offense are taken into account (Hagan 1974; Kleck 1981). Other studies find that race does matter even when these factors are considered, with African Americans and Latinos receiving harsher sentences than whites (Kempf and Austin 1986; Petersilia 1985; Spohn and Holleran 2000; Steffensmeier and Demuth 2001); this racial disadvantage tends to be greatest among young males. In general, racial discrimination is found more often for the decision to imprison (the "in/out" decision) than for actual sentence length of those imprisoned, and for less serious crimes than for more serious crimes. Racial discrimination also appears in homicide and rape cases when the race of the victim is taken into account: defendants accused of killing or raping white victims tend to receive harsher sentences than those accused of killing or raping African American victims. In effect, the legal system (as in some studies of arrest noted earlier) considers crimes with white victims to be more serious and important than crimes with African American victims.

Walker et al. (2007:280) conclude that the evidence on race and sentencing is mixed and that racial discrimination in sentencing "is not universal but is confined to certain types of cases, certain types of settings, and certain types of defendants," a pattern they call "contextual discrimination." They continue, "It thus appears that although flagrant racism in sentencing has been eliminated, equality under the law has not been achieved. . . . Racial minorities who find themselves in the arms of the law continue to suffer discrimination in sentencing."

Although evidence of racial discrimination in police practices and in sentencing is mixed and complex overall, evidence of racial discrimination in the legal "war" against drugs is clear and compelling. Crack cocaine and powder cocaine are pharmacologically similar. Since the mid-1980s, the war against drugs has focused much more on crack cocaine than on powder cocaine. Because of this focus, African Americans have been arrested much more often than whites for drug use and received much harsher sentences than whites, even though they do not use illegal drugs more often than whites. A former president of the American Society of Criminology called the war on drugs "a major assault on the black community" and said, "What is particularly troublesome . . . is the degree to which the impact has been so disproportionately imposed on non-whites. There is no clear indication that the racial differences in arrest truly reflect different levels of activity or of harm imposed" (Blumstein 1993:4–5). Agreeing with this sentiment, a former criminal justice official likened the war against drugs to a "search and destroy" mission against African Americans (Miller 1996).

Overall, then, the racial profiling, arrest, and sentencing evidence is mixed and methodologically difficult to interpret. However, several well-designed studies do find evidence of racial discrimination in all three areas that tends to support the law and inequality thesis. The legal war against drugs of the last two decades also manifests the use of law against people of color far out of proportion to their actual use of illegal drugs.

The law has contributed to racial and ethnic inequality in at least one other way by disenfranchising felons. Because felons are disproportionately African American and Latino, several million people from these backgrounds have lost the right to vote, not only while they are imprisoned but, in several states, even after they serve their term and return to society. Only two states, Maine and Vermont, permit a convicted felon to vote while imprisoned. As of the time of this writing, seven states take the right to vote away from felons forever, even after they serve their sentence and finish any period of parole that may be required of them. Another 35 states prohibit released felons from voting while they remain on parole. At any one time, about 4.7 million felons, still imprisoned or released back to society, are not allowed to vote. This figure equals more than 2 percent of all American adults and 13 percent of African American men nationwide. In some states, as many as 25 percent of African American men are disenfranchised because of their felony records (The Sentencing Project 2007).

Felony disenfranchisement probably affected the presidential election in 2000, when 600,000 ex-felons in Florida were not allowed to vote. Most of them were African American and probably would have voted for Al Gore, the Democratic Party candidate, over George W. Bush, the Republican Party candidate. Bush eventually was considered the winner of the disputed election because of his margin of 537 votes in Florida; had felons there been allowed to vote, Gore almost certainly would have become president. Scholars of felony disenfranchisement estimate that it has also affected the results of at least seven U.S. Senate elections (Manza and Uggen 2006). Because voting is perhaps the most fundamental right in a democracy and because Republican Party candidates, who do not normally favor the various social programs that would benefit the African American community, have won elections they would have lost if felons had been allowed to vote, the law has contributed to racial inequality by contributing to felony disenfranchisement.

In sum, the United States has moved far beyond the days of slave law and of "Jim Crow" racism in which the law in the South upheld segregation in all walks of life and virtually guaranteed conviction and imprisonment any time an African American was accused of a crime. At the same time, racial discrimination continues to affect the operation of the criminal justice system in several complex respects, and the criminal justice system in turn has disenfranchised substantial numbers of African Americans and Latinos. All these dynamics provide ample evidence for the law and inequality thesis in regard to race and ethnicity.

Gender

We saw earlier that law buttressed women's subordination in many ways in the American past. Since the 1960s, however, the law has changed dramatically with respect to women's equality, and a large body of federal and state legislation and court rulings now bans overt discrimination against women in the workplace, family, and other social realms. Perhaps most notably, Title VII of the 1964 Equal Rights Act banned employment discrimination on the basis of gender, and the 1965 Equal Pay Act required that the two sexes receive equal pay for the same jobs. Although gender inequality continues to exist, it is not so clearly reinforced by the law as it was before the 1960s.

Civil Justice. This overall conclusion notwithstanding, it remains true that the law, as interpreted by the U.S. Supreme Court, is allowed to treat women and men differently to some extent. According to the Court's **intermediate scrutiny standard**, such differential treatment is permitted as long as it is "substantially related" to the achievement of important governmental goals. This standard takes into account the fact that women and men are biologically different and thus have different roles with respect to child rearing and other responsibilities. Based on this standard, women and men may be treated differently in some circumstances but not in other circumstances. As political scientist Suzanne Samuels (2006:296) observes, "The fact that men and women have different roles in human reproduction underlies many U.S. laws that treat women differently than men."

The Supreme Court developed and applied the intermediate scrutiny standard to gender in *Craig v. Boren* (429 U.S. 190 [1976]), in which it struck down an Oklahoma law that prohibited the sale of 3.2 percent beer to males under 21 but allowed it to be sold to women older than 18. The Court ruled that the law discriminated against males 18–20 because there was no sufficient reason in its legal judgment to treat the sexes differently with respect to the sale of 3.2 beer. The intermediate scrutiny standard in this case, then, was used to end a particular example of differential gender treatment.

In a 2001 case, *Tuan Anh Nguyen et al. v. Immigration and Naturalization Service* (533 U.S. 53), the Court used the standard to support differential gender treatment. In this case, plaintiff Nguyen was born in Vietnam to a Vietnamese woman and American man who were not married and was raised by his father in the United States. After Nguyen was convicted of a sex offense in Texas at age 22, the Immigration and Naturalization Service (INS) took steps to deport him. Nguyen's claim that he was a U.S. citizen because his father was a citizen was rejected because federal law requires that a child born abroad out of wedlock can be a U.S. citizen only if the mother is a citizen; for purposes of the child's citizenship, the father's citizenship is irrelevant. Nguyen appealed the INS ruling, but a federal appellate court and then the Supreme Court both ruled against him. Following the intermediate scrutiny standard, the Court reasoned in part that the state needs to determine that a biological relationship between a parent and child exists, and that the mother's status is evident from the birth itself while the father obviously does not even have to be present while the child is being born and may not even know that the child was born. As the Court put it, "Because fathers and mothers are not similarly situated with regard to proof of biological parenthood, the imposition of different rules for each is neither surprising nor troublesome from a constitutional perspective." Partly for this reason, the Court concluded that the mother's U.S. citizenship is sufficient to establish the child's citizenship, while the father's U.S. citizenship is not sufficient.

Gays and lesbians continue to experience discrimination in many realms of life because the law does not prohibit such discrimination. In many locations, workplace discrimination on the basis of sexual orientation is permitted under the law and denies full equal employment opportunity to gays, lesbians, and bisexual and transgendered individuals. In February 2007, the American Civil Liberties Union, the Human Rights Campaign, and the National Gay and Lesbian Task Force announced stepped-up efforts to have Congress pass the Employment Non-Discrimination Act, which would prohibit employment discrimination on the basis of sexual orientation. The House of Representatives passed the act in November 2007 (*Washington Post* 2007), but the Senate was still considering it at the time of this writing.

Another area of discrimination on the basis of sexual orientation concerns marriage. Except for Massachusetts, no state as of mid-2007 permitted same-sex couples to legally marry. Three states—Connecticut, New Jersey, and Vermont—recognized civil union status for same-sex couples that grants them the legal rights and responsibilities that spouses enjoy. California, Hawaii, and Maine provided some spousal rights and responsibilities to same-sex couples, while the remaining 43 states granted no legal protection to such couples (Human Rights Campaign 2007).

Criminal Justice. Men account for about 91 percent of all prison and jail inmates in the United States (Sabol, Minton, and Harrison 2007). Does this gender disparity reflect gender discrimination in arrest, conviction, and sentencing, or does it reflect actual gender differences in rates of criminal behavior? Criminologists agree that this gender disparity almost entirely reflects actual gender differences in criminal behavior (Chesney-Lind and Pasko 2004; Muraskin 2007). Males commit more serious crime than females for many reasons: they are socialized to be assertive and dominant; they are given more opportunity during adolescence to be away from home and thus to get into trouble; they feel less attached to their families and other social institutions; and they have more friends, other males, who break the law. For all these reasons, it is not surprising that the vast majority of prison and jail inmates are men, far out of proportion to their representation in the general population.

That said, it is still possible that some gender discrimination does exist in arrest and sentencing that increases the number of one sex or the other who end up incarcerated. Here, it is helpful to consider arrest and sentencing separately, and within each stage to then consider juveniles and adults separately. Regarding arrest and juveniles, the evidence indicates that girls are more likely than boys to be arrested or otherwise brought to the attention of juvenile authorities for *status offenses* like running away from home or frequent sexual behavior, as it is felt that girls need more protection than boys during the teenage years (Chesney-Lind and Pasko 2004). Regarding arrest and adults, the evidence is more complex. Female prostitutes are more likely than their male customers to be arrested, representing a form of gender discrimination that gets little public attention. For other types of offenses, female and male suspects appear to be at equal risk for arrest overall, but older women and women who act "femininely" appear to be at lower risk for arrest. In an interesting racial difference, African American women appear to as likely as African American men to be arrested, but white women appear less likely than white men to be arrested (Visher 1983). Despite popular perceptions, there do not appear to be gender differences in the chances of receiving a traffic ticket for speeding (Schmitt, Langan and Durose 2002).

The evidence for sentencing partly reflects what is found for juveniles. Once they are arrested or otherwise referred to juvenile justice authorities, girls receive harsher legal treatment than boys for status offenses, but they receive somewhat more lenient treatment for more serious offenses (Chesney-Lind and Pasko 2004). Turning to adults, women appear to be treated more leniently than men at the sentencing stage. Overall, they are 10 to 25 percent less likely than men convicted of similar crimes to be sent to prison or jail, although the two sexes receive similar terms of incarceration once the decision is made to put them behind bars (Daly 1994).

Several reasons appear to explain the somewhat more lenient treatment that women offenders receive in the decision to incarcerate (Daly 1994). Judges and prosecutors believe that women pose less threat to society than men convicted of the same offenses, that women are more likely than men to have significant family and childcare responsibilities, and that women are less responsible than men for the crimes they commit. Some scholars believe that women do deserve more lenient treatment in incarceration because of these reasons, while others feel that such leniency still constitutes gender discrimination in the legal system.

Summary

1. Much historical evidence supports a law and inequality thesis. After the Civil War, the labor movement grew rapidly and posed a significant threat to management. Because it was a crime in most locations for workers to go on strike, many workers were arrested for striking and then prosecuted. Because juries often acquitted striking workers, management obtained injunctions that permitted workers to be jailed without benefit of jury trial. In another area, economic downturns after the Civil War led to massive unemployment, and during the 1870s hundreds of thousands of workers took to the rails to find employment. This led most states to enact "tramp acts" that made it illegal to travel without visible means of support. These acts were used in Buffalo and elsewhere during the 1890s to arrest unemployed and striking workers. According to leftist scholars, the modern police force, which arose during the 1830s and 1840s, was generally used to suppress the poor and, in particular, was used to suppress labor strikes in many cities in many years.
2. The history and aftermath of slavery provide much evidence for the law and inequality thesis in the American past. Slavery was endorsed in the U.S. Constitution and was supported by federal, state, and local law. After the Civil War, the South established the Black Codes to subjugate African Americans so much that they mirrored the antebellum slave codes. Although Reconstruction ended these laws, the end of Reconstruction in 1877 led to the rise of Jim Crow laws, followed by the U.S. Supreme Court's *Plessy v. Ferguson* decision that upheld "separate but equal" racially segregated public facilities. Jim Crow laws and segregation continued in the South into the 1960s until segregation finally ended thanks to the civil rights movement.
3. Law also played a key role in the historical subjugation of Native Americans. Beginning in the early colonial period and continuing well after the Civil War, white settlers and eventually U.S. troops waged a virtual war on Native Americans and, in particular, forced them to sign treaties that ceded control of much of their land to whites and confined them to reservations. The criminal justice system also discriminated against Native Americans in early America. During the colonial period, for example, Native Americans living in white towns had to obey curfews and other limits on their behavior, and whites who killed Native

Americans generally received more lenient sentences than Native Americans who committed the same type of killing.

4. The Chinese, too, were victims of legal discrimination in the American past. Chinese immigrants began to move into California and other western areas in large numbers to try to strike it rich after the 1849 California Gold Rush and to work on the burgeoning railroad construction industry. Anti-Chinese sentiment grew rapidly and led to the 1882 Chinese Exclusion Act that banned Chinese immigration for ten years and then evolved into a permanent ban in 1902.

5. The origins of several U.S. drug laws reflect racial prejudice and discrimination. Prohibition came about in part because of prejudice against Catholic immigrants; opium was banned partly because of prejudice against Chinese immigrants; cocaine was banned partly because of prejudice against African Americans; and marijuana was banned partly because of prejudice against Mexican-American immigrants.

6. The law and inequality thesis finds ample historical support in the legal treatment of women, as women were second-class citizens in the eyes of the law for most of U.S. history. They could not vote before 1920, and employment discrimination against women was lawful until it was prohibited by the 1964 Civil Rights Act. Underlying the law's support for gender discrimination was the widespread view that women are naturally different from and subordinate to men, and, in particular, biologically destined to be mothers and homemakers.

7. In the contemporary era, social class continues to affect legal outcomes. In the civil justice system, the "haves" come out ahead of the "have-nots" for several reasons, including the fact that they tend to be repeat players instead of one-shotters and simply have greater financial resources to employ skilled legal counsel and exploit other advantages. In the criminal justice system, the poor face several obstacles even though they officially enjoy several legal protections under the Constitution. Because they cannot afford legal counsel, they are typically represented by public defenders or assigned counsel, yet many reports indicate that their legal representation is substandard. Although the poor's crimes do much less harm overall than white-collar crime, their crimes are punished much more severely.

8. Because they are disproportionately poor, people of color are at a disadvantage in the civil justice system compared to their white counterparts. They are also at a disadvantage in the criminal justice system, even if it is true, as many scholars think, that racial and ethnic discrimination in the criminal justice system is not systematic. Instead, such discrimination varies from one context to another and depends on a variety of circumstances. In particular, it is found more often for less serious offenses than for more serious offenses and for the incarceration decision than for sentence lengths. It is also found when the race of the victim in serious crime is considered, as crimes involving white victims receive harsher legal treatment than those involving people of color as victims. Regarding police, the racial profiling evidence is complex overall but does indicate that some racial profiling exists. Some but not all studies find some racial discrimination in the decision to arrest. Clear evidence of differential racial treatment is found in the legal war against drugs because the penalties for crack cocaine are much more severe than those for powder cocaine, leading to harsher legal treatment of African Americans than of whites. Felony disenfranchisement has also disproportionately affected people of color and in this way contributed to their inequality.

9. Since the 1960s, federal and state legislation and court rulings have banned overt discrimination against women in the workplace, family, and other social realms. Even so, the law does allow women and men to be treated differently to some extent under some circumstances, as the Supreme Court's intermediate scrutiny standard permits such differential treatment as long as it is substantially related to the achievement of significant governmental objectives. Many kinds of discrimination continue to exist under the law on the basis of sexual orientation. Workplace discrimination is lawful in many locations, and the vast majority of states still do not provide legal rights and responsibilities to same-sex couples. In the criminal justice system, adult women receive somewhat more lenient treatment than their male counterparts, while girls receive harsher treatment than boys for status offenses but more lenient treatment for more serious offenses.

Key Terms

Coverture
Dangerous classes
Intermediate scrutiny standard
One-shotters
Repeat players
Separate sphere
Social stratification
Tramp acts

CHAPTER 8

The Legal Profession

Q. What's the difference between a dead snake in the road and a dead lawyer in the road?
A. Skid marks in front of the snake.

Q. Why are research labs using lawyers instead of rats in their experiments?
A. There are some things a rat just won't do.

Q. What do you call 6,000 lawyers at the bottom of the sea?
A. A good start.

Source: Law professor Marc Galanter, quoted in Shinkle 2005: C6

Lawyer jokes like these abound on the Internet and elsewhere. They reflect our ambivalent view that lawyers are important for many reasons but also engage in practices that many of us find distasteful. As the public has become more concerned about a possible litigation "crisis" (see Chapter 6), lawyer jokes have arguably become more common and harsher in tone (Galanter 2006). It is an exaggeration to say that the legal profession is under attack, but it is not an exaggeration to say that the public often takes a dim view of the legal profession. This negative view persists despite the fact, or perhaps because of the fact, that the legal profession wields enormous power and influence in so many aspects of our society. Because the study of the legal profession is, unsurprisingly, an important part of the study of law and society, sociolegal scholars have explored many aspects of the practice of law. This chapter reviews this body of work to shed light from a social science perspective on lawyers, lawyering, and legal education.

IMAGES AND PERCEPTIONS OF LAWYERS

Perhaps ever since lawyers first practiced law, they have endured a negative public image. The Bible says, "Woe unto you also, ye lawyers! for ye lade men with burdens grievous to be borne, and ye yourselves touch not the burdens with one of your fingers" (Luke 11:46). Contemplating a transformation to an idyllic society, a character in Shakespeare's *Henry VI, Part II* (Act IV, scene 2, line 59), declares, "The first thing we do, let's kill all the lawyers." Plato referred to the "small and unrighteous souls" of lawyers, and the poet John Keats observed, "I think we may class the lawyer in the natural history of monsters" (quoted in Vago 2006:365).

This negative image is reflected in public opinion surveys of random samples of Americans nationwide. In a survey commissioned by the American Bar Association (2002), only 19 percent of respondents said they were "extremely" or "very" confident in the legal profession and lawyers. This percentage ranked above only that for the media in a list of 10 institutions and far behind the leading institution, the medical profession/doctors, in which 50 percent of respondents said they were extremely or very confident. Almost three-fourths of the sample agreed that lawyers "are more interested in winning than in seeing that justice is served" and that they "spend too much time finding technicalities to get criminals released." Almost 70 percent said that lawyers "are more interested in making money than in serving their clients," and half said "we would be better off with fewer lawyers."

Reasons for the Negative Image

Why do such negative perceptions and jokes about lawyers exist? The major reason lies in the nature of the work that lawyers do, as several aspects of lawyering lend themselves almost automatically to negative views.

Involvement in Heated Disputes. Lawyers often represent people involved in heated disputes. Just as we often "blame the messenger," as the saying goes, for bringing bad news, so do we dislike lawyers because they take sides as they represent a disputant. As Benjamin Civiletti, a former U.S. Attorney General, once observed, "I don't think anyone's ever going to love lawyers. After all, it's our lot to get involved in society's most heated disputes between people, to be buffers. It may be that it's also our lot to get kicked and gouged, like a bumper on an automobile" (quoted in Smith 1985:22).

"Making Things Worse." A related reason for the negative public image of lawyers stems from the apparently widespread belief that lawyers complicate matters with their penchant for argumentation and turgid prose. Recall from Chapter 4 that adjudication may aggravate disputes, in part because attorneys try to secure a victory for their clients rather than to look for points of compromise and reconciliation. Lawyers are thus perceived as contentious advocates who may even aggravate the ill-feelings between the disputants in their efforts to best represent their clients. Any readers who might have read a lease or other legal contract or parts of the tax code also know how wordy and unclear the language of legal documents can often be. The idea that lawyers "make things worse" is the basis for many jokes about them.

Defense of Unsavory Clients. Lawyers also defend unsavory, dislikeable clients: robbers, rapists, murderers, and other criminals. In a democratic society, it is essential that every defendant, no matter how despicable, enjoy legal representation. It is also perhaps inevitable that such representation will stigmatize the lawyers who offer it and, more generally, lead the public to hold the legal profession in low regard.

Use of Legal Stratagems. Related to this, lawyers are bound by their professional ethics to vigorously represent every client to the best degree possible. Under this *total commitment* model of lawyering, they must do everything possible within the bounds of the law to ensure that a criminal defendant is not found guilty or that a civil client wins a case. This means they must also use every legal stratagem and take advantage of every legal technicality to win their cases. As a former dean of Cornell University Law School once commented, "Under this model, the lawyer is obliged to do everything for the client that the client would do for himself if he had the lawyer's skill and knowledge. That means virtually anything short of breaking the law, but much that is immoral or undesirable is not illegal" (quoted in Smith 1985:22). Yet, the obligation to follow this model wins lawyers few friends outside their own profession, especially because they are seen as doing their best to win freedom for dangerous criminals through legal maneuvering. As former attorney general Civiletti noted, "In the popular vernacular, that becomes getting the defendant off on a technicality. But what people forget is that those technicalities are the basic rights of freedom, the detailed procedural and due process rights that protect us against tyranny" (quoted in Smith 1985:22).

"Ambulance Chasing." Another aspect of lawyering involves seeking fair compensation for clients who have suffered various personal injuries and, sometimes, great tragedy. Because lawyers are paid for doing this, they are often seen as little better than vultures who greedily take advantage of their clients' misfortune for their own personal gain. This is the familiar image of the lawyer as an "ambulance chaser," someone who follows an ambulance after a traffic accident or other incident and then tries to represent the ill or injured person inside. This image came to mind and was probably reinforced in December 1984 when a Union Carbide storage tank in Bhopal, India, leaked 27 tons of poisonous gas that killed more than 3,700 Indians and permanently disabled thousands more. In the weeks after the tragedy, many U.S. lawyers flew to Bhopal to try to represent the families of the dead in multimillion dollar lawsuits against Union Carbide. They were hoping to be able to try the cases in American courts so that they could collect contingency fees amounting to about one-third of any money they won for their clients. The lawyers' rush to India led an American newspaper columnist to remark, "One thing already seems clear in the wake of the tragic poisonous gas leak at a Union Carbide plant in Bhopal, India: The reputation of American lawyers is about to sink even lower than it ever has lurked before" (Greene 1984:C5). An editorial cartoon at the time depicted a vulture flying to India while carrying a legal briefcase.

Work on Behalf of the "Haves." Some lawyers, as we shall see below, work for corporations and other wealthy clients. Sometimes their work on behalf of the "haves" pits them against the "have-nots" as they help Goliath to defeat David. In his classic work *Democracy in America,* Alexis de Tocqueville (1994 [1835]) observed that lawyers in the United States are members of the aristocracy and tend to support established interests and the status quo. Their work on behalf of the "haves" has contributed historically to their negative public image.

The Positive Image: Lawyer as Hero

Lawyers have certainly endured a negative image over the years for all the reasons just discussed, but one positive image, the lawyer as a hero for the innocent or the oppressed, is extolled in film and fiction (Asimow and Mader 2004). In Harper Lee's (1960) majestic novel, *To Kill a Mockingbird,* made into a 1962 film starting Gregory Peck, Alabama lawyer Atticus Finch defends an African American man falsely accused of raping a white woman in a racist town, even though Finch realizes that doing so might harm his practice and endanger his life. In the 1982 film "The Verdict," lawyer Frank Galvin, played by Paul Newman, takes on a large law firm in a medical malpractice case involving a woman who went into a coma while having a baby; against all odds, he wins the case. The popular books and TV show about Perry Mason depict an attorney who works vigorously to defend innocent clients. In real life, attorneys such as the famed Clarence Darrow are heralded for taking on unpopular defendants and unpopular causes. Like Atticus Finch, attorneys who represented civil rights movement clients in the South during the 1960s risked their practice and personal safety (Barkan 1985). Lawyers may be villains for all the reasons discussed in the previous section, but some have also been, and continue to be, heroes to the less fortunate among us by offering them legal assistance and the hope for a better society. Our later section on "cause lawyering" returns to this theme.

HISTORY OF THE LEGAL PROFESSION

As some of the preceding discussion indicated, the negative image of lawyers is at least partly rooted in their working for established interests. The history of the legal profession in the United States provides ready evidence of this dynamic, and it also provides ready evidence of a profession that for much of its history systematically excluded or relegated to its lower ranks people of color, women, and Catholics and Jews. Although these racial, gender, and religious barriers formally ended some decades ago, the bar remains stratified along these lines. A brief discussion of the history of the legal profession will provide a backdrop for understanding these issues.

Early Origins of Lawyers

The concept of the *attorney* or legal professional with which we are all familiar—someone who has a specialized knowledge of the law, who gives legal advice to clients, and who represents the client in court and other venues (Rueschemeyer 1973)—is a relatively modern development. Preliterate societies, both those that existed thousands of years ago and those studied since the nineteenth century by anthropologists, typically lacked such third-party representation (Schwartz and Miller 1964). Individuals who became involved in what we might call a legal proceeding (keeping in mind that law in these societies differs from the modern model most familiar to us) would represent themselves or perhaps have family members or friends speak out on their behalf. No one specialized in knowing more about the rules or law of the society than anyone else.

It was not until ancient Rome that the legal profession began to develop (Crook 1984). Early on, some individuals represented others but were best considered orators rather than attorneys, as they were trained in rhetoric (public speaking) rather than law. Later, about the time of Cicero (106 BC–43 BC), certain individuals did acquire some knowledge of law and gave legal advice, but they did not know much about the law and gave their advice only as a part-time activity. A full-fledged legal profession, with individuals practicing law full-time, finally appeared during the Imperial Period (about the first five centuries AD) out of necessity, as Roman law by that time had become extremely detailed. After Rome collapsed by the sixth century AD and the Middle Ages began, the Church's control of continental Europe ensured that religion became the law of the land. Accordingly, secular law and the legal profession collapsed as well, only to reemerge with the beginning of capitalism in the mercantile period (see Chapter 6).

To help understand the history of the U.S. legal profession to which we will soon turn, it is helpful to trace changes in the legal profession in England during the eighteenth century (Abel 1988; Duman 1980; Lemmings 2000). Before this period, most English lawyers learned their law as apprentices to existing lawyers, and no special training or expertise was required to practice law. By the end of the eighteenth century, however, the professional English bar had begun to emerge. Formal legal education became a growing expectation, and licenses became required to provide many types of legal services. These new emphases were promoted by professional societies of attorneys, the Inns of Court, which established ethical codes and sought to limit the provision of legal services to persons formally trained and licensed in the law. Although one goal of this limitation was to increase the quality of the work lawyers provided, another was to restrict the number of people allowed to practice law and thus to raise their legal fees (owing to simple supply and demand considerations) and income. This attempt to monopolize the practice of law to ensure greater wealth and influence is called the *market control* dynamic (Abel 1988) and was repeated in the United States, to which we now turn.

Origin and Development of the American Legal Profession

While the bar was emerging and monopolizing the practice of law in England during the eighteenth century, America was still in its colonial period. Although almost half of the 56 signers of the Declaration of Independence, including John Adams and Thomas Jefferson, were lawyers, the legal profession was actually held in low esteem during the early colonial years, when lawyers were so "distinctly unwelcome" that several colonies banned people from practicing law for money (Friedman 1985:94). An individual named Thomas Morton came to Plymouth in 1624 or 1625 and was probably the first lawyer in Massachusetts colony; he was soon jailed and then banished from the colony for practicing law.

Various writings and statements by the colonists illustrate their antipathy toward lawyers. One person wrote that in Pennsylvania, "They have no lawyers. Everyone is to tell his own case, or some friend for him . . . 'Tis a happy country." A North Carolinian called lawyers "cursed hungry Caterpillars" and complained that their fees "eat out the very Bowels of our Common-wealth" (quoted in Friedman 1985:94, 96).

Some colonists translated words into action. Mobs rioted against lawyers in New Jersey in 1769 and 1770. Shortly after the birth of the new nation, the famous Shays' Rebellion began in Massachusetts in 1786. Daniel Shays was a farmer who helped lead an armed farmers' revolt against the severe debt they had incurred after the American Revolution. Their aim was to free farmers from debtors' prisons and to disrupt court proceedings against debtors. The farmers especially disliked lawyers, who, they felt, aided the efforts of banks to foreclose on and take away their farms when they could not pay their mortgages (Richards 2002).

The colonists disliked lawyers for at least two reasons (Auerbach 1978). Recalling Chapter 4's discussion, they distrusted law itself: they thought its adversary nature violated their Christian beliefs and disrupted the social harmony characterizing their small communities. Not surprisingly, they easily translated this distrust of law into distrust and dislike of lawyers. As noted above, many colonists also disliked lawyers for representing banks and other established interests. This attitude was the key motivator for the rioting against lawyers in New Jersey in 1769 and 1770 and in Massachusetts sixteen years later.

As Chapter 4 explained, law became more important as the colonies grew in population and as their mercantile economies expanded during the eighteenth century. Lawyers perforce became more important and influential as a "necessary evil" (Friedman 1985:96). Thus, it was not surprising that so many signed the Declaration of Independence and played a key role in the eventual ratification of the Constitution and other significant events in the life of the new nation.

Professionalization of the American Bar. During the colonial period, one could practice law without any formal legal training or any legal training at all. Typically, however, lawyers learned law, as in England, by apprenticing themselves to existing lawyers, while a few traveled back to England to learn law there. The apprenticeship model continued well into the nineteenth century and is reflected in the legal training of Abraham Lincoln. As Friedman (2004a:166) recounts,

> For most of the nineteenth century, it was easy to become a lawyer. Most lawyers learned their trade as apprentices—they were gofers in the offices of established lawyers; here they picked up scraps of information, read law books, copied documents, and made themselves generally useful. After doing this sort of thing for a year or two, the fledgling would usually go to a local judge, answer a few questions, and that was that.

Because it was so easy to become a lawyer, many people who did become lawyers were poorly trained, creating concern in legal circles about the quality of the bar. At the same time, their sheer numbers also created concern that the plethora of attorneys was keeping legal fees and lawyers' incomes much lower than they would be if there were a smaller number of attorneys. These concerns led the American legal profession to undertake the same market control effort that its English counterpart achieved a century earlier (Abel 1989). This effort was led by the many professional associations of elite attorneys, white men who worked for the many corporations and smaller businesses that stimulated the postbellum Industrial Revolution. In 1870, these attorneys formed a city bar association in New York City, which was followed by the formation of several other city and state bar associations during the rest of that decade. In 1878, some 75 attorneys, many of them again corporate lawyers, from almost two dozen states met in Saratoga Springs, New York, and formed the American Bar Association, the nation's first national association of attorneys.

Thanks in large part to the new bar associations' efforts, law school became an increasing requirement for the practice of law by the early twentieth century. Before the Civil War, relatively few law schools existed, and no state required a law degree or even any college education to practice law. By 1900, many more law schools existed, but less than half required even a high-school degree for admission, and about one-third did not require any law schooling at all. As the twentieth century progressed, more law schools were established and required at least some college education for admission, and states began to require a law school degree and/or full-fledged, lengthy entrance exam for admission to the bar. Today, of course, these are requirements throughout the nation. (The *Law School and Legal Education* section below discusses the evolution of legal education further.)

This massive market control effort by the ABA and state bar associations achieved its intended effects. First, it improved the training of new attorneys and the quality of the bar. Second, it restricted the number of lawyers and thus raised the fees that established attorneys could charge and the incomes they could earn. By monopolizing the practice of law, the American bar, like its English counterpart about a century earlier, ensured that its power and wealth would not be challenged by upstart entrants into the legal profession.

The ABA also ensured that its ranks would remain free of the non-elite members of society. As Friedman (1984:245–246) bluntly puts it, the ABA began as a "club for white males" and "has a rather shameful history of snobbery and bias." The founders and members of the ABA and of the many state bar associations were typically white Protestants and, especially in the ABA, graduates of elite law schools such as Harvard and Yale. Women, Catholics, Jews, and African Americans and other people of color were excluded from ABA membership at its founding and well into the twentieth century, and they also had little hope during that time of joining the corporate law firms that stood at the pinnacle of the legal profession.

A notorious example of the ABA's exclusion occurred in 1912, when the ABA executive committee accidentally admitted three African American lawyers as members, but canceled their membership when it discovered its "error." The chair of the ABA membership committee wrote that the issue of admitting African Americans raised the "question of keeping pure the Anglo-Saxon race" (quoted in Auerbach 1978:66). The entire membership then considered the matter and decided to permit the three attorneys to remain in the ABA but stipulated that racial identification should be part of any future applications. As Auerbach (1978:66) characterizes this outcome, the ABA "thereby committed itself to lily-white membership for the next half-century. It had elevated racism above professionalism." Although Catholics were gradually admitted into bar associations as the twentieth century progressed, Jews were not admitted into the ABA and hired by corporate law firms until after World War II when the Holocaust came to light; women and African Americans and other people of color were generally not admitted or hired until the 1960s and 1970s. As we shall discuss below, the bar is still stratified today along these sociodemographic grounds.

The ABA's elite orientation was also in evidence in 1908 when it adopted its Canons of Professional Ethics, a document that outlined ethical standards guiding the practice of law. Although adoption of the Canons was spurred by a denunciation three years earlier by President Theodore Roosevelt of corporate attorneys for helping big business to violate government regulation, the Canons instead focused on the activities of attorneys with working-class clients in urban areas. For example, the Canons prohibited attorneys from advertising their services, a practice that corporate attorneys hardly needed to do but one that struggling urban lawyers depended on to find clients. Similarly, the Canons prohibited attorneys from actively soliciting clients (i.e., "ambulance chasing"). Again, this was a practice that urban attorneys needed to find clients but one for which corporate attorneys had no need at all. Not coincidentally, many of the urban lawyers most affected by the Canons were Catholic or Jewish immigrants who did not come from the advantaged backgrounds enjoyed by ABA members. These urban lawyers' behavior was now considered "unethical because established Protestant lawyers said it was" (Auerbach 1978:50).

THE U.S. LEGAL PROFESSION TODAY

This brief history provides a context for understanding the nature of the legal profession today. We will look first at its growth over the years, then at the type of work lawyers do, and then at an activity, cause lawyering, that is the focus of recent work by law and society scholars, before moving on to some other issues.

Growth and Demographics of the Legal Profession

The number of attorneys in the United States today far exceeds what the founders of the bar associations in the 1870s and 1880s could have ever imagined. Since the nineteenth century, the legal profession has grown tremendously both in absolute numbers and as a percentage of the population (Abel 1989; American Bar Foundation 2007). In 1850, about 24,000 attorneys were in practice, equivalent to 1 lawyer for every 969 individuals in the national population. By 1900, the number of attorneys was about 109,000, or

FIGURE 8.1 Growth in Number of United States Practicing Attorneys, 1950–2006

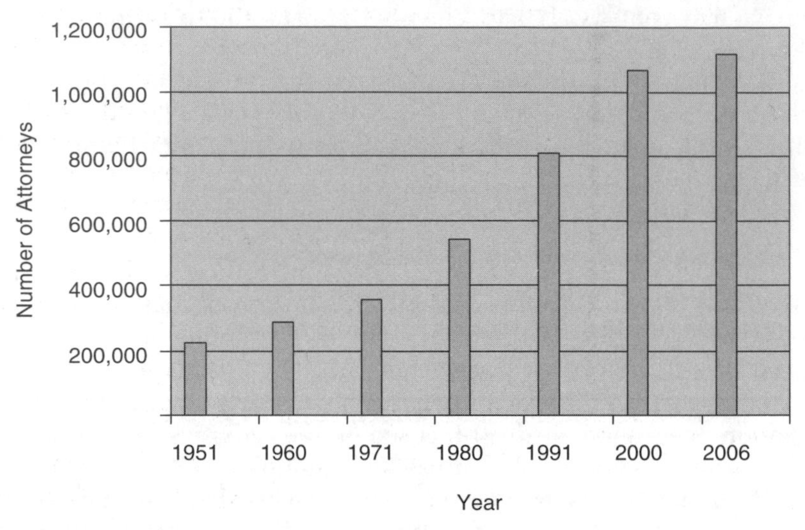

Source: American Bar Foundation 2007

1 for every 696 individuals. The number doubled to about 213,000 by 1950, while the lawyer-to-population ratio remained about the same. It reached 286,000 in 1960 and 355,000 in 1970, and then soared to 542,000 in 1980, with the lawyer-to-population ratio rising to 1 for every 418 individuals. The number of lawyers continued to rise dramatically, reaching more than 800,000 lawyers in 1991 and more than 1 million in 2000. By 2006, the number of attorneys had risen to more than 1.1 million, a ratio of about 1 for every 264 individuals (see Figure 8.1).

Thus, two related trends are notable during the past half century. First, the number of attorneys in the United States has almost quadrupled since 1960. Second, this increase is much greater than the population increase during this time period, as the lawyer-to-population ratio rose from 1 lawyer for every 627 individuals in the 1960 national population to 1 for every 264 in 2006.

Interestingly, the United State has more lawyers per capita than other nations. A few years ago, when the U.S. ratio was 1 lawyer for every 320 Americans, the corresponding rates for other nations were: England, 1 for every 694; France, 1 for every 2,461; Japan, 1 for every 8,195 (Stonehill 2007). We have seen in previous chapters that law plays a greater role in interpersonal and national affairs in the United States than in most other nations, and international comparisons of the number of lawyers per capita support this conclusion.

Certain other numbers yield a demographic profile of the bar (American Bar Foundation 2007). In particular, they suggest that although the legal profession is no longer the exclusive "white men's club" that it was in the nineteenth century, it still does not reflect the diversity of the American citizenry. In 2006, women comprised 30.2 percent of all lawyers, a percentage that is obviously much lower than their approximately 50 percent share of the population. Only 4.2 percent and 3.4 percent of U.S. attorneys in 2000 were African American and Latino, respectively, even though both these groups comprise about 12–13 percent of the U.S. population. Law school enrollments, as we shall see, exhibit in some respects greater diversity than the bar itself. In addition, the lack of racial/ethnic diversity in the legal profession may reflect social inequality in the larger society and not just any continuing discrimination in the bar. Although the bar has obviously changed, it is still evident that the U.S. legal profession has not yet escaped its historical legacy of exclusion.

The Stratification and Social Organization of the Bar

The U.S. legal profession is *stratified*: some attorneys earn much more money than others and enjoy more respect and influence than others. These disparities depend largely on the kind of work they do and the settings in which they perform their work. To help understand the **stratification** of the American bar, we first present some income statistics and then turn to the nature and venues of legal work in the United States.

Although the legal profession by and large is a wealthy one, income data gathered by the U.S. Department of Labor show that some lawyers are especially wealthy while some attorneys are relatively unwealthy, at least by legal profession standards (Bureau of Labor Statistics 2007). In 2006, the median annual income (half earning above this and half earning below it) of attorneys was $102,470 (equivalent to more than $105,000 in 2008 dollars), almost 3 times higher than the median income for the overall population. One-fourth of all attorneys earned more than $145,600, while one-fourth earned less than $69,910, with the remaining 50 percent of attorneys obviously earning between these two figures.

These income differences reflect the type of work attorneys do and the types of settings in which they practice law. About three-fourths of all lawyers are in private practice (see Figure 8.2), while 8 percent each work in government (federal, state, or local) or for industry, with most of these "industry" attorneys working as "house counsel" for corporations. Another 3 percent work as judges, while the remainder ("other" in Figure 8.2) either teach law, work in legal aid or as public defenders, or are retired. The many attorneys in private practice are commonly distinguished according to the size of their practice (see Table 8.1). Almost half of private practice attorneys work by themselves as solo attorneys, while the remainder work in law firms ranging in size from as few as 2 attorneys to more than 100 attorneys.

The attorneys in the large law firms, whose clients are typically corporations, and those who work as house counsel for corporations occupy the pinnacle of the legal profession (Heinz et al. 2005; Rostain 2004). These

FIGURE 8.2 Settings for Legal Practice, 2000 (percent of all attorneys in each setting)

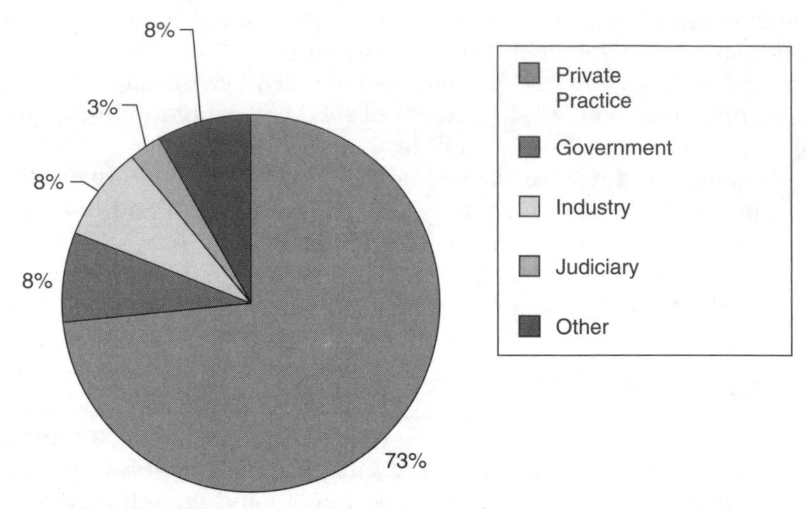

Source: American Bar Foundation 2007

corporate attorneys earn the highest incomes, with many partners of the large firms earning more than $1 million yearly, and they occupy key positions in the American Bar Association and other important legal bodies. They typically come from wealthy families of origin and have attended the elite national law schools at Harvard, Yale, Columbia, Stanford, and other such universities. In contrast, solo attorneys and those who work in small firms (fewer then ten lawyers) represent clients in criminal cases, divorces, real estate transactions, personal injury cases, and wills and other estate planning. As **personal services attorneys**, they enjoy much less wealth and

TABLE 8.1 Percentage of Private Practice Attorneys Employed by Size (number of attorneys) of Practice, 2000

Size of Practice (# of attorneys in the practice)	Percent of Attorneys Employed in This Practice
1 (solo practice) lawyer	48
2–5 lawyers	15
6–10 lawyers	7
11–20 lawyers	6
21–50 lawyers	6
51–100 lawyers	4
101+ lawyers	14

Source: American Bar Foundation 2007

status than their large firm counterparts, come from less wealthy families of origin, and much less often have degrees from the elite national law schools. As a recent review characterizes this stratification of the bar, "Research has established a dramatic divide in the American private bar between lawyers who represent corporations, and who enjoy greater economic rewards and social standing, and those who represent individuals, who are the lower end of the economic and social hierarchy of the bar" (Rostain 2004:147–48).

Corporate and personal services attorneys thus occupy two "hemispheres" of the bar (Heinz and Laumann 1982), and they differ in other ways as well. In particular, corporate attorneys spend little time in the courtroom and instead practice their trade in their offices and conference rooms. Personal services attorneys spend somewhat more of their time (but by no means most of it) in the courtroom. In another difference, corporate lawyers are necessarily involved with law that is very complex; in contrast, the law practiced by personal services attorneys is much less complex and rather routine (Rostain 2004).

Important differences also exist within the personal services segment of the bar (Mather, McEwen, and Maiman 2001; Seron 1996). In particular, some personal services attorneys are generalists, in that they handle all the matters—divorces, real estate transactions, personal injury cases, and so forth—that were listed just above. In this respect, they parallel physicians who are general practitioners or the classic "family doctor." Other personal services attorneys are more specialized. For example, some primarily handle divorce cases, while others handle personal injury cases. These attorneys obviously parallel physicians who have a specialty like cardiology or radiology. Specialist personal services attorneys are generally more knowledgeable than generalist attorneys of the areas in which they specialize and often have higher incomes than their generalist counterparts.

Whether they are generalists or specialists, many personal services attorneys see themselves as problem solvers who talk extensively with their clients and actually listen to what they have to say (Rostain 2004). They view their role as one that not only offers clients legal advice and assistance but also helps them when appropriate with other aspects of their lives. At the same time, they typically have to balance their clients' legal problems and personal needs with their own sense of what is practical to achieve through the law. This balance is particularly difficult to achieve in divorce cases, which often are very emotional for both spouses. Divorce lawyers thus often find themselves having to urge their clients to be less emotional and more reasonable in their expectations of what they will get in the eventual divorce settlement (Mather, McEwen, and Maiman 2001; Sarat and Felstiner 1995). Personal injury lawyers similarly have to persuade their clients to lower their expectations for financial damages to a more reasonable level (Baker 2001).

The personal services attorneys just described are called *traditional* attorneys because they emphasize getting to know their clients, talking with them, and treating them as individuals. Other personal service attorneys operate in a more *entrepreneurial* fashion that emphasizes the provision of legal services to many clients in the least expensive and most efficient manner possible (Seron 1996). Many work in small- or medium-sized law firms, including franchise firms that are like chain stores in that they have many offices within a region or even in different states (Van Hoy 1997). Entrepreneurial attorneys take a business approach to their practice of law. According to Tanina Rostain (2004:159), they place "great emphasis on marketing and on organizing their practices so that they could handle a large volume of similar cases. To market their services, they pioneered the use of large targeted advertising campaigns in electronic and print media and organized their practices around selling prepackaged 'one size fits all' legal services."

Whatever type of law practiced, the stratification of the legal profession into large firms, medium and small firms, and one- or two-person firms (to use a simple ranking) represents a notable change since the professionalization of the American bar began in the late nineteenth century, when almost all attorneys worked by themselves or at most with only one other partner, and large firms were unknown (Heinz et al. 2005). In 1872, for example, only four law firms in the entire nation had at least five attorneys. The growing importance of corporations noted earlier and corresponding rise of the large corporate law firm slowly but surely changed the social organization of the bar. Today, about 18 percent of U.S. lawyers work in firms with more than fifty attorneys (see Table 8.1).

John P. Heinz and colleagues (2005) argue that the American bar is now somewhat less unified and coherent than it had been in the 1970s. Growing gender, racial/ethnic, and religious diversity in the bar means that attorneys no longer all come from the same social backgrounds and congregate in the same social networks. Because lawyers have become more specialized in the work they do, they are also less likely to share the same understanding of what practicing law is all about. Since the 1970s, the large law firms have become even larger, and many are now not just local firms (situated in only one city) but rather have branches in several cities and even in other nations. Further, the great growth of the bar since the mid-1900s has weakened social bonds among the many attorneys now in practice, just as people in a large city are less apt to know each other than those in a small village. As Heinz et al. (2005:8) put it,

> [A]s the numbers grow, the probability of chance transactions between any given pair or any given sets of lawyers decreases. Since individual lawyers' circles of acquaintance are unlikely to expend to the same extent as the growth of the bar, there will be an increasing number of their fellow lawyers with whom they have no ties. Thus, communication is likely to be restricted to more narrow slices of the whole. The bar, therefore, has become more diverse and less well integrated.

The international "branching out" of U.S. law firms has led to some interesting problems because of several differences between American lawyers and those overseas (Tagliabue 2007). In recent years, an increase in corporate mergers and acquisitions in Europe has prompted an increase in the demand for corporate law. To meet this demand, U.S. firms have increasingly either bought or merged with European firms or established less formal working relationships with them. This development has caused tension for several reasons. First, many American lawyers are unfamiliar with continental Europe's civil law tradition and corresponding legal procedure. Second, many also lack fluency in French, German, and the other native languages of the continent. Third, and reflecting a larger cultural difference between the United States and Europe, American lawyers and their legal practice tend to be more aggressive than their European counterparts. Citing an example of this latter difference, a recent news report commented, "French law practice has been by tradition modest and discreet. Even some of the larger firms . . . have no Web site, considering it unprofessional" (Tagliabue 2007:C1).

Cause Lawyering

Private practice attorneys, whether they rank at the pinnacle of the legal profession in corporate law firms or at the bottom in one-person firms, work for clients with various legal problems and needs. They offer their legal assistance in a neutral manner, do not necessarily agree with any aims the client may have, and certainly are not trying to change society through the legal help they render. In contrast, a small segment of the bar practices law with the direct goal of improving society. Because they are working for a principle or social change goal or for a cause like civil rights, environmental protection, gay rights, or women's rights, they are called **cause lawyers**, and the law they practice is called **cause lawyering**. Perhaps because the work these lawyers do is so uncommon in the field of law but also very important, cause lawyering has received much attention during the last decade or two (Heinz, Paik, and Southworth 2003; Sarat and Scheingold 2006; Scheingold and Sarat 2004).

Cause lawyers, of course, existed long before social scientists began writing about them. Clarence Darrow, mentioned earlier, began his legal career working for a railroad corporation but soon switched to the practice of labor law and represented clients, including notable socialist and labor leader Eugene V. Debs, who had been arrested for going on strike and engaging in other labor protest (Tierney 1979). Darrow later represented teacher John Scopes in the celebrated 1925 "monkey trial" that challenged a Tennessee law prohibiting the teaching of evolution.

Darrow handled the case on behalf of the American Civil Liberties Union (ACLU), which was founded in 1920 during an era when the federal government was suppressing the civil liberties of socialists and labor radicals (Finan 2007). Today, the ACLU is probably the best known association of cause lawyers. Its Web site (www.aclu.org) says that its mission is to "preserve" several "protections and guarantees," including: (a) the several freedoms listed in the First Amendment; (b) equal treatment under the law without regard to race, sex, religion, and national origin; (c) due process; and (d) the right to privacy. Another national organization of cause lawyers is the National Lawyers Guild, which was founded in 1937 and, according to its Web site (www.nlg.org), is "dedicated to the need for basic change in the structure of our political and economic system." Since its founding, says the Web site, the Guild "has been an important part of the American people's struggle for real democracy, for economic and social justice, and against oppression and discrimination based on race, ethnicity, immigration status, class, gender or sexual orientation."

Dozens of other cause lawyers and cause lawyer groups in the past and present could also be cited, but the heyday of cause lawyering was no doubt the period of the 1960s and 1970s, when attorneys in the South represented civil rights activists and attorneys elsewhere represented Vietnam antiwar activists, various "new left" causes, and black power advocates (Black 1971; Ginger 1972; James 1973). The many other social movements that began during this time, including the environmental movement and gay and women's rights movements, all involved attorneys whose help in civil and criminal cases alike was invaluable.

Many of these cause lawyers and others before and since have represented unpopular clients and unpopular causes, often at great expense, both personally and financially. Their clients often have little money to pay them, and taking on their cases may mean that the lawyers must decline other, more lucrative cases. Even more ominously, cause lawyers have sometimes risked their practice and even their personal safety, at least in the 1960s and 1970s and before. Lawyers back then who took on unpopular clients and causes risked the loss of more conventional clients, who might not wish to be associated with such attorneys. Some lawyers also became targets of violent reprisal by individuals who disagreed with the work the lawyers were doing.

These problems were paramount in the South during the early 1960s, when thousands of people were arrested during civil rights protests and for other activities related to their involvement in the civil rights movement. Rejecting their goal of desegregation or fearing for their practice, few attorneys were willing to represent civil rights clients. Those who did suffered various costs. As one study of this issue summarized what happened, they "faced threats of contempt and disbarment in court, and possibilities of physical attack, loss of business, and social ostracism outside of court" (Barkan 1984:555).

One of the few African American attorneys in the South during this time was Chevenne Bowers (C.B.) King, who began his practice in Albany, Georgia, in 1954 at the dawn of the civil rights movement and later defended countless numbers of civil rights protesters. In the early 1960s, he went to a local jail to see a civil rights activist who was beaten by his white cellmates after he was arrested. The local sheriff told him to leave and, when King declined to do so, shouted at him, "You goddamn Black sonofabitch, you still here?" The sheriff then smashed King's forehead and neck with a walking cane (James 1973:296). One summer, King had a white law student working with him. When the two went to a county jail to see some clients, the sheriff said within the student's earshot, "Lower than a goddamn dog helping that n——, C.B." When the student asked whether that was for the record, the sheriff replied, "You goddamn right. C.B. tell that goddamn white boy I'll rip his ass aloose" (James 1973:298).

Cause lawyers today in the United States do not risk their safety or life as C.B. King and other lawyers in the South once did. But they still sacrifice the earnings they could realize from conventional legal practice because they want to use the law to achieve what they regard as noble goals. Cause lawyers who work for conservative causes may fare better financially overall than their liberal and radical counterparts thanks to funding from conservative foundations and think tanks, business organizations, and other sources, but salaries and resources still vary widely among such lawyers (Heinz, Paik, and Southworth 2003). Whatever their political leanings, cause lawyers and the work they do remind us that law does not exist in a vacuum and instead is often used for political ends. Cause lawyering thus illustrates one of the many connections between law and society that are highlighted in the social scientific understanding of law.

Women in the Law

Recall (from Chapters 2 and 7) the U.S. Supreme Court ruling in 1873 (*Bradwell v. Illinois* [83 U.S. 130]) that women were not entitled to practice law. Arguing that women were biologically unsuited to practice law, a concurring opinion said in part, "The natural and proper timidity and delicacy which belongs to the female sex evidently unfits it for many of the occupations of civil life. . . . The paramount destiny and mission of woman are to fulfill the noble and benign offices of wife and mother. This is the law of the Creator."

These words will sound ludicrous to many and probably most readers, but they reflected widespread expert and public opinion when they were written. Women have obviously come a long way in the many years since, both in the law and in other spheres of life, but they still have a long way to go before achieving full equality. Women comprised about 8 percent of all American attorneys in 1980 and, as we saw earlier, about 30 percent in 2006. Although the latter figure represents a significant increase during the last quarter-century, it still lags behind women's 50-percent share of the population. The numbers look more favorable if we focus just on law school enrollments. As Figure 8.3 illustrates, women's law school enrollment soared during the 1970s and rose steadily through the early part of this decade before a recent slight decline.

FIGURE 8.3 Women's Percentage of Total J.D. Enrollment, Academic Year 1951–1952 to 2006–2007

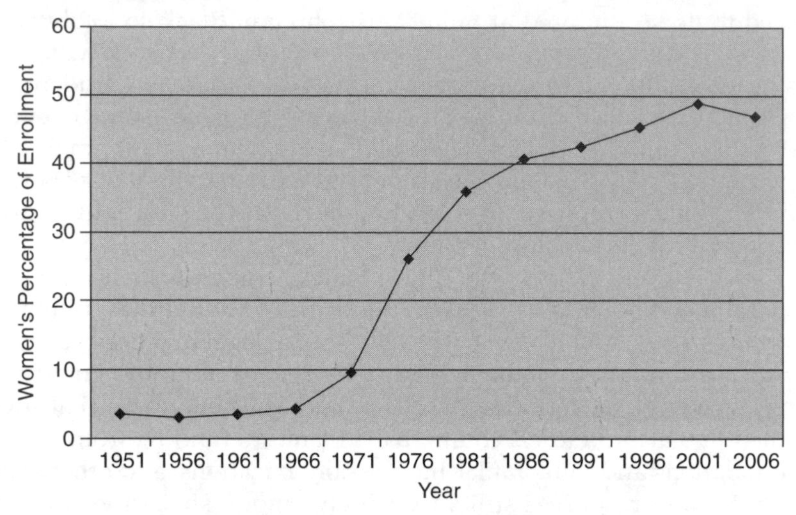

Source: http://www.abanet.org/legaled/statistics/charts/stats%20-%206.pdf

Women comprised only 3.5 percent of law school students in 1951–1952, but today almost half (46.9 percent) of law school students are women. These figures suggest that the female percentage of attorneys should also continue to rise and may one day match the male percentage.

These numbers provide a somewhat encouraging picture of women in law, but they provide only part of the picture. Although the number of women law students and lawyers has been growing, the upper echelon of the bar in terms of wealth and prestige remains very much a male domain, as a "glass ceiling" continues to exist (Rhode 2001; Rostain 2004). Compared to men, women are more likely to be solo attorneys or in small firms and less likely to be in the large firms, and they are paid less than men even when size of firm, family situation, work hours, and other relevant factors are taken into account (Bernat 2006; Chambliss and Uggen 2000; Heinz et al. 2005). The reasons for this disparity are complex, and scholars do not agree on which reason(s) make the most sense.

The dispute over these reasons reflects disputes about the reasons for the lack of women in other traditionally male-dominated professions. Some scholars say that the lack of women in law's upper echelon (as in other traditionally male professions) reflects the choices women make in and after law school (Carroll and Brayfield 2007). Because they tend to be more liberal than men and because they tend to be more family oriented, they tend not to apply for corporate law jobs (Abel 1989); for similar reasons, women who do end up in large firms tend to leave these jobs more often than their male counterparts.

While conceding this "choice" argument may have some merit, other scholars instead attribute the lack of women in law's upper echelon to patterns of discrimination and exclusion (Bernat 2006; Rhode 2001). One major problem is a hostile and discriminatory work environment. Demeaning comments and sexual harassment continue to exist in law firms as they do in other workplaces, and women are less likely than men to be assigned lucrative and prestigious corporate work instead of more pedestrian work on estates, wills, and other matters (Bernat 2006). These problems were arguably worse during the 1970s than they are today (Epstein 1981), but to the extent a hostile, discriminatory work environment does exist for these reasons, it limits women's entry and advancement in the legal profession. In a related problem, recruitment and advancement in the legal profession still occurs through an "old boys' network" that excludes women. The large law firms also expect very long workdays and work weeks for their new attorneys and do not provide the type of more flexible schedule that women often prefer. As one scholar observes, "For women, the extraordinary time demands of elite practice are inconsistent with bearing primary child care responsibilities—a role they are still by and large expected to assume" (Rostain 2004:155). Regardless of the exact reasons for women's inequality in the legal profession, it does not appear that the favorable numbers of women law students will soon translate into full equality in the legal profession itself once the law degree is achieved.

Legal Ethics and Corruption

We noted at the outset of this chapter that lawyers sometimes tread a thin line between the most aggressive representation of their clients and what is legally permissible and/or morally defensible. To defend a client, should a criminal defense lawyer point to evidence that someone else did the crime even though the lawyer does not believe that this other person committed the offense? This was a common tactic on the 1997–2004 ABC show *The Practice,* whose lawyers labeled the tactic "Plan B." For better or worse, the tactic often worked, as the defense lawyers were able to raise enough doubt about their clients' guilt that the jury would return a not guilty verdict. This is a fictitious example, but real-life attorneys often must decide how aggressive their legal representation should be vis-à-vis moral standards and the rules of legal procedure.

Recall that the ABA's 1908 Canons of Professional Ethics condemned certain practices as unethical, including advertising and "ambulance chasing," and that these practices were, and still are, typically those of the solo or small-firm legal services attorneys who take on working-class clients and rank at the bottom of the bar's hierarchy. Reflecting the elite, corporate backgrounds of the ABA membership at the time, the Canons were silent on other ethical issues that might have been raised, such as whether attorneys should defend corporations that knowingly kept dangerous workplaces that harmed their workers' health and even killed them (Auerbach 1978). The Canons also established other principles for legal practice that still apply today to lawyers throughout the bar; these principles include an inviolable commitment to lawyer–client confidentiality, a ban on the knowing use of perjured testimony, and a ban on knowingly helping a client commit fraud.

Several decades ago, sociolegal scholars began studying the extent and nature of ethical violations and financial corruption (typically the failure to render legal services after a fee is paid) by the legal profession (Carlin 1962, 1966; Davis and Elliston 1986; Handler 1967; Shaffer and Shaffer 1991; Zitrin and Langford 1999). Reflecting the Canons' emphasis, these studies typically focus on the behavior of solo and small-firm legal services attorneys, who are asked to indicate in interviews or in *self-report surveys* certain ethical violations they may have committed. This line of research finds a rate of ethical violation that is thought to be much higher than in the world of corporate attorneys (Smigel 1969), but also a pattern of weak enforcement and punishment by the bar. Drawing on his classic studies, Jerome Carlin (1962, 1966) estimated that only about 2 percent of unethical attorneys are investigated, and that fewer than 0.2 percent of unethical attorneys are penalized.

One problem in dealing with attorney misconduct is that most ethical violations remain unknown unless a client complains to the police or appropriate bar association. However, many clients are either unaware of lawyer's malpractice or unfamiliar with the ethical standards for legal practice (Abel 1989). Moreover, many clients who suspect malpractice may not know where to report it or may simply not want to endure the personal difficulties that their complaint might initiate. Some clients also do not complain about malpractice with which they are aware because the malpractice benefits their case. Lawyers who know of other lawyers' ethical misconduct often decline to report it (Abel 1989). When complaints of ethical violations do reach the appropriate bar association, many do not get investigated. Of those that do get investigated, very few lead to disbarment, the most serious professional penalty a bar association may render. As Richard L. Abel (1989:145) characterizes this problem. "Starting from a population of complaints that already overlooks most misconduct, the disciplinary process then displays extraordinary lenience."

Professions typically rely on self-regulation to enforce codes of behavior, but research on the professions shows that self-regulation is often lax and ineffective, akin to the proverbial fox guarding the chicken coop (Abel 1989). Critics say that professional ethical codes serve mainly to help a profession to put on a good public face. As Abel (1989:143) explains this concern in his highly regarded study of the American bar, "The suspicion that professional associations promulgate ethical rules more to legitimate themselves in the eyes of the public than to engage in effective regulation is strengthened by the inadequacy of enforcement mechanisms." For better or worse, the lax regulation by the legal profession reflects a problem found in the professions more generally.

Compounding the ethical problems of the legal profession, several studies find that lawyers simply do not know some of the rules governing their behavior. To illustrate this ignorance, Abel (1989:143) discussed the example of fee-splitting, which involves a lawyer compensating another lawyer for referring a client. Although most states prohibit this practice, a survey of 600 attorneys found that two-thirds either thought that fee-splitting was allowed or did not know one way or the other. Other surveys of attorneys, Abel (1989) reported, similarly find them unfamiliar with some ethical standards and/or quite willing to violate them even if they are familiar with them. Abel (1989:156) concluded twenty years ago, "Despite periodic scandals and persistent public criticism, there is little evidence that the legal profession has engaged in more effective regulation of misconduct or incompetence in recent years." Two decades later, many scholars would say that conclusion still applies.

Satisfaction with Legal Careers

How satisfied are lawyers with their careers? What affects their level of satisfaction? These questions have been addressed by a number of studies on *lawyer satisfaction*. These studies ask lawyers how satisfied they are with aspects of their jobs and careers, including the work they do, the salaries they enjoy, and their relationships with their colleagues. Several findings emerge from these studies (Dinovitzer and Garth 2007; Hagan and Kay 2007; Heinz et al. 2005).

First, lawyers as a whole are generally highly satisfied with their jobs and careers. Research on people in other occupations also finds a relatively high degree of job satisfaction (Firebaugh and Harley 1995), so lawyers are little different in this regard.

Second, the degree of overall satisfaction is about the same for women and men lawyers and for white lawyers and for those of color. Put another way, gender and race do not appear to affect how satisfied lawyers are with their jobs and careers and with their decision to enter law. Because, as we have seen, female and minority lawyers experience workplace problems and rank in the lower echelon of the bar, the fact that they do not have lower levels of satisfaction has been considered something of a paradox. Investigation of this paradox among women lawyers finds that although their overall satisfaction with their jobs is the same as men's, they are somewhat less satisfied than men with specific aspects of their jobs, including their level of responsibility, recognition for their work, and chances for advancement (Hull 1999). Research also shows that even though women lawyers report high overall satisfaction, they are also "substantially more likely to report feelings of depression or despondency about their lives" (Hagan and Kay 2007:68–69). These feelings stem partly from women's concern that having children will impede their career path in the legal profession. They internalize this concern into depression rather than let it affect satisfaction with the legal work they do.

Third, lawyers with higher incomes are more satisfied with their jobs than those with lower incomes, but lawyers in large law firms are less satisfied with their jobs than those in small firms or solo practice. In a related finding, lawyers working in government or in public interest law are more satisfied than those in private practice. In general, "private practice in large firm settings is the least satisfying type of practice, even though it is the most lucrative" (Dinovitzer and Garth 2007:7). The high stress and time-consuming nature of large firm careers appear to explain this apparent paradox.

Fourth, graduates of elite law schools exhibit less satisfaction with their decision to enter law than graduates of low-tier law schools, even when only lawyers working in large law firms are considered (Dinovitzer and Garth 2007). To turn that around, graduates of low-tier schools are happier with their decision to enter law than are graduates of elite schools. Because the former tend to come from lower socioeconomic backgrounds, they may see law as an especially attractive career path and thus become more satisfied that they entered it.

LAW SCHOOL AND LEGAL EDUCATION

Law school today is the gateway to the practice of law, but this was not always the case. As our historical overview of the legal profession indicated, before the twentieth century most people who began legal practice did so without formal legal schooling. The growing importance of the law school toward the end of the nineteenth century and especially during the first half of the twentieth century was an intrinsic part of the growing professionalization of the bar discussed earlier. This section discusses the history of legal education further and reviews critiques and defenses of the education that law students receive today.

The Growth of the American Law School

We saw earlier that the growth and growing importance of law schools beginning in the late nineteenth century increased the quality of the bar, but that it also reduced the supply of attorneys and thus made legal practice more lucrative for the reduced numbers that were able to enter and graduate from law school. Because these reduced numbers were largely white Protestant males from relatively wealthy backgrounds, the increasing prominence of law school for entry into the legal profession contributed to the bar's historic exclusion of people who did not fit this sociodemographic profile. As we shall soon see, other developments in growth of law schools reinforced this exclusion (Abel 1989; Auerbach 1978; Friedman 1985).

The first U.S. law school was established in Litchfield, Connecticut, in 1784. Students heard lectures about commercial law for 1½ hours per day, with an exam every Sunday, over a span of fourteen months (with a two-month vacation) before getting their degree. The Litchfield school dissolved in 1833; despite its appearance during the first decade of the new nation, neither its example nor model of legal education caught on, and apprenticeship or self-education remained the normal avenue for entry into the legal profession.

A key development in the history of the law school and thus of the legal profession itself was the establishment in 1829 of a law school at Harvard University. This was the first American law school at a university; although there are many independent law schools today, most law schools, and certainly the most prominent law schools, are part of universities. Harvard's example soon led to the rise of other law schools, most at universities: there were 15 law schools in 1850, 21 in 1860, 31 in 1870, 51 in 1880, 61 in 1890, and 102 in 1900. The number of law students in the nation rose concomitantly, from fewer than 900 in 1850, to 1,600 in 1870, and then 7,600 by the mid 1890s (Freidman 1985).

During this time, a law school degree required only one or, increasingly, two years of courses; students typically did not need a college degree to be admitted into law school; and most students were taught by part-time teachers whose main occupation was as a practicing lawyer or judge. This model began to change after the appointment of Christopher C. Langdell as dean of Harvard Law School in 1870. Langdell instituted many changes at Harvard that soon became the model at law schools throughout the country and remain the model today. He required either a college degree or passing grade on an entrance exam for admission into Harvard; he increased the tenure for a law degree from 1 to 2 years in 1871 and then to 3 years in 1876; he hired full-time law professors; and he established a required sequence of courses and mandated final exams.

Most importantly, Langdell introduced the **case study method** (also just called the *case method*) of teaching law. Before this development, most law students learned law by listening to lectures and by reading textbooks that discussed law. The case study method introduced by Langdell relied on the reading of casebooks, collections of actual cases and court opinions designed to illustrate various legal principles. With the case study method came the **Socratic method** of teaching law. Rather than lecture on law, professors engaged students in a vigorous question-and-answer process about the cases (with which students were expected to be very familiar by reading their casebooks) that was meant to help them recognize the legal principles at stake. Langdell considered this a more scientific way of teaching law, one that emphasized law as logic and separate from society and thus the *traditional view of law* discussed in Chapter 1. For better or worse, his new model "purged from the curriculum whatever touched directly on economic and political questions" (Friedman 1985:617).

The twin case study–Socratic method was initially unpopular after Langdell introduced it; students cut his classes and enrollment declined. As a history of Harvard Law school recounted, "To most of the students, as well as to Langdell's colleagues, it was abomination" (quoted in Friedman 1985:615). Within a few years later, however, this new model of legal education became the norm at Harvard and gradually became the norm at other law schools. Today, of course, it is the standard way of teaching and learning law throughout the nation.

As law schools grew in number from the late nineteenth century well into the next century, a two-tier system of law schools materialized.

Although many law schools were at universities, independent law schools also emerged that catered to people with day jobs who thus took their classes at night. As Jerold S. Auerbach (1978:74) summarized this development, "A law-school hierarchy emerged in which Harvard Law School and its emulators trained aspirants to the professional elite, while night law schools prepared members of ethnic minority groups for careers in business, politics, and in the professional underclass," and, we should add, for careers as personal services attorneys in the lower echelon of the bar. Because the university law schools were largely limited to white male Protestants from wealthy backgrounds, poorer individuals and Catholic and Jews were forced to attend the night law schools. In this manner, the growth of the American law school contributed to the hierarchy of the bar and to its historic exclusion of people who were not wealthy white male Protestants.

From the late nineteenth century into the twentieth century, states increasingly began to require a law school degree for admission to the bar (Abel 1989). In 1879, only about one-third of the states required any formal legal study, but by 1904 this proportion had grown to two-thirds and by 1928 to more than 80 percent. In 1935, only 9 states required a degree from an accredited law school; in 1984, 48 states had such a requirement. In 1949, only 62 percent of practicing lawyers in the United States had a law school degree; by 1970, 93 percent had a law school degree.

Law school enrollments reflect the growing importance of law school degree (see Figure 8.4). The most noticeable growth in law school enrollment came during the 1960s and 1970s, with several reasons accounting for this growth. First, the post–World War II baby boom came of age, and the sheer numbers of young people drove up not only college admissions but also law

FIGURE 8.4 Law School Enrollments, 1900–2006

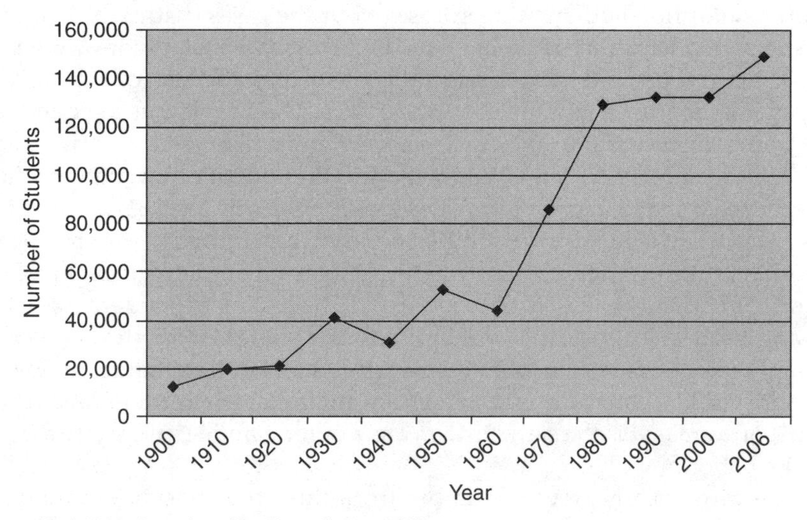

Sources: Abel 1989; American Bar Foundation 2007

school admissions. Second, the legal changes that helped end racial segregation and other cause lawyering developments during this time (see earlier discussion) inspired many young people to look to a legal career so that they, too, might be able to "make a difference." Third, there was probably increasing recognition that legal careers can be very lucrative, and the rising salaries of Wall Street and other corporate attorneys a generation ago no doubt contributed to greater interest in a legal career. Fourth, and as we have seen, women began to attend law school in increasing numbers, and their growing presence in law schools accounted for much the growth in law school enrollments during the period in question.

Critique and Defense of Legal Education Today

As noted above, the case study and Socratic method dominates law school teaching today. This method has been criticized since Langdell introduced it in 1870, and criticism mounted during the 1960s and 1970s as a new generation of more left-leaning lawyers and sociolegal scholars arose. Three major criticisms exist (Abel 1989; Auerbach 1978).

First, legal education is said to rely so heavily on the traditional view of law that it shuns social science and neglects important considerations of race, class, gender, and other social variables (Ansley 1991; Kassam 2003). Michael Lowy and Craig Haney (1980), an anthropology and psychology professor, respectively, went to Stanford Law School, where they attained a law degree and observed how law was taught. Given their background in the social sciences, they were alarmed by what they saw: "As social scientists we were struck by the extent to which law and law school was consciously, explicitly, and vehemently anti-social science" (p. 36). They continued, "Many law professors savagely attack and purge alternative explanations of human behavior. Control over students, many of whom were social science majors in college, is established quickly." Students became passive as a result and did not question the nonlegal aspects and implications of what they were taught.

As an example of the exclusion of social science considerations, Lowy and Haney discussed the classroom treatment of a New York case involving a woman who sued the police for not protecting her after she sought their protection from an ex-boyfriend who had threatened her. The boyfriend eventually hired someone to throw lye in the woman's face, an act that damaged her vision permanently. The court dismissed the lawsuit because it said it did not have the authority to tell the police how they should spend their time. Class discussion guided by the professor developed the legal principle at stake, that some questions lie beyond a court's purview. Lowy and Haney noted that this discussion omitted real-life questions such as: How many similar threats occur? How busy were the police in their other activities? Were some other people being protected by the police? In short, the case was discussed in a social and political vacuum. In another example, they wrote, a criminal law professor mentioned the disproportionate arrest of minority males for felonies without mentioning the possibility of racial bias in such arrests and the heavy involvement of whites in white-collar crime. Although a generation has passed since Lowy and Haney went to law school, it is fair to say that social science considerations are still lacking in the law school curriculum.

A second and related criticism of legal education is that it socializes students to be more politically conservative than when they began law school and, more specifically, orients them to the needs of corporations and other established interests rather than to the needs of the poor and people of color (Erlanger and Klegon 1978; Granfield 1992; Kennedy 2004). According to this critique, the required curriculum in most law schools is much more oriented to the needs of the wealthy than to the needs of the poor in regard to the specific courses that are required and the examples used in these courses. David N. Rockwell (1971) referred to this problem as "the education of the capitalist lawyer." Just as the legal profession is heavily oriented toward the interests of corporations and other "haves" in society, Rockwell wrote, so is legal education oriented toward these interests. As a result, law schools have a bias "toward the solution of corporate and property problems before human problems" (p. 95). By focusing on court opinions, casebooks omit questions of racial prejudice, sexist processes, and other real-life matters.

A third and again related criticism of legal education is that it is brutal for students and particularly alienating for women (Garner 2000). The grueling nature of legal education is vividly portrayed in books like *One L*, Scott Turow's (1977) account of the first year at Harvard Law School, and the 1973 film *The Paper Chase*. In both these works, some students suffered tremendously from the strain of their studies even to the point of suicide. To the extent that legal education is brutal in real life, the casebook study and Socratic methods are largely responsible. From the outset, students are required to read legal opinions in casebooks but typically have much trouble understanding them despite having excelled in college. Using the Socratic method, some professors humiliate students who do not understand the cases adequately (Lowy and Haney 1980). Although such humiliation has arguably lessened or disappeared in the years since *One L* was written and *The Paper Chase* produced, students still spend hour upon hour reading relatively few pages of court opinions and must demonstrate their knowledge in class under intense questioning by their professors. Some evidence even indicates that law students suffer more than their fair share of psychological distress as a result (Benjamin et al. 1986).

Law school faculty and other observers defend legal education from all these criticisms. They say that law must be taught as logic and in a value-free manner because that is what law is all about. They concede the orientation of law school courses and topics toward established interests, but they say that most law students desire corporate law careers where this sort of knowledge will be essential; empirical studies find that law students do in fact want such careers (McGill 2006). Legal education defenders also concede that law school is tough, but they defend the case study and Socratic methods as necessary because students must learn to be quick on their feet and to "think like a lawyer."

Supporting this latter point, Carol E. Rinzler (1979) wrote of her experience in Yale Law School, which she entered at age 35 after several years working as a book editor. She recounted that for the first day of classes she had to read 70 pages in casebooks, a task she thought would be easy since as an editor she often read an entire book manuscript in an evening. It took her twelve hours to finish the first 40 pages in the casebooks, and she could not understand what she had read. "For the first several weeks," she wrote, "I was convinced there had been a summer orientation session to which they did not invite me" (p. 102). Although she found life in law school grueling, she wrote that she was "unqualifiedly" glad she went to law school. She added, "In fact, every intelligent person should spend at least a term in law school. I think I think a lot better than I did when I went in. Learning to 'think like a lawyer,' means learning to go straight to the heart of a matter, while simultaneously seeing 50 possibly issues on both sides. It's sort of like sending your mind to the gym" (p. 102).

What should we make of the debate over legal education? Which side is correct? There is no certain answer to this latter question, and our ideological leanings no doubt affect which side we might favor in the debate. It does appear that law students become more conservative during their time in law school and more oriented toward corporate careers and less interested in public interest careers (Erlanger and Klegon 1978; Granfield 1992). Their education in law school is also truly demanding. Both sides to the debate therefore make valid points, but, if a social scientific understanding of law is important, as this book obviously assumes, it is unfortunate that this understanding is largely lacking in the law school curriculum today. Because law school graduates do not practice law in a social and political vacuum, a greater understanding of the social scientific aspects of law would enhance both their practice and their public citizenship.

Summary

1. Lawyers have a negative image for several reasons. First, they often represent people involved in heated disputes. Second, they are thought to complicate matters with their penchant for argumentation and turgid prose. Third, they defend criminals and other unsavory clients. Fourth, they sometimes use legal stratagems as they vigorously represent their clients to the best degree possible. Fifth, when they represent people who have suffered personal injuries, they are seen as "ambulance chasers" or as vultures who greedily take advantage of their clients' misfortune for their own personal gain. Finally, many lawyers work for corporations and other wealthy clients and are disliked for this reason by "have-nots."

2. Lawyers also have a positive image as heroes for the innocent or oppressed. Fictitious examples include Perry Mason and Atticus Finch in *To Kill a Mockingbird*. A real-life example includes famed attorney Clarence Darrow.
3. Lawyers are a relatively modern development that were lacking in preliterate societies. Part-time lawyers appeared in ancient Rome, and full-time lawyers appeared in Rome's Imperial Period. Although the advent of the medieval period in Europe weakened the importance of law and lawyers, these twin phenomena again became important in early capitalism.
4. During the eighteenth century, the English legal profession became increasingly professionalized. Formal legal education became a growing expectation, and licenses became required to provide many types of legal services. This professionalization enhanced the quality of legal services and restricted the number of people allowed to practice law. This latter consequence raised lawyers' incomes.
5. Lawyers were very unpopular during the colonial period of the United States, but became increasingly important as the break with England neared owing to growing commerce. The colonists disliked lawyers for at least two reasons. First, they distrusted law and thus lawyers because they thought its adversary nature violated their Christian beliefs and disrupted the social harmony characterizing their small communities. Second, they disliked lawyers for representing banks and other established interests.
6. The American bar professionalized during the late nineteenth century after a period of many decades in which apprenticeship was a prime path into the legal profession. Because it was so easy to become a lawyer, many people who did become lawyers were poorly trained, creating concern in legal circles about the quality of the bar. At the same time, their sheer numbers also created concern that the plethora of attorneys was keeping legal fees and lawyers' incomes much lower than they would be if there were a smaller number of attorneys. These concerns led bar associations to lobby for stricter requirements for entry into the bar.
7. These bar associations tended to be composed of elite attorneys, almost all of them white male Protestants from wealthy backgrounds who primarily practiced corporate and business law. The bar associations excluded women, people of color, and Catholics and Jews for many decades. Their professionalization effort succeeded in improving the quality of the bar and also in restricting entry into it (and thus in increasing lawyers' incomes).
8. The U.S. legal profession has grown tremendously since the late nineteenth century and now numbers more than 1.1 million attorneys. The United States has more lawyers per capita than most other Western nations. Although the U.S. legal profession is no longer the exclusive

"white men's club" that it was in the nineteenth century, it still does not reflect the diversity of the American citizenry, as women and people of color continue to be underrepresented in the bar.

9. The U.S. legal profession is stratified, with corporate attorneys enjoying much greater incomes and prestige than personal services attorneys. About three-fourths of all lawyers are in private practice, while 8 percent each work in government (federal, state, or local) or for industry. Another 3 percent work as judges, while the remainder either teach law, work in legal aid or as public defenders, or are retired.

10. The many attorneys in private practice are commonly distinguished according to the size of their practice. Almost half of private practice attorneys work by themselves as "solo" attorneys, while the remainder work in law firms ranging in size from as few as two attorneys to more than 100 attorneys.

11. The American bar is now somewhat less unified and coherent than it had been in the 1970s. Growing gender, racial/ethnic, and religious diversity in the bar means that attorneys no longer all came from the same social backgrounds. Because lawyers have become specialized in the work they do, they are less likely to share the same understanding of what practicing law is all about. Further, the great growth of the bar since the mid-1900s has weakened social bonds among the many attorneys now in practice.

12. Cause lawyers attempt to use the law to change society and can be liberal/radical or conservative in their political orientation. The heyday of cause lawyering was the period of the 1960s and 1970s, when attorneys in the South represented civil rights activists and attorneys elsewhere represented Vietnam antiwar activists, various "new left" causes, and black power advocates. The many other social movements that began during this time also involved attorneys whose help in civil and criminal cases alike was invaluable.

13. Women have made significant advances in the legal profession during the past few decades, but still face a "glass ceiling" that limits their advancement and full equality with their male counterparts. Women are less likely than men to work in large law firms, and they are paid less than men even when size of firm, family situation, work hours, and other relevant factors are taken into account. Possible reasons for this disparity include (1) women are less likely than men to desire stressful corporate law careers; (2) women in corporate law settings face a hostile and discriminatory work environment; and (3) recruitment and advancement in the legal profession still occurs through an "old boys' network" that excludes women.

14. Research finds a good deal of ethical violations among personal services attorneys. Lawyerly misconduct usually remains unknown unless a client complains. However, many clients are either unaware of lawyer's malpractice or unfamiliar with the ethical standards for legal practice. Many of the clients who suspect malpractice may also not know where to report it or may simply not want to endure the personal difficulties that their complaint might initiate. Some clients also do not complain about malpractice with which they are aware because the malpractice benefits their case. Cases of lawyerly misconduct that are investigated rarely lead to disbarment, and self-regulation by the legal profession is often lax and ineffective.

15. The number of law schools grew rapidly after the Civil War and into the twentieth century. A key development in the growth of the American law school was the introduction at Harvard Law School in 1870 of the case study and Socratic methods, which eventually became the model for legal education across the nation. A surge in law school enrollment occurred during the 1960s and 1970s, thanks partly to the baby boom coming of age and to growing interest in the use of law to achieve social change.

16. Legal education in the past and today has been criticized on several grounds. First, it shuns social science and neglects important considerations of race, class, gender, and other social variables. Second, it socializes students to be more politically conservative than when they began law school and, more specifically, orients them to the needs of corporations and other established interests rather than to the needs of the poor and people of color. Third, legal education is said to be brutal for students and particularly alienating for women. Defenders of legal education say that law needs to be taught as a system of logic, that law students want corporate careers, and that the case study and Socratic methods are necessary to train students to "think like a lawyer."

Key Terms

Case study method
Cause lawyering
Cause lawyers
Corporate attorneys
Personal services attorneys
Socratic method
Stratification

CHAPTER 9

Courts and Juries

Chapter Outline

This last chapter of the book examines two further features of the legal system that law and society scholars have studied: courts and juries. Many books discuss the structure and functioning of courts and juries in minute detail, but our purpose here is simply to sketch some of their most important sociolegal aspects and to review the social scientific evidence regarding certain issues and controversies about them. Our discussion will focus their operation in the criminal justice system.

CRIMINAL COURTS IN THE UNITED STATES

One of the key social functions of law in a democracy, to recall some material from Chapter 1, is to preserve individual freedom by protecting civil rights and civil liberties. In the criminal justice system, this function is primarily the province of the criminal courts, at least in theory. Once arrests are made, prosecutors and judges must decide whether the evidence is sufficient to bring charges against defendants and to bring the case to trial if necessary, and judges preside over certain hearings and the trial if it occurs. Defense attorneys ideally represent defendants to the best of their ability. If the court system works as intended in our democracy, the guilty will be found guilty, the innocent will not be convicted, and no outcome will depend on extralegal factors such as the defendant's social class, race/ethnicity, or gender.

Yet, as we saw in earlier in the book, the criminal courts do not always work as intended. As Chapter 7 discussed, extralegal factors do affect court case outcomes, and the poor and people of color (with the findings on gender more complex) are sometimes more at risk for prosecution, conviction, and incarceration because of their lack of money and/or race and ethnicity. Innocent people are found guilty, as an estimated 1 percent of all felony convictions, or about 10,000 of the roughly 1 million convictions each year in the United States, are of innocent people (Huff 2002). Innocent people have probably also been executed, with at least a dozen innocent defendants estimated to have lost their lives between the 1970s and the end of the 1990s (Bohm 2007). Chapter 1 indicated that a central theme of law and society scholarship is that how law actually works is often very different from how it should ideally work. We have thus already seen that the criminal courts sometimes illustrate this theme, for better or worse, and will see more evidence of this theme in the sections that follow.

Understanding the Criminal Court

Certain *models* of the criminal courts help us understand how they should work in theory and perhaps how they actually work in practice. These models also point to some of the problems and dilemmas that the courts face as they try to meet the various demands placed on them.

Adversarial and Consensual Models. The most familiar model is undoubtedly the **adversarial model** (also called the *combat model* or the *adversary system model*) (Abadinsky 2008; Kagan 2001). According to this model, the prosecutor and defense attorney "fight it out" by vigorously contesting the evidence before a judge acting as a neutral referee over the proceedings. Through this process, the truth about a defendant's guilt or innocence will emerge and justice will be done. This model of the courts is integral to law school curricula and illustrated in almost every novel, film, or TV show about lawyers. Perry Mason, the hero of the Erle Stanley Gardner novels and of the 1957–1966 TV series bearing his name and starring Raymond Burr, is perhaps the epitome of the adversary model lawyer.

Although the adversarial model does characterize the criminal courts to some degree, and more so than for those in other Western nations, social scientists say the adversarial model is largely a myth for the bulk of criminal cases. In most cases, a **consensual model**, characterized by a good deal of cooperation between prosecution and defense, operates. According to this conception, the prosecutor and defense attorney share understandings of typical crimes and of the typical punishments for these crimes. Because "fighting it out" over any one case slows down the processing of all cases, their main goal is to expedite the mass of cases that come into the courts as quickly and efficiently as possible (Abadinsky 2008). For this reason, the adversarial model largely operates for only two types of criminal cases: (1) *celebrated* cases, in which the crime is particularly grisly and/or the defendant is well-known (for example, O.J. Simpson) or unusual in some other respect and (2) *serious felonies*, that is, many homicides, robberies, and rapes committed by strangers. In many of these two types of cases, the prosecutor and defense attorney do "fight it out" in real life as on TV or in film or fiction, and culminate their combat with a criminal trial (Walker 2006). We return to the consensual model in our discussion of the courtroom workgroup below.

Crime Control and Due Process Models. As noted earlier in the book, the U.S. response to the terrorism of 9/11 involved greatly increased powers of the FBI and other law enforcement agencies and, for many suspects and defendants, a lessening of the rights traditionally afforded the accused in a democracy like the United States. According to critics, this response has involved "sacrificing civil liberties in the name of national security," as the title of one critical book put it (Cole and Dempsey 2006). As the controversy generated by the U.S. response to 9/11 illustrates, a key dilemma faced by the criminal courts and other aspects of the legal system is to balance the need for public safety with the need for civil liberties and individual freedom.

This dilemma reflects the tension between two competing models of the criminal courts, as outlined by law professor Herbert Packer (1964) four decades ago. These were the *crime control* and due process models, respectively. The essential goal of the **crime control model**, as its name implies, is to ensure public safety. This model assumes that most criminal suspects are guilty and thus emphasizes the need to expedite their cases through the courts and other branches of the criminal justice system. Packer wrote that the legal process under this model is similar to a factory line or conveyer belt, with criminal cases passing by on the conveyer belt and the people at the belt having to attend to each case as quickly and efficiently as possible. Under this model, public safety takes precedence over the rights of suspects and defendants and, more generally, over civil liberties and individual freedom. As should be apparent, the crime control and consensual models complement each other, since the former model emphasizes the need to convict defendants as quickly as possible to ensure public safety and the latter emphasizes a similar need to ensure the smooth operation of the criminal courts.

In contrast, the essential goal of the **due process model**, as its name also implies, is to protect the individual from possible government abuse of power. This model assumes that some suspects and defendants may be innocent of the crimes of which they are accused and that, in any event, both the innocent and the guilty in a democracy deserve to have their rights observed in order to preserve individual liberty. Packer wrote that the legal process under this model is similar to an obstacle course, as the Constitution and Bill of Rights make it relatively difficult (at least compared to the situation in dictatorships) for the government to arrest, convict, and incarcerate someone accused of a crime. As should be evident, the due process and adversarial models complement each other, since the former model emphasizes the need to protect defendants' rights and the latter emphasizes the need for vigorous contesting of the evidence.

The Discretionary Model. After the police decide to arrest a suspect, a prosecutor must decide whether to bring charges and, if so, what charges to bring. A judge must decide whether there is sufficient evidence to hold a suspect and whether to release the suspect on bail. If the case goes to trial, a judge or jury must decide whether to convict or to acquit the defendant. If the verdict is conviction, the judge must decide whether and how long to incarcerate the defendant. It should be clear from this brief and admittedly simple summary that the sequence of any case in the criminal justice system involves a series of many decisions by legal personnel. The term **discretionary model** helps us understand this important dimension of the criminal courts and other stages of the criminal justice process (Abadinsky 2008; Albonetti 1987; Kadish and Kadish 1973).

Discretion is essential to the operation of the legal system, but it also opens the door for arbitrariness and injustice (Engen et al. 2003; Songer and Unah 2006). No two defendants are exactly alike. Although they may commit the same crime, say robbery, the ways in which the two robberies happen will not be identical. The two defendants will also differ in their backgrounds: their age, their upbringing, their prior criminal record, their potential for future criminality, and so forth. It thus makes sense for prosecutors and judges to exercise at least some discretion as they make their many decisions. At the same time, if they do not treat similar defendants identically, they may end up treating some defendants (e.g., people of color) more harshly than other defendants. Even if we grant that this disparate treatment is entirely or largely unconscious, the legal treatment that results may still be arbitrary and unfair and smack of racial or ethnic prejudice.

In this regard, recall from Chapter 7 that some criminal suspects and defendants do receive harsher treatment because of their race, ethnicity, and other extralegal characteristics. A fundamental dilemma of the criminal courts, then, is to strike the right balance between the need to exercise reasonable discretion and the need to ensure that similar suspects and defendants are treated similarly. If discretion were eliminated altogether, that would ensure that race and other extralegal characteristics of the defendant would not affect what happens to defendants, but it would also open the door for other sorts of unfairness, precisely because no two defendants are exactly alike.

Normal Crimes and the Courtroom Workgroup

The consensual and discretionary models form the backdrop for a further understanding of the criminal courts rooted in work by sociologists and political scientists in the 1960s and 1970s, respectively. To a large degree, this work reflected the perspectives of the legal realism and sociological jurisprudence schools from earlier in the century. Recall from Chapter 2 that both these schools emphasized that how law actually worked in practice differed in many respects from how it was supposed to work in theory. In particular, Roscoe Pound (1997 [1930]), dean of Harvard Law School and a key figure in sociological jurisprudence, emphasized that the criminal courts were settings in which the poor received little justice, and he called for reforms to address this problem. "If democracy involves regard for the interests of the mass of humanity," he wrote, criminal "justice in our cities falls far short of its demands" (p. 199).

After three decades of largely neglecting Pound's emphasis on the poor in the criminal courts, social scientists returned to this issue beginning in the 1960s amid a wave of concern about the poor and research on the problems they face in many spheres of life (Bagdikian 1964; Caplovitz 1963; Harrington 1962). Quite naturally, sociolegal scholars began to address the problems the poor faced in the civil and criminal courts (Carlin, Howard, and Messenger 1966; Carlin 1966; Wald 1965). Abraham S. Blumberg (1967b) likened defense lawyers' representation of poor and working-class clients to a "confidence game" in which the lawyers' main goal is to collect their fees as quickly as possible. This meant that they spent as little time as possible on any one case while fooling their clients into think they spent much time. To expedite their handling of cases and thus their collection of fees, defense lawyers cooperated with prosecutors to have their clients plead guilty instead of defending their clients vigorously according to the adversarial model's assumptions. Because defense lawyers in effect pretend to help their clients but in fact let them down, Blumberg called them "double agents." In effect, said Blumberg, lawyers were selling out their clients.

In a study that helped formulate the consensual model discussed earlier, David Sudnow (1965) explained how the expeditious processing of criminal cases occurs. Sudnow said that prosecutors, defense attorneys, and judges share common understandings of typical crimes and common understandings of the appropriate punishment for these crimes. These understandings draw on several criteria, including the strength of the evidence, the amount of harm (physical or financial) caused by the crime, and the nature of the relationship between the offender and victim. If a given offense seems typical of other crimes like it when considering all these criteria, courtroom actors deem it a **normal crime** and almost immediately have the same understanding of what its punishment should be. This understanding allows prosecutors and defense attorneys to quickly decide on the punishment a defendant should expect in return for a guilty plea. Because judges share these understandings, they almost always agree with the punishment that the prosecutor recommends based on the agreement reached with the defense attorney.

The studies by Blumberg and Sudnow were among the first in flurry of works by social scientists and journalists on the lack of justice for the poor in the urban criminal courts (Clarke and Koch 1976; Mather 1973). These courts were portrayed as "assembly lines" in which defendants, even those who were innocent, were spending only seconds with their assigned counsel or public defender before agreeing at their lawyer's behest to plead guilty. Defense attorneys were depicted as caring more for their fees and other professional needs than for their clients' legal welfare. To quote the titles of two popular journalistic accounts at the time, this situation amounted to "justice denied" (Downie 1972) and "injustice for all" (Strick 1977). One of these accounts charged that "chaos, injustice, and cynical indifference" characterized urban courts and that urban courts were settings for "vagrants sleeping in the corridors, incompetent lawyers and bail bondsmen swarming like vultures, and hack political appointees clothed in the robes of justice destroying lives through prejudice, whim, and limited legal ability" (Downie 1972:7, 16).

Later in the 1970s, political scientists began to examine decision making in the criminal courts and took a somewhat more nuanced view than the courts' severest critics of the fate of poor defendants (Eisenstein and Jacob 1977). These scholars developed the concept of the **courtroom workgroup** to help explain how and why criminal cases are processed. The workgroup consists of the prosecutor, defense attorney, and judge. All three parties are well aware of the heavy caseloads that burden the system generally and each of the parties individually. They thus also recognize the need to process these cases quickly and efficiently, lest the system break down. In this regard, guilty pleas are obviously much faster and efficient than criminal trials. For this reason, the courtroom workgroup favors guilty pleas over trials, and guilty pleas account for more than 90 percent of all convictions in many criminal courts (Abadinsky 2008; Smith 2005).

At the same time, these scholars emphasized that prosecutors do not proceed with a case unless they are fairly sure that a jury would convict the defendant if the latter demanded a jury trial as guaranteed by the Constitution. Prosecutors thus dismiss up to half of all felony arrests if the evidence is too weak and/or if the victim or witnesses indicate they will not want to testify. Prosecutors also drop cases for other reasons, for example, if the victim is of less than upstanding character and not very believable. After prosecutors complete this winnowing process, only the strongest cases, where the defendant is almost certainly guilty, remain for plea bargaining or possible trial.

This understanding led the new scholarship on courtroom workgroups to draw some significant conclusions (Heumann 1978; Rosett and Cressey 1976). First, because the defendants in the cases that remain after the winnowing process are probably guilty, plea bargaining does not do them a disservice and even benefits them for reasons discussed just below. Second, plea "bargaining" does not really occur because the courtroom workgroup, as Sudnow (1965) explained, already knows the punishment for a given "normal" crime. Third, sentences given to American defendants are fairly strict (and indeed are generally harsher than those given to defendants in other Western nations; Kappeler and Potter 2005). Although the new scholarship challenged the 1960s' and early 1970s' view about injustice for the poor in the criminal courts, it still suggested that the courts are only rarely settings for the vigorous contesting of the evidence assumed by the adversarial model.

Plea Bargaining: Evil, Necessary Evil, or Not Evil?

The considerations discussed so far help to understand one of the most controversial issues today concerning the criminal courts, **plea bargaining**. Leonard Downie, Jr. (1972:23), the author of one of the critical journalistic accounts of the courts in the 1970s, condemned the practice for harming defendants: "A lawyer who knows next to nothing about is client or the facts of the crime with which he is charge barters away a man's right to a trial, and, along with it, the presumption that a defendant is innocent until proved guilty." He also wrote that serious offenders sometimes receive sentences via plea bargaining that are too lenient and that plea bargaining thus harmed society in two ways: "In fact, nobody can be certain that innocent persons are not being convicted or, more frequently, that habitual criminals are not being let off lightly" (p. 30).

The work of the political scientists and other sociolegal scholars in the 1970s suggested that plea bargaining was more benign than this critique suggested and also necessary for the criminal court system to function at all. More recent studies agree and emphasize that plea bargaining benefits all parties in a case (Emmelman 1996; Fisher 2003; Piehl and Bushway 2007). Because of these benefits, each of these parties has several reasons for desiring plea bargains over actual trials (Abadinsky 2008).

From the prosecutor's standpoint, the major benefit of a plea bargain is that it ensures a conviction and guarantees that a defendant will be punished, and, if convicted of a serious offense, be incarcerated. The outcome of any case that goes to a jury is something of a gamble, as a jury may acquit a defendant even if the evidence for conviction is very strong. Deadlocked ("hung") juries are also possible. Because unanimous verdicts are required in jury trials, even one juror can deadlock a jury, forcing the prosecutor to have to try the case a second time. A second trial obviously costs more money and takes up more time, and the elapse of additional weeks and even months can mean that witnesses' memories may fade. Because the outcome of any case tried by a jury (and perhaps also by a judge) is at least somewhat uncertain, a prosecutor who declines to negotiate a plea with the defense knows the defendant might end up not being convicted. Prosecutors may also wish to spare certain kinds of victims, including rape and child abuse victims, the obvious stress that would result from having to testify in a trial of their accused offender.

Prosecutors desire the certainty of conviction from plea bargains to help protect public safety, but their personal career interests are also at stake. An important component of prosecutors' job performance is their **conviction**

rate, the percentage of all their cases that end in a conviction. When this rate is determined, a conviction counts as a conviction regardless of whether it results from a guilty plea or from a conviction after trial. Prosecutors who decline to negotiate a plea risk lowering their conviction rate and thus the ranking of their job performance, with obvious negative implications for their salary, career advancement, and other professional considerations. In celebrated cases that receive heavy media attention, the prosecutor's decision to accept or decline a plea bargain often becomes quite difficult, with the possible impact of the case on the prosecutor's career playing no small role (Walker 2006). A plea bargain will ensure a conviction and avoid the negative publicity if the defendant is acquitted after trial, but might also be perceived as allowing the defendant a more lenient sentence than might have resulted if the case had gone to trial. A celebrated case that goes to trial will by definition receive a lot of media publicity, and a conviction at the end of such a trial can benefit the prosecutor's career in many ways; by the same token, an acquittal can have the opposite career effect.

Prosecutors desire plea bargains for one last reason: *caseload pressure*. As indicated earlier, all members of the courtroom workgroup must deal with heavy caseloads. They simply cannot afford to spend too much time (and also money) on any one case unless, perhaps, the case falls into the celebrated or serious felony categories. It is not an exaggeration to say that any prosecutor's office that had to handle even a small increase in the number of jury trials would barely be able to function without an infusion of more attorneys and other staff and of a major increase in its budget.

Before proceeding, it should be noted that some scholars question the importance of caseload pressure in understanding why plea bargaining is so common (Heumann 1978). Plea bargaining was also common a century ago, they say, when caseloads were much lower than now and thus caseload pressure could not have been a very important factor (Alschuler 1995; Smith 2005). These scholars instead highlight the informal pressures of courtroom workgroup functioning. Workgroup members spend so much time with each other that personal ties develop. Although these ties are usually not very close, they still lead the workgroup members to want to cooperate with one another. They also realize that if they "rock the boat" by refusing to plea bargain in a particular case, other workgroup members might not help them out, and might even make things very difficult, in a future case.

While the actual importance of caseload pressure remains in dispute, it still plays a role, as we have seen, in prosecutors' preferences for plea bargains. Judges also desire and even expect plea bargains because they realize that trials slow down their dockets. Any one trial may tie up a judge's time and attention for hours, days, or weeks at a time. Even now, with fewer then 10 percent of felony cases nationwide going to trial, it can take many weeks for a case to come to trial. Judges almost always agree to any plea bargain and recommended punishment that result from discussions between the prosecutor and defense attorney.

Why might defendants prefer plea bargains over trials when a plea bargain guarantees conviction and a trial offers at least some hope of acquittal? Several reasons seem to matter (Abadinsky 2008; Kramer, Wolbransky, and Heilbrun 2007; www. nolo. com). First, plea bargains allow defendants to know the sentence and thus to avoid the uncertainty of what the sentence might be if they were convicted after a trial. Because good evidence exists that sentences after trial convictions are more severe than those resulting from plea bargaining (Jost 1999), defendants often opt for plea bargains rather than take the gamble of a trial. Second, plea bargains also help them avoid having to wait several months in jail (since most cannot afford bail) before their trials could occur; this is a particular benefit for defendants whose potential sentence likely includes no jail or prison time. Third, for those relatively few defendants represented by private counsel (for which they pay), a plea bargain will save a good deal of money in lawyer fees over a trial. Fourth, any case is a source of stress for defendants and their families; because a plea bargain ends a case sooner than a trial, it reduces the amount of stress they experience. Fifth, pleading guilty to lesser charges, say to a misdemeanor instead of a felony, yields a less unfavorable prior record and may leave the door open for potential employment opportunities that a felony conviction would have precluded. Finally, a plea bargain results in less media publicity than a trial; some defendants wish to accept a plea to avoid the publicity a trial would bring.

Although, as all these reasons suggest, plea bargaining may not be the sin that scholarly work before the late 1970s implied, it still raises some troubling questions in a democracy. Law professor John H. Langbein (1992) says these questions stem from the fact that plea bargaining greatly reduces the jury's role in dispensing justice. This reduced role means the fate of defendants is left largely to the state and its represented officials rather than to the jury as our democratic system of government intended. As Langbein puts it,

> We are accustomed to celebrating the Bill of Rights. With it, the American constitution-makers opened a new epoch in the centuries-old struggle to place effective limits on the abuse of state power. Not all of the Bill of Rights is a success story, however. While we are celebrating the Bill of Rights, we would do well to take note of that chapter of the Bill of Rights that has been a spectacular failure: the Framers' effort to embed jury trial as the exclusive mode of proceeding in cases of serious crime (p. 119).

Langbein criticizes plea bargaining for additional reasons. First, its use means that justice takes place behind closed doors rather than in public via a trial. Second, because sentences after trials are often more severe than sentences from guilty pleas, the defendant in effect is being coerced to plead guilty and thereby waive the constitutional right to a jury trial: "A legal system that comes to depend upon coercing people to waive their supposed rights is by definition a failed system. The system can no longer function by adhering to its own stated principles" (p. 124). Many other critics echo this view, and legal historians point out that American and English judges before the twentieth century often cautioned defendants against accepting plea bargains for this reason (Alschuler 1995). Finally, says Langbein, plea bargaining is dishonest. In effect, the legal system pretends that a man who committed a rape only committed the physical assault to which he pleads guilty (to use a hypothetical example) or that a man who committed a burglary only committed the act of trespassing to which he pleads plead guilty. This dishonesty means that conviction records have little credibility.

For better or worse, plea bargaining is entrenched in American courts. Many people think it lets hardened criminals back on the streets too quickly, while many other people think it results in innocent defendants being convicted and denies them important constitutional rights. An understanding of the courtroom workgroup provides a context for this controversy and suggests that plea bargaining may not be as evil as critics on either side of the controversy claim. Whatever position one takes about plea bargaining, its dominance in the American courts reinforces the conclusion that the adversary system is largely a myth for the bulk of criminal cases in the United States (Abadinsky 2008; Smith 2005).

THE JURY IN DEMOCRATIC SOCIETY

If plea bargaining is so dominant, why has the jury seemingly always been considered important? Why has there been so much debate about the jury over the years, and why have scholars been so interested in it? Several reasons help explain the American jury's importance.

First, the jury is historically significant for acting as a "bulwark against grave official tyranny" (Kalven and Zeisel 1966:296). We will recount some of the jury's history shortly, but its significance in the founding of the United States merits mention here. The framers of the Constitution and Bill of Rights guaranteed the right to trial by jury in Article III, Section 2 of the Constitution and in Amendment VI of the Bill of Rights. This guarantee reflected the importance of the jury during the colonial period that is discussed below. When the Constitution and Bill of Rights were written, this importance was fresh in the framers' minds. The American jury's historic significance helps account for the continuing interest in it today.

Second, the jury has long been regarded as a quintessential democratic institution, one that in effect allows the public to render important decisions in the courtroom and public views to be represented in the courtroom (Abramson 1994). In this sense, the jury is said to serve an educational function for the public. Alexis de Tocqueville (1994 [1835]), the great nineteenth century French observer of American democracy, was one of the first observers to discuss this function. In his majestic *Democracy in America*, he wrote, "The jury is both the most effective way of establishing the people's rule and the most effective way of teaching them how to rule.... [The jury] should be regarded as a free school which is always open and in which each learns his rights . . . and is given practical lessons in the law.... I think the main reason for the political good sense of the Americans is their long experience with juries ..." Modern evidence shows that jury service does indeed have some of the educational effects Tocqueville predicted. In particular, people who serve on juries that reach verdicts are more likely to vote in later elections than people chosen for juries that do not reach a verdict because the case was dismissed or for some other reason (Gastil, Deess, and Weiser 2002).

Third, although the jury is only rarely used in criminal cases, the possibility of a jury trial shapes the decision-making of prosecutors in many ways. Prosecutors negotiate pleas precisely because they fear a jury acquittal or deadlocked verdict if the defendant were to demand a jury trial and because they do not wish to incur the extra time and cost of jury trials.

Fourth, when jury trials do occur, the jury's verdict often receives much attention from the news media and other interested parties. As Valerie P. Hans and Neil Vidmar (2004:198) point out, jury verdicts "send messages to potential wrongdoers and victims, to negligent corporations, to lawyers and judges, about likely outcomes in criminal and civil cases." For this reason, they say, "juries cast a long shadow, having an impact far beyond their decisions in specific cases." This long shadow is a fundamental reason for the jury's contemporary importance despite the fact that jury trials are uncommon.

To understand the jury today, it is important to appreciate its past. A brief history of the jury follows.

History of the Jury

The exact origins of the jury are fairly murky, but it is traced back to England during the thirteenth century when a group of neighbors would meet in a criminal case because they had witnessed some of the events surrounding the case. Their role was both to testify to what they had seen or heard and, based on their special knowledge of the case, to decide the verdict. Over the next few centuries this witnessing function gave way as the jury's role became more limited to interpreting the facts of a case and the law governing a case and, based on its interpretation, to reaching a verdict (Dwyer 2002; Green 1985; Moore 1973).

During this time, however, a jury could be punished by a judge for returning a verdict that the judge disliked. This possibility meant that juries were not truly free to decide their verdicts, and it also meant that jury trials were of little use for defendants. For this reason, one of the most significant developments in the history of the jury occurred in 1670 in *Bushell's Case*. Two Quakers, William Penn (who later founded Pennsylvania) and William Mead had been arrested for preaching about Quakerism in public. In previous years, other Quakers had been arrested for a similar "crime," and juries would acquit them and then be fined by the judge for doing so. When four of the twelve jurors in Penn and Mead's case voted to acquit these two defendants, the judge threw them in jail, starved them, and fined them. One of the four jurors, Edward Bushell, filed a writ of habeas corpus to contest his punishment and won his case. The court ruling established the principle, so fundamental to the role of the jury, that juries in criminal cases could not be punished for any verdict they reached.

Juries used their new power to acquit with impunity in England in the ensuing two centuries to protect two kinds of defendants: those accused of seditious libel against the English government and those accused of a myriad of petty offenses for which death was the punishment (Barkan 1983). In so doing, wrote political scientist James P. Levine (1992:24), "the harshness of British law was mitigated by jury's willingness to stretch the facts to spare the accused." In the seditious libel cases, the jury was, according to law, only to consider whether the allegedly seditious material was in fact published; it was not to consider whether the material was seditious. Rejecting this limited view of their role, many juries acquitted defendants who had clearly published the material the prosecutor accused them of publishing; evidently, the juries either decided that the material was not seditious or that the material was seditious but truthful. The death penalty cases arose from the fact that England had about 230 capital offenses by the early 1800s, including forgery and the publishing of defamatory letters (Hay 1975). Evidently thinking that death was too harsh a punishment for these offenses, juries often acquitted defendants accused of them. For this reason, the jury "played a major role in the gradual abolition of the death penalty in England" (Kalven and Zeisel 1966:49).

During the American colonial period, the jury protected colonists accused of violating English trade law. Many colonists were arrested for violating the Navigation Acts, which among other provisions required that all goods exported from America be channeled through England with the payment of heavy shipping fees and that such fees be paid even for goods sent from one colony to another. To protect English farmers, related laws also made it very difficult for grain to be imported into England from the colonies. Other laws also prohibited the exporting of hats and other manufactured goods from the colonies. Colonists found ways of violating all these laws, and smuggling was not only extensive but also vital to the colonial economy. Arrests were common, but so were acquittals because the defendants' juries were their peers, that is, other colonists, who sympathized with the defendants' behavior and found them not guilty despite clear evidence of their guilt. In response to these acquittals, England established courts of vice-admiralty to try smuggling offenses without juries. England's abolition of the jury was one of the grievances listed in the Declaration of Independence: "For depriving us in many cases, of the benefit of Trial by Jury."

One particular case from the colonial period deserves special mention. In 1735, John Peter Zenger, the publisher of a New York colony newspaper, was arrested and tried for seditious libel after publishing material critical of the colony's royal governor. Like English juries at the time, Zenger's jury was to decide only whether he had published the material, not whether the material was seditious or whether what it said was true. Zenger's lawyer told the jury that his client had indeed published material but asked the jury to acquit him because what he wrote was true: "And this I hope is sufficient to prove that jurymen are to see with their own eyes, to hear with their own ears, and to make use of their own consciences" (quoted in Goodell 1973:23). Zenger's jury did acquit him; its verdict reduced colonial prosecutions for seditious libel and is celebrated for contributing to the freedom of the press (Finkelman 1981; Garrow 2006).

The jury acquittals during the colonial period of Zenger and smuggling defendants reflected the prevailing view at the time that the jury's role included both interpreting the law and finding the facts (Alschuler and Deiss 1994). This dual role continued well into the nineteenth century and again was reflected in jury acquittals of controversial defendants. During the abolitionist era preceding the Civil War, many people were arrested for violating the 1850 Fugitive Slave Law that made it a crime to help fugitive slaves escape or to obstruct their capture and return. Many Northern juries acquitted these defendants even though the evidence was clear that they had in fact violated the law. In 1851, for example, several people broke into a Boston to rescue a slave. President Millard Fillmore urged that the lawbreakers be prosecuted, and three members of the group eventually went to trial. A judge called them "beyond the scope of human reason and fit subjects either of consecration or of a mad-house" (quoted in Friedman 1971:38). Despite clear evidence of their guilt, the jury acquitted them. Historians commend the jury acquittals in the Fugitive Slave Law cases for helping the antislavery movement (Pease and Pease 1974).

The jury again came to the aid of controversial defendants two decades later during the height of the postbellum labor movement. Workers in many locations were arrested for going on strike and for other labor activities, but juries often acquitted them anyway. Judges began issuing injunctions against strikes; when workers were arrested for violating the injunction, they were placed in jail for contempt of court without benefit of a jury trial (Frankfurter and Greene 1930).

By the end of the nineteenth century, the prevailing view of the American jury's role in legal system no longer included interpreting the law. Instead, the jury's role was now restricted to fact-finding, that is, determining whether the evidence was sufficient to support a conviction beyond a reasonable doubt, the prevailing view of its role today. At least two reasons account for this restriction. First, judges had become better trained and more versed in the law. One reason the jury historically had been expected to interpret the law was that judges were often poorly trained in it (Alschuler and Deiss 1994). As judges became more knowledgeable, "the perceived value of the untutored members of the jury began to decrease, and restrictions on what the jury could do were put into place" in England during the eighteenth century and in the United States a century later (Hans and Vidmar 2004:198). Second, the American jury toward the end of the nineteenth century was perceived by corporations and their allies in the upper echelon of the legal profession as too sympathetic to workers and too hostile to corporations. Juries were not only acquitting workers for striking and other anti-management offenses, as noted above; they were also ruling in favor of plaintiffs who filed damage suits against corporations. According to one writer, juries had thus "developed agrarian tendencies of an alarming character" (quoted in Yale Law Journal 1964:192).

The historic importance of the American jury notwithstanding, jury verdicts, as we have seen, decide fewer than 10 percent of all criminal cases today. Although this is a very low number, the performance of the jury has long concerned many observers who believe that juries often ignore the facts in a case and convict or acquit as a result of their respective hostility or sympathy toward a particular defendant (Adler 1994b; Frank 1949; Penrod and Heuer 1998); there is probably more concern that juries acquit guilty defendants than that they convict innocent defendants. Saying that juries are "out of control" in civil cases, other observers want to limit or abolish jury trials for civil matters; although used in United States, parts of Canada, and Australia, civil juries are virtually unknown elsewhere in the world (Hans and Vidmar 2004). The criticism of the American jury from so many circles has prompted social scientists to study its performance. The next section summarizes the findings of this research.

How Well Does the Jury Perform?

The general answer to this question is "very well," but this simple response obviously needs some elaboration. Little elaboration is needed here, though, on the performance of civil juries, as Chapter 4 reviewed the many studies of their performance. The conclusion of these studies was that jury verdicts and awards in civil cases are reasonable and not excessive despite the heavy attention the news media and social critics give to the occasional jury award that is highly excessive. If, as most observers assume, judges' verdicts and awards are held as the reasonable standard with which to compare jury decisions, jury verdicts and awards are strikingly similar overall to judges' verdicts and awards (Eisenberg et al. 2002, 2006). If judges are not "out of control" in civil cases, then neither are juries.

A similar conclusion can be drawn for criminal juries (Devine et al. 2001; Diamond and Rose 2005; Hans and Vidmar 2004; Vidmar and Hans 2007). These juries have been studied in several ways, including surveys of criminal court judges who are asked to comment on jury verdicts in cases over which they presided, and simulated jury research involving mock juries (usually composed of college students). The classic study remains the Chicago Jury Project initiated by Harry Kalven, Jr., and Hans Zeisel in 1953 and reported in their 1966 book, *The American Jury* (Kalven and Zeisel 1966).

This project observed jury trials and interviewed jurors after their trials ended, analyzed data on jury verdicts, observed jury trials, and conducted experiments with mock juries. The most important and influential phase of their research, however, involved surveys completed by 555 judges who reported on 3,576 criminal jury trials over which they had presided.

The judges were asked a number of questions. Perhaps the most significant question was whether they had agreed or disagreed with the jury's verdict. Judges agreed with the jury's verdict in 78 percent of the cases; in 19 percent, the judge would have convicted while the jury acquitted; and in 3 percent, the judge would have acquitted while the jury convicted. Jury scholars interpret the 78 percent figure as indicating a high level of judge–jury agreement and interpret this high level as evidence (assuming judges' verdicts are again the reasonable standard with which to compare juries' verdicts) that the criminal jury is performing satisfactorily (Dwyer 2002; Hans and Vidmar 2004). More recent studies also find that judges and juries tend to agree much more often than they disagree (Hannaford, Hans, and Munsterman 2000; Heuer and Penrod 1994).

Although the juries in Kalven and Zeisel's 1966 study were more lenient than judges, studies of verdicts in the 1980s found that juries were somewhat *more* likely than judges to convict (Levine 1983; Myers 1981; Roper and Flango 1983). To the extent that juries became less lenient during the intervening decades, rising U.S. crime rates may have been the reason (Levine 1992). In any event, if juries are in fact more likely than judges to convict criminal defendants, claims that juries "let criminals go free" appear to be unfounded.

What about the 19 percent of cases, almost one-fifth, in the Kalven and Zeisel study in which the jury proved more lenient than the judge would have been? The judges in the survey were asked to indicate why they thought the jury's verdict in these cases was different from what their own verdict would have been. Their responses permitted Kalven and Zeisel to determine whether these "disagreement" verdicts were evidence of "out of control" juries that cannot be trusted to convict guilty defendants or, instead, whether these verdicts were reasonable even if the judge disagreed with them. Significantly, judges indicated that evidence problems (e.g., the evidence of guilt was not overwhelmingly strong) accounted at least partly for the vast majority of the disagreements. This finding indicated that the jury's acquittals in cases in which the judge would have convicted were not unreasonable. As a later review of this evidence stated, "[T]he bottom line is that *The American Jury* concluded that it was not jury incompetence that caused disagreement between judge and jury" (Hans and Vidmar 1986:118).

Other kinds of evidence also indicate competent jury performance. If juries are performing adequately, then the strength of evidence should play a large role in their verdicts. Studies of mock and actual juries find that this is indeed the case (Devine et al. 2001). When the evidence is strong, juries in these studies are much more likely to convict than when the evidence is weak.

Testimony by corroborating witnesses considerably increases the chances of jury convictions, as it should if the jury is taking evidence into account. A study of 201 actual criminal jury verdicts in Indianapolis, Indiana, found that verdicts were based primarily on the evidence; in the relatively few times in which juries reached verdicts contrary to the evidence, they did so out of a sense of fairness and justice and in this manner adhered to the historic role of the jury (Myers 1979). In another sign of jury competence, studies also find that juries are generally able to understand complex evidence (Diamond and Rose 2005).

Anecdotal evidence also paints a positive picture of the jury. Federal judge William L. Dwyer (2002:134–35) writes, "In the hundreds of trials I have seen as a practicing lawyer and judge, the jury nearly always has returned what I believed to be a just and fair verdict. This holds true in cases involving scientific or technical evidence, and in those with racial overtones." Law professor James Gobert served on several juries and later wrote,

> It was an eye-opening experience. On the juries on which I served were men and women from all walks of life. . . . What united us was our common pursuit of the right verdict in the case on which we had been brought together. . . . All of our members contributed to the deliberations and all played a meaningful role in the decision-making process. What most impressed me was the seriousness of purpose with which my fellow jurors approached their task. (quoted in Dwyer 2002:139)

Other kinds of evidence present a more conflicting picture of juries' performance. In mock jury studies, jurors' demographic characteristics, including their gender, income, and race or ethnicity, "have only modest or inconsistent impact on verdicts" (Hans and Vidmar 2004:203), and the same conclusion can be drawn regarding defendants' demographic characteristics (Devine et al. 2001). Taken in isolation, then, neither jurors' nor defendants' demographic characteristics appear to play a large role in jury verdicts. However, a *similarity bias* does exist: "jury demographic factors interact with defendant characteristics to produce a bias in favor of defendants who are similar to the jury in some salient respect" (Devine et al. 2001:673). Jurors are at least somewhat less likely to convict defendants who are of the same race/ethnicity or socioeconomic status as the jurors. Significantly, this effect appears only when the strength of the evidence is moderate (i.e., there is some evidence of guilt, but this evidence is not very strong); it does not appear when there is strong evidence for either guilt or innocence, respectively (Devine et al. 2001). In more ominous findings, juries are more likely to prefer death sentences in capital cases when the defendant is African American and the victim is white.

In interpreting the evidence of jury racial bias, recall from Chapter 7 that judges' sentencing of people of color is somewhat harsher than their sentencing of whites (Steffensmeier and Demuth 2001). Given such evidence, it is unclear whether any racial/ethnic biases jurors exhibit are any worse than those that judges exhibit. If, as argued earlier, judges' decisions are a reasonable standard for assessing juries' decisions, the evidence of jurors' racial/ethnic bias does not necessarily indicate that juries' performance is worse than judges' performance. This conclusion may constitute faint praise for the jury in regard to such bias, but it does indicate that juries are at least no worse in this regard than the judges with whom they are typically compared. The fact that strength of the evidence typically plays a much larger role in jury deliberations than any racial/ethnic bias certain jurors may have (Diamond and Rose 2005; Garvey et al. 2004) also contributes to a conclusion of jury competence.

This brief review only sketches the plethora of research on jury deliberations (Devine et al. 2001; Vidmar and Hans 2007). An overall conclusion from this research is that "most of the negative claims about the criminal and civil jury systems have not stood up to careful empirical tests" (Hans and Vidmar 2004:207). Juries are not perfect, but neither do they appear to be the out-of-control, irrational decision-makers that their critics claim. Levine (1992:192-93) wrote that if juries make mistakes, these are "democratic" mistakes:

> Just as allowing people to elect their leaders on occasion puts demagogues and charlatans in government, accepting as inevitable a certain number of verdicts that seem unthinkable may be necessary. . . . Jurors make critically important decisions, sometimes life and death decisions. They use their heads; they use their hearts; they use their souls. They become politicians in the best sense of that word: they make public decisions with the interests of the public in mind. They deserve our congratulations.

The Issue of Jury Nullification

Jury nullification in criminal trials occurs when juries reach verdicts that contradict the evidence. In practice, discussions of jury nullification concern the acquittals of defendants despite clear evidence of their legal guilt. The earlier discussion of the jury's history included several examples of jury nullification: the refusal of English juries to convict defendants accused of seditious liberal and capital offenses; the refusal of colonial juries to convict colonial defendants accused of smuggling; and the acquittals of John Peter Zenger, abolitionists, and striking workers. Jury nullification has thus played a very important role in English and American history and is a major reason that the jury is regarded historically, as noted earlier, as a "bulwark against grave official tyranny" (Kalven and Zeisel 1966:296) and as a "palladium of liberty" (Moore 1973).

Recall that during and after the colonial era and well into the nineteenth century, the prevailing view was that the jury should both interpret the law and assess the evidence, and through both processes, decide whether a conviction was warranted. This meant that the jury was free to decide whether the law that a defendant had allegedly violated was a law that should be enforced. John Adams wrote of the juror in 1771 that "it is not only his right, but his duty . . . to find the verdict according to his own best understanding, judgment, and conscience, though in direct opposition to the direction of the court" (quoted in Yale Law Journal 1964:173). Alexander Hamilton wrote in 1804 that a jury should disregard the judge's instructions and acquit a defendant "if exercising their judgment with discretion and honesty they have a clear conviction that the charge of the court is wrong" (quoted in Sax 1968:487).

For the first half of the nineteenth century, judges typically instructed juries that they were free to disregard the judge's view of the law. As the jury's role restricted by the end of the nineteenth century for the reasons discussed earlier, judges no longer gave juries this instruction as various court rulings and new state legislation declared that juries should only decide the facts. The matter became settled in 1895, when the U.S. Supreme Court ruled in *Sparf v. U.S.* (156 U.S. 51) that defense teams could not tell juries that they had the power to disregard the law and acquit with impunity and that judges were not obliged to inform juries of this power (Rubenstein 2006).

During the Vietnam war, many antiwar protesters who were arrested and tried wished to have juries informed of their power to nullify the law and acquit, and sociolegal scholars wrote several pieces arguing for the right of defense teams to have juries so informed (Sax 1968; Scheflin 1972; Van Dyke 1970; Yale Law Journal 1964). Because juries were sometimes sympathetic to antiwar defendants, more acquittals might have resulted if the juries had been allowed to be told that they could disregard the law (Barkan 1983). As one juror in an antiwar trial that ended in conviction commented, "I had great difficulty sleeping that night after the summing-up arguments. I sympathized very strongly with the defendants; I detest the Vietnam war.... But it was put so clearly by the judge. It was a law violation.... The jury can't say, 'Was he justified in violating the law?' If the judge had said, 'If you find they were justified, find them not guilty,' it would have been beautiful" (quoted in Mitford 1969:224). In this regard, modern mock jury research shows that juries are indeed more likely to disregard the law and acquit defendants if they are reminded by the judge of their power to do so (Devine et al. 2001).

The issue of jury nullification reflects a paradox: juries have the historic power to nullify the law and acquit defendants with impunity if they do not like the law or feel the punishment is too severe, but they are not allowed to be told or reminded of this power. Instead, judges' instructions typically order juries instead to follow the law as the judge explains it to them. Defense teams often try to play to a jury's sympathies but can only hope that juries will be more flexible in their deliberations than the judge's instructions allow them to be.

Although jury nullification became an issue during the Vietnam war, it continues to be controversial today (Leipold 1996; Ostrowski 2001; Rubenstein

2006; Weinstein 1993). Many observers call for juries to be informed of their power to disregard the law if their sense of justice leads them to do so, in line with their historic role before the late nineteenth century, while critics say this information would lead to jury anarchy. As one former federal judge explained, informing juries of their power to nullify the law would create a "lawless society, a society without law, without regulations. That is a monstrosity" (quoted in Van Dyke 1970:278). These critics add that although jury nullification has historically yielded positive contributions (e.g. helping the antislavery movement), it has also led to far less desirable results. For example, white juries in the South during the 1960s civil rights movement routinely acquitted white defendants accused of murdering or assaulting civil rights activists. Critics of jury nullification worry that similar unjust verdicts would result if juries were reminded of their power to nullify the law (King 2003).

Calling for judges' instructions to juries to become less stringent regarding their need to follow the law without question, a recent law review article commented,

> Jury nullification, though controversial, is an important part of the jury's role in a criminal trial. It supports democratic and anti-tyrannical values and can assist the disempowered in resisting majoritarian control. While nullification is a tool that can be used for undesirable purposes, when properly regulated its benefits substantially outweigh its detriments. . . . Jury nullification can protect liberty, democracy, and equality—the very values that the jury trial was intended to secure. (Rubenstein 2006:993)

This is obviously not the final word on the subject, and no doubt the controversy over informing juries of their power to nullify the law will continue for some years to come. As this controversy suggests, the jury remains a vital part of the American legal system even if jury trials are the exception rather than the rule.

Summary

1. Models of criminal courts help understand how they should work in theory and also how they work in practice. The adversarial model assumes the prosecutors and defense attorneys vigorously contest the evidence, while the consensual model assumes that they instead cooperate in order to expedite case processing. The crime control model assumes that most criminal suspects are guilty, while the due process model assumes that some are innocent and emphasizes the need to protect their various legal rights. The discretionary model emphasizes the many decisions that legal official must make before and after a suspect is arrested. The discretion that legal actors exercise opens the door

for arbitrariness and injustice but it also helps theoretically ensure that defendants with different backgrounds and different needs are treated differently.

2. Normal crimes are those that fit preconceived notions of the typical crime held by prosecutors, defense attorneys, and judges. These parties form the courtroom workgroup and recognize the need to process heavy caseloads quickly and efficiently.
3. Plea bargaining continues to be a controversial procedure, but research suggests it benefits all parties in a case. Prosecutors prefer plea bargains because they ensure convictions and usually incarceration and because they guarantee high conviction rates that benefit a prosecutor's career. The importance of caseload pressure for prosecutors' interest in plea bargaining has been contested. Judges desire plea bargains to ensure that dockets do not proceed too slowly as jury trials would guarantee.
4. Defendants typically desire plea bargains for several reasons. First, plea bargains allow defendants to avoid the uncertainty of what the sentence might be if they were convicted after a trial and, in particular, to receive a more lenient sentence than they would receive if convicted at trial. Second, plea bargains help them avoid having to wait several months in jail before trial. Third, plea bargains save money for the relatively few defendants represented by private counsel. Fourth, because plea bargains end a case much sooner than if it went to trial, they reduce the stress for defendants and their families. Fifth, plea bargains often involve pleading guilty to lesser charges and thus yield a less unfavorable criminal record were a plea bargain not to occur. Finally, plea bargains are less publicized than trials.
5. Despite the benefits of plea bargaining, critics worry that it deprives defendants of their constitutional right to jury trials and thus is antidemocratic. They also say that plea bargaining constitutes "secret justice" and that it in effect coerces defendants into pleading guilty because they realize that a more severe sentence awaits if they request a jury trial.
6. The jury is important for several reasons. Historically, it has protected defendants against government abuse of power. It is also regarded as a quintessential democratic institution that educates the public. The possibility of a jury trial shapes prosecutorial decision-making in many ways, and jury verdicts send a message to many parties about the outcomes they can expect in civil and criminal cases.
7. The jury began in thirteenth-century England as a group of neighbors with special knowledge of a criminal case would meet to testify about their knowledge and to decide the verdict. Over the next few centuries this witnessing function gave way as the jury's role became more limited to interpreting the facts of a case and the law governing a case. The 1670 *Bushell's Case* established the right of juries to return a verdict without fear of legal punishment. Juries used their new power to

acquit with impunity in England in the ensuing two centuries to protect two kinds of defendants: those accused of seditious libel against the English government and those accused of a myriad of petty offenses for which death was the punishment.

8. During the American colonial period, the jury protected colonists accused of violating English trade law. In response to jury acquittals, England established courts of vice-admiralty to try smuggling offenses without juries. England's abolition of the jury was one of the grievances listed in the Declaration of Independence. In a celebrated case from the colonial period, a jury acquitted John Peter Zenger, a New York colony newspaper publisher, of seditious libel despite clear evidence that he had published the offending material. During the nineteenth century, juries also acquitted abolitionists who helped fugitive slaves and workers arrested for striking.
9. By the end of the nineteenth century, the prevailing view of the American jury's role in legal system had restricted this role to fact-finding, At least two reasons account for this restriction. First, judges became better trained and more versed in the law. Second, the jury was perceived by corporations and their allies in the upper echelon of the legal profession as too sympathetic to workers and too hostile to corporations.
10. Research on jury deliberations and functioning is abundant. The classic Chicago Jury Project by Kalven and Zeisel concluded that judges agree with criminal jury verdicts in 78 percent of all cases. Later studies have also found a high level of judge–jury agreement. As further evidence of jury competence, studies of mock and actual juries find that strength of evidence plays a key role in jury verdicts. Evidence of racial/ethnic bias by jurors for or against defendants who share their racial/ethnic background may be troubling, but such bias by juries does not appear to be greater than that exhibited by the judges with whom juries are compared.
11. Jury nullification occurs when juries reach verdicts that contradict the evidence. Jury acquittals of colonial smugglers, antislavery activists, and labor protesters during the colonial period and nineteenth century were examples of jury nullification. During the Vietnam war, many antiwar protesters who were arrested and tried wished to have juries informed of their power to nullify the law and acquit, and more acquittals might have resulted if the juries had been allowed to be told that they could disregard the law.
12. The issue of jury nullification reflects a paradox: juries have the historic power to nullify the law and acquit defendants with impunity if they do not like the law or fee the punishment is too severe, but they are not allowed to be told or reminded of this power. Many observers call for juries to be informed of their power to disregard the law if their sense of justice leads them to do so, in line with their historic role before the latter nineteenth century, while critics say this information would lead to jury anarchy and undermine the rule of law.

Key Terms

Adversarial model
Consensual model
Conviction rate
Courtroom workgroup
Crime control model
Discretionary model
Due process model
Jury nullification
Normal crime
Plea bargaining

REFERENCES

Abadinsky, Howard. 2008. *Law and Justice: An Introduction to the American Legal System*. Upper Saddle River, NJ: Prentice Hall.

Abel, Richard L. 1973. "A Comparative Theory of Dispute Institutions in Society." *Law & Society Review* 8:217–347.

———. 1988. *The Legal Profession in England and Wales*. New York: Blackwell Publishing.

———. 1989. *American Lawyers*. New York: Oxford University Press.

Abramson, Jeffrey. 1994. *We The Jury: The Jury System and the Ideal of Democracy*. New York: Basic Books.

Adler, Jeffrey S. 1994a. "The Dynamite, Wreckage, and Scum in Our Cities: The Social Construction of Deviance in Industrial America." *Justice Quarterly* 11:33–49.

Adler, Stephen J. 1994b. *The Jury: Trial and Error in the American Courtroom*. New York: Times Books. Alaya, Ana M., and David Kinney. 2004. "Leaving Children in Hot Cars Becomes Life and Death Issue." *Newark Star-Ledger* May 26:1.

Albonetti, Celesta A. 1987. "Prosecutorial Discretion: The Effects of Uncertainty." *Law & Society Review* 21:291–313.

Alschuler, Albert W. 1995. "Plea Bargaining and Its History." Pp. 138–160 in *The Law & Society Reader*, edited by Richard L. Abel. New York: New York University Press.

Alschuler, Albert W., and Andrew G. Deiss. 1994. "A Brief History of the Criminal Jury in the United States." *University of Chicago Law Review* 61:867–928.

American Bar Association. 2002. *Public Perceptions of Lawyers*. Chicago: American Bar Association.

American Bar Foundation. 2007. "http://www.abanet.org/marketresearch/resource.html.".

Anderson, Ellen Ann. 2005. *Out of the Closets and into the Courts: Legal Opportunity Structure and Gay Rights Litigation*. Ann Arbor: University of Michigan Press.

Ansley, Frances Lee. 1991. "Race and the Core Curriculum in Legal Education." *California Law Review* 79:1511–1597.

Armstrong, David. 1999. "U.S. Lagging on Prosecutions." *The Boston Globe November* 16:A1.

Asimow, Michael, and Shannon Mader. 2004. *Law and Popular Culture: A Course Book*. New York: Peter Lang.

Associated Press. 1995. "Maine High Court Rules Patient Missed Deadline for Malpractice Suit." *The Boston Globe* October 5:G5.

———. 2006. "Israeli Supreme Court Curbs Powers of Rabbinic Courts." April 6: http://news.findlaw.com/ap/o/51/04-07-2006/ddbf000acb99effd.html.

Auerbach, Jerold S. 1978. *Unequal Justice: Lawyers and Social Change in Modern America*. New York: Oxford University Press.

———. 1983. *Justice Without Law*. Oxford: Oxford University Press.

Austin, James. 1986. "Using Early Release to Relieve Prison Crowding: A Dilemma in Public Policy." *Crime and Delinquency* 32:391–403.

Austin, James, John Clark, Patricia Hardyman, and D. Alan Henry. 1999. *Three Strikes and You're Out: The Implementation and Impact of Strike Laws*. Washington, DC: National Institute of Justice, U.S. Department of Justice.

Austin, John. 1995 (1832). *The Province of Jurisprudence Determined.* Cambridge: Cambridge University Press.

Ayad, Moustafa. 2006. "Surrogate Mom Loses Triplets: Court Awards Boys to Biological Father." *Pittsburgh Post-Gazette* April 22:A1.

Babbie, Earl. 2007. *The Basics of Social Research.* Belmont, CA: Wadsworth.

Baer, Judith A. 2002. *Women in American Law: The Struggle Toward Equality from the New Deal to the Present.* New York: Holmes & Meier.

Bagdikian, Ben H. 1964. *In the Midst of Plenty: The Poor in America.* Boston: Beacon Press.

Bailey, William C. 1998. "Deterrence, Brutalization, and the Death Penalty: Another Examination of Oklahoma's Return to Capital Punishment." *Criminology* 36:711–733.

Bailis, Daniel S., and Robert J. MacCoun. 1996. "Estimating Liability Risks with the Media as Your Guide: A Content Analysis of Media Coverage of Civil Litigation." *Law and Human Behavior* 20:419–429.

Baker, Tom. 2001. "Blood Money, New Money, and the Moral Economy of Tort Law in Action." *Law & Society Review* 35:375–320.

Balbus, Isaac. 1982. *Marxism and Domination.* Princeton: Princeton University Press.

Baldus, David C., George Woodworth, and Charles A. Pulaski. 1990. *Equal Justice and the Death Penalty: A Legal and Empirical Analysis.* Boston: Northeastern University Press.

Barkan, Steven E. 1983. "Jury Nullification in Political Trials." *Social Problems* 31:28–45.

———. 1984. "Legal Control of the Southern Civil Rights Movement." *American Sociological Review* 49:552–565.

———. 1985. *Protesters on Trial: Criminal Prosecutions in the Southern Civil Rights and Vietnam Antiwar Movements.* New Brunswick, NJ: Rutgers University Press.

———. 1996. "The Social Science Significance of the O.J. Simpson Case." in *Representing O.J.: Murder, Criminal Justice and Mass Culture*, edited by Gregg Barak. Albany, NY: Harrow and Heston.

———. 2006a. "Criminal Prosecution and the Legal Control of Protest." *Mobilization: An International Journal* 11:181–194.

______. 2009. *Criminology: A Sociological Understanding,* 4th ed. Upper Saddle River, NJ: Prentice Hall.

Barkan, Steven E., and George Bryjak. 2003. *Fundamentals of Criminal Justice.* Boston: Allyn & Bacon.

Barkan, Steven E., and Steven F. Cohn. 2005. "On Reducing White Support for the Death Penalty: A Pessimistic Appraisal." *Criminology & Public Policy* 4:39–44.

Barstow, David. 2003. "U.S. Rarely Seeks Charges for Death in Workplace." *The New York Times* December 22:A1.

Barton, Roy F. 1969(1919). *Ifugao Law.* Berkeley: University of California Press.

Baxi, Upendra. 1974. "Comment—Durkheim and Legal Evolution: Some Problems of Disproof." *Law & Society Review* 8:645–652.

Beirne, Piers. 1979. "Empiricism and the Critique of Marxism on Law and Crime." *Social Problems* 26:373–385.

Benjamin, G. Andrew H., Alfred Kaszniak, Bruce Sales, and Stephen B. Shanfield. 1986. "The Role of Legal Education in Producing Psychological Distress Among Law Students and Lawyers." *Law & Social Inquiry* 11:225–252.

Berenson, Alex. 2007. "Lilly Settles with 18,000 OverZyprexa." *The New York Times* January 5:C1.

Bergen, Raquel Kennedy. 2006. *Marital Rape: New Research and Directions*. Harrisburg, PA: National Resource Center on Domestic Violence.

Bernat, Frances. 2006. "The Practice of Law in the Twenty-First Century." Pp. 133–146 in *Women, Law, and Social Control*, edited by Alida V. Merlo and Joycelyn M. Pollock. Boston: Allyn & Bacon.

Bishop, Donna. 2000. "Juvenile Offenders in the Adult Criminal Justice System." *Crime and Justice: A Review of Research* 27:81–167.

Black, Bernard, Charles Silver, David A. Hyman, and William M. Sage. 2005. "Stability, Not Crisis: Medical Malpractice Claim Outcomes in Texas, 1988–2002." *Journal of Empirical Legal Studies* 2:207–259.

Black, Donald J. 1976. *The Behavior of Law*. New York: Academic Press.

———. 1980. "The Social Organization of Arrest." Pp. 151–162 in *Police Behavior: A Sociological Perspective*, edited by Richard J. Lundman. New York: Oxford University Press.

Black, Jonathan. 1971. *Radical Lawyers: Their Role in the Movement and in the Courts*. New York: Avon.

Blankenburg, Erhard. 1994. "The Infrastructure for Avoiding Civil Litigation: Comparing Cultures of Legal Behavior in The Netherlands and West Germany." *Law & Society Review* 28:789–808.

Blumberg, Abraham S. 1967a. *Criminal Justice*. Chicago: Quadrangle Books.

———. 1967b. "The Practice of Law as a Confidence Game: Organizational Cooptation of a Profession." *Law & Society Review* 1:15–39.

Blumstein, Alfred. 1993. "Making Rationality Relevant—The American Society of Criminology 1992 Presidential Address." *Criminology* 31:1–16.

Blumstein, Alfred, and Joel Wallman (Eds.). 2006. *The Crime Drop in America*. Cambridge: Cambridge University Press.

Bohm, Robert M. 2007. *Deathquest: An Introduction to the Theory and Practice of Capital Punishment in the United States*. Cincinnati: Anderson Publishing Company.

Boyd, Susan B. 2004. "Legal Regulation of Families in Changing Societies." Pp. 255–270 in *The Blackwell Companion to Law and Society*, edited by Austin Sarat. Malden, MA: Blackwell Publishing.

Boyle, Elizabeth Heger. 2000. "Is Law the Rule? Using Political Frames to Explain Cross-national Variation in Legal Activity." *Social Forces* 79:385–418.

Bracey, Dorothy H. 2006. *Exploring Law and Culture*. Long Grove, IL: Waveland Press.

Brandwin, Pamela. 2000. "Slavery as an Interpretive Issue in the Reconstruction Congresses." *Law & Society Review* 34:315–366.

Bravin, Jess. 2002. "Bush Says Taliban, Al Qaeda Fighters Aren't POWs Under Geneva Conventions." *The Wall Street Journal* A20.

Brinkley, Joel. 2006. "Bush Budget Would Slash Bolivia Military Aid." *The New York Times* February 9:A10.

Brown, Dorothy A. 2003. *Critical Race Theory: Cases, Materials, and Problems*. St. Paul, MN: Thomson/West.

Brown, M. Craig, and Barbara D. Warner. 1995. "The Political Threat of Immigrant Groups and Police Aggressiveness in 1900." Pp. 82–98 in *Ethnicity, Race, and Crime: Perspectives Across Time and Place*, edited by Darnell F. Hawkins. Albany, NY: State University of New York Press.

Brown-Nagin, Tomiko. 2005. "Elites, Social Movements, and the Law: The Case of Affirmative Action." *Columbia Law Review* 105.

Bumiller, K. 1988. *The Civil Rights Society: The Social Construction of Victims*. Baltimore: Johns Hokpins University Press.

Bureau of Labor Statistics. 2007. "Occupational Outlook Handbook, 2006–2007 Edition, Lawyers." http://www.bls.gov/oco/ocos053.htm.

Burns, W. Haywood. 1998. "Law and Race in Early America." Pp. 279–284 in *The Politics of Law: A Progressive Critique*, edited by David Kairys. New York: Basic Books.

Burros, Marian. 2006. "KFC Is Sued Over the Use of Trans Fats in Its Cooking." *The New York Times* June 14:A16.

Burstein, Paul. 1994. *Equal Employment Opportunity: Labor Market Discrimination and Public Policy*. New York: Aldine de Gruyter.

Cain, Maureen, and Alan Hunt (Eds.). 1979. *Marx and Engels on Law*. New York: Academic Press.

Campo-Flores, Arian. 2005. "The Legacy of Terri Schiavo." *Newsweek* April 4:22–29.

Cantor, Norman F. 1997. *Imagining the Law: Common Law and the Foundations of the American Legal System*. New York: HarperCollins Publishers.

Caplovitz, David. 1963. *The Poor Pay More: Consumer Practices of Low-Income Families*. New York: Free Press.

Carlin, Jerome E. 1962. *Lawyers on Their Own: A Study of Individual Practitioners in Chicago*. New Brunswick, NJ: Rutgers University Press.

———. 1966. *Lawyers' Ethics: A Survey of the New York City Bar*. New York: Russell Sage Foundation.

Carlin, Jerome, Jan Howard, and Sheldon Messenger. 1966. "Civil Justice and the Poor." *Law & Society Review* 1:9–90.

Carrasco, Miguel, Edward D. Barker, Richard E. Tremblay, and Frank Vitaro. 2006. "Eysenck's Personality Dimensions as Predictors of Male Adolescent Trajectories of Physical Aggression, Theft and Vandalism." *Personality & Individual Differences* 41:1309–1320.

Carrington, Paul D. 1984. "Of Law and the River." *Journal of Legal History* 34: 222–228.

Carroll, Catherine, and April Brayfield. 2007. "Lingering Nuances: Gendered Career Motivations and Aspirations of First-Year Law Students." *Sociological Spectrum* 27:225–255.

Catalano, Shannan M. 2006. *Criminal Victimization, 2005*. Washington, DC: Bureau of Justice Statistics, U.S. Department of Justice.

Chamallas, Martha. 2003. *Introduction to Feminist Legal Theory*. New York: Aspen Publishers.

Chambliss, Elizabeth, and Christopher Uggen. 2000. "Men and Women of Elite Law Firms: Reevaluating Kanter's Legacy." *Law and Social Inquiry* 25:41–68.

Chambliss, William J. 1964. "A Sociological Analysis of the Law of Vagrancy." *Social Problems* 12:67–77.

———. 1967. "Types of Deviance and the Effectiveness of Legal Sanctions." *Wisconsin Law Review* Summer:703–719.

Chambliss, William, and Robert Seidman. 1982. *Law, Order, and Power*. Reading, MA: Addison-Wesley Publishing Company.

Chamlin, Mitchell B. 1991. "A Longitudinal Analysis of the Arrest-Crime Relationship: A Further Examination of the Tipping Effect." *Justice Quarterly* 8:187–199.

Chesney-Lind, Meda, and Lisa Pasko. 2004. *The Female Offender: Girls, Women, and Crime*. Thousand Oaks, CA: Sage Publications.

Chiricos, Ted, Kelly Welch, and Marc Gertz. 2004. "Racial Typification of Crime and Support for Punitive Measures." *Criminology* 42:359–389.

Chiricos, Theodore G., and Gordon P. Waldo. 1975. "Socioeconomic Status and Criminal Sentencing: An Assessment of a Conflict Proposition." *American Sociological Review* 40:753–772.

Christie, Jim. 2005. "California Sets Standard for Lesbian Parents." *The Boston Globe* August 23:A3.

Clark, Blue. 1999. *Lone Wolf v. Hitchcock: Treaty Rights and Indian Law at the End of the Nineteenth Century*. Lincoln, NE: University of Nebraska Press.

Clark, Walter Van Tilburg. 1940. *The Ox-Bow Incident*. New York: Random House.

Clarke, James W. 1990. *On Being Mad or Merely Angry: John W. Hinckley, Jr., and Other Dangerous People*. Princeton: Princeton University Press.

Clarke, Stevens H., and Gary G. Koch. 1976. "The Influence of Income and Other Factors on Whether Criminal Defendants Go to Prison." *Law & Society Review* 11:57–92.

Cochran, John K., Mitchell B. Chamlin, and Mark Seth. 1994. "Deterrence or Brutalization? An Impact Assessment of Oklahoma's Return to Capital Punishment." *Criminology* 32:107–134.

Coglianese, Cary. 2001. "Social Movements, Law, and Society: The Institutionalization of the Environmental Movement." *University of Pennsylvania Law Review* 150:85–118.

Cohen, Thomas H. 2005. *Punitive Damage Awards in Large Counties, 2001*. Washington, DC: Bureau of Justice Statistics, U.S. Department of Justice.

Cohen, Thomas H., and Steven K. Smith. 2004. *Civil Trial Cases and Verdicts in Large Counties, 2001*. Washington, DC: Bureau of Justice Statistics, U.S. Department of Justice.

Cole, David, and James X. Dempsey. 2006. *Terrorism and the Constitution: Sacrificing Civil Liberties in the Name of National Security*, 3rd ed. New York: W.W. Norton.

Coleman, James William. 2006. *The Criminal Elite: Understanding White-Collar Crime*. New York: Worth Publishers.

Collins, Hugh. 1982. *Marxism and Law*. Oxford: Clarendon Press.

Collins, Randall. 1994. *Four Sociological Traditions*. New York: Oxford University Press.

Costanzo, Mark. 1997. *Just Revenge: Costs and Consequences of the Death Penalty*. New York: St. Martin's Press.

Cotterrell, Roger. 2004. "Law in Social Theory and Social Theory in the Study of Law." Pp. 15–29 in *The Blackwell Companion to Law and Society*, edited by Austin Sarat. Malden, MA: Blackwell Publishing.

Crook, John A. 1984. *Law and Life of Rome*. Ithaca, NY: Cornell University Press.

Cross, F. 1997. "Political Science and the New Legal Realism." *Northwestern University Law Review* 92:251–326.

Cullen, Francis T., William J. Maakestad, and Gray Cavender. 2006. *Corporate Crime Under Attack: The Fight to Criminalize Business Violence*. Cincinnati: Anderson Publishing Company.

Cunningham, David. 2004. *There's Something Happening Here: The New Left, the Klan, and FBI Counterintelligence*. Berkeley: University of California Press.

D'Alessio, Stewart J., and Lisa Stolzenberg. 1993. "Socioeconomic Status and the Sentencing of the Traditional Offender." *Journal of Criminal Justice* 21:61–77.

———. 1998. "Crime, Arrests, and Pretrial Jail Incarceration: An Examination of the Deterrence Thesis." *Criminology* 36:735–761.

Daly, Kathleen. 1994. *Gender, Crime, and Punishment*. New Haven: Yale University Press.

Dammer, Harry R., and Erika Fairchild. 2006. *Comparative Criminal Justice Systems*. Belmont, CA: Wadsworth Publishing Company.

Daniels, Stephen, and Joanne Martin. 1995. *Civil Juries and the Politics of Reform*. Evanston, IL: Northwestern University Press.

Davenport, Anniken U. 2006. *Basic Criminal Law: The U.S. Constitution, Procedure, and Crimes*. Upper Saddle River, NJ: Prentice Hall.

Davey, Monica. 2006. "Iowa's Residency Rules Drive Sex Offenders Underground." *The New York Times* March 15:A1.

Davis, Michael, and Frederick A. Elliston. 1986. *Ethics and the Legal Profession*. Buffalo: Prometheus Books.

Death Penalty Information Center. 2006. *The Death Penalty in 2006: Year End Report*. Washington, DC: Death Penalty Information Center.

Decker, Scott, and Carol Kohfeld. 1985. "Crimes, Crime Rates, Arrests, and Arrest Ratios: Implications for Deterrence Theory." *Criminology* 23:437–450.

Delaney, David. 2003. *Law and Nature*. New York: Cambridge University Press.

Delgado, Richard, and Jean Stefancic. 2001. *Critical Race Theory: An Introduction*. New York: New York University Press.

Demillo, Andrew. 2006. "Judge Halts Construction to Protect Woodpecker." *The Boston Globe* July 21:A5.

Devine, Dennis J., Laura D. Clayton, Benjamin B. Dunford, Rasmy Seying, and Jennifer Pryce. 2001. "Jury Decision Making: 45 Years of Empirical Research on Deliberating Groups." *Psychology, Public Policy, and Law* 7:622–727.

Diamond, Shari Seidman, and Mary R. Rose. 2005. "Real Juries." *Annual Review of Law and Social Science* 1:255–284.

Dinovitzer, Ronit, and Bryant G. Garth. 2007. "Lawyer Satisfaction in the Process of Structuring Legal Careers." *Law & Society Review* 41:1–50.

Doob, Anthony N., and Cheryl Marie Webster. 2003. "Sentence Severity and Crime: Accepting the Null Hypothesis." *Crime and Justice: A Review of Research* 30:143–195.

Dowd, Nancy E., and Michelle S. Jacobs (Eds.). 2003. *Feminist Legal Theory: An Anti-Essentialist Reader*. New York: New York University Press.

Downie, Leonard, Jr. 1972. *Justice Denied: The Case for Reform of the Courts*. Baltimore: Penguin Books.

Dror, Yehezkel. 1969. "Law and Social Change." Pp. 90–99 in *Sociology of Law: Selected Readings*, edited by Vilhelm Aubert. Harmondsworth, England: Penguin Books.

Duman, Daniel. 1980. "Pathway to Professionalism: The English Bar in the Eighteenth and Nineteenth Centuries." *Journal of Social History* 13:615–628.

Durkheim, Emile. 1933 (1893). *The Division of Labor in Society*. London: The Free Press.

———. 1962 (1895). *The Rules of Sociological Method*. New York: Free Press.

Durose, Matthew R., and Patrick A. Langan. 2007. *Felony Sentences in State Courts, 2004*. Washington, DC: Bureau of Justice Statistics, U.S. Department of Justice.

Dwyer, William L. 2002. *In the Hands of the People: The Trial Jury's Origins, Triumphs, Troubles, and Future in American Democracy*. New York: Thomas Dunne Books.

Echeverria, John D. 2003. "Standing and Mootness Decisions in the Wake of Laidlaw." *Widener Law Review* 10:183–204.

Eckhoff, Torstein. 1969. "The Mediator and the Judge." Pp. 171–181 in *Sociology of Law: Selected Readings*, edited by Vilhelm Aubert. Harmondsworth, England: Penguin Books.

Edgerton, Robert. 1976. *Deviance: A Cross-Cultural Perspective*. Menlo Park, CA: Cummings Publishing Company.

Egelko, Bob. 2005. "Court Grants Equal Rights to Same-Sex Parents: Breaking Up Partnerships Doesn't End Parental Obligations." *The San Francisco Chronicle* August 23:A1.

Ehrensaft, Miriam K., Patricia Cohen, and Jeffrey G. Johnson. 2006. "Development of Personality Disorder Symptoms and the Risk for Partner Violence." *Journal of Abnormal Psychology* 115:474–483.

Ehrlich, Eugen. 1936 (1913). *Fundamental Principles of the Sociology of Law*. Cambridge: Harvard University Press.

Eisenberg, Theodore, Neil LaFountain, Brian Ostrom, David Rottman, and Martin T. Wells. 2002. "Juries, Judges, and Punitive Damages: An Empirical Study." *Cornell Law Review* 87:743–782.

Eisenberg, Theodore, Paula L. Hannaford-Agor, Michael Heise, Neil LaFountain, G. Thomas Munsterman, Brian Ostrom, and Martin T. Wells. 2006. "Juries, Judges, and Punitive Damages: Empirical Analyses Using the Civil Justice Survey of State Courts 1992, 1996, and 2001 Data." *Journal of Empirical Legal Studies* 3:263–295.

Eisenstein, James, and Hebert Jacob. 1977. *Felony Justice: An Organizational Analysis of Criminal Courts*. Boston: Little, Brown and Company.

Eller, Jack David. 2006. *Violence and Culture: A Cross-Cultural and Interdisciplinary Approach*. Belmont, CA: Thomson/Wadsworth.

Ellickson, Robert C. 2000. "Trends in Legal Scholarship: A Statistical Study." *Journal of Legal Studies* 29:517–543.

Elliott, Janet. 2005. "Gay Marriage Ban Put in Texas Constitution." *The Houston Chronicle* November 9:A1.

Elliott, Michael. 1999. "Telling the Difference: Nineteenth-Century Legal Narratives of Racial Taxonomy." *Law and Social Inquiry* 24:611–636.

Elon, Menachem. 1994. *Jewish Law: History, Sources, Principles*. Philadelphia: Jewish Publication Society.

Ely, John Hart. 1993. *War and Responsibility: Constitutional Lessons of Vietnam and Its Aftermath*. Princeton: Princeton University Press.

Emmelman, Debra S. 1996. "Trial by Plea Bargain: Case Settlement as a Product of Recursive Decisionmaking." *Law & Society Review* 30:335–360.

Engel, David M. 1984. "The Oven Bird's Song: Insiders, Outsiders, and Personal Injuries in an American Community." *Law & Society Review* 18:551–582.

Engel, David M., and Frank W. Munger. 2003. *Rights of Inclusion: Law and Identity in the Life Stories of Americans with Disabilities*. Chicago: University of Chicago Press.

Engel, Robin Shephard, and Jennifer M. Calnon. 2004. "Examining the Influence of Drivers' Characteristics During Traffic Stops with Police: Results from a National Survey." *Justice Quarterly* 21:49–90.

Engels, Friedrich. 1926. *The Peasant War in Germany*. New York: International Publishers.

———. 1993 (1845). "The Demoralization of the English Working Class." Pp. 48–50 in *Crime and Capitalism: Readings in Marxist Criminology*, edited by David F. Greenberg. Philadelphia: Temple University Press.

Engen, Rodney L., Randy R. Gainey, Robert D. Crutchfield, and Joseph G. Weis. 2003. "Discretion and Disparity Under Sentencing Guidelines: The Role of Departures and Structured Sentencing Alternatives." *Criminology* 41:99–130.

Epstein, Cynthia Fuchs. 1981. *Women in Law*. New York: Basic Books.

Erikson, Kai T. 1976. *Everything in Its Path: Destruction of Community in the Buffalo Creek Flood*. New York: Simon and Schuster.

Erlanger, Howard S., and Douglas A. Klegon. 1978. "Socialization Effects of Professional School: The Law School Experience and Student Orientations to Public Interest Concerns." *Law & Society Review* 13:11–35.

Eskirdge, William M., Jr. 2001. "Channeling: Identity-Based Social Movements and Public Law." *University of Pennsylvania Law Review* 150:419–525.

Evan, William M. 1965. "Law as an Instrument of Social Change." Pp. 285–293 in *Applied Sociology: Opportunities and Problems*, edited by Alvin W. Gouldner and S. M. Miller. New York: The Free Press.

Ewick, Patricia, and Susan S. Silbey. 1998. *The Common Place of Law: Stories from Everyday Life*. Chicago: University of Chicago Press.

Fagan, Jeffrey, and Garth Davies. 2001. "Street Stops and Broken Windows: Terry, Race, and Disorder in New York City." *Fordham Urban Law Journal* 28:457–504.

Farole, Donald J., Jr. 1999. "Reexamining Litigant Success in State Supreme Courts." *Law & Society Review* 33:1043–1057.

Farrington, David P., Rolf Loeber, and Magda Stouthamer-Loeber. 2003. "How Can the Relationship Between Race and Violence be Explained?" Pp. 213–237 in *Violent Crime: Assessing Race and Ethnic Differences*, edited by Darnell F. Hawkins. Cambridge: Cambridge University Press.

Fass, Simon M., and Chung-Ron Pi. 2002. "Getting Tough on Juvenile Crime: An Analysis of Costs and Benefits." *Journal of Research in Crime and Delinquency* 39:363–399.

Federal Bureau of Investigation. 2007. *Crime in the United States, 2006*. Washington, DC: Federal Bureau of Investigation.

Feest, Johannes. 1968. "Compliance with Legal Regulations: Observations of Stop Sign Behavior." *Law & Society Review* 2:447–462.

Felstiner, William L. F. 1974. "Influences of Social Organization on Dispute Processing." *Law & Society Review* 9:63–94.

Felstiner, William L. F., Richard L. Abel, and Austin Sarat. 1980–1981. "The Emergence and Transformation of Disputes: Naming, Blaming, Claiming." *Law & Society Review* 15:631–654.

Festinger, Leon. 1957. *A Theory of Cognitive Dissonance*. Stanford: Stanford University Press.

Finan, Christopher M. 2007. *From the Palmer Raids to the Patriot Act: A History of the Fight for Free Speech in America*. Boston: Beacon Press.

Finkelman, Paul. 1981. "The Zenger Case: Prototype of a Political Trial." Pp. 21–42 in *American Political Trials*, edited by Michal Belknap. Westport, CT: Greenwood Press.

Firebaugh, Glenn, and Brian Harley. 1995. "Trends in Job Satisfaction in the United States by Race, Gender, and Type of Occupation." *Research in the Sociology of Work* 5:87–104.

Firestone, David. 1999. "Alabama Acts to Limit Huge Awards by Juries." *The New York Times* June 2:A16.

———. 2001. "Defense System in Georgia Needs Overhaul, Lawyers Say." *The New York Times* July 31:A8.

Fisher, George. 2003. *Plea Bargaining's Triumph: A History of Plea Bargaining in America*. Stanford: Stanford University Press.

Fisher, Karla, Neil Vidmar, and René Ellis. 1993. "The Culture of Battering and the Role of Mediation in Domestic Violence Cases." *Southern Methodist University Law Review* 46:2117–2174.

Footlick, Jerrold K. 1977. "Too Much Law?" *Newsweek* January 10:40–47.

Forbath, William E. 1991. *Law and the Shaping of the American Labor Movement*. Cambridge, MA: Harvard University Press.

Frank, Jerome. 1930. *Law and the Modern Mind*. New York: Brentano's.

———. 1949. *Courts on Trial: Myth and Reality in American Justice*. Princeton: Princeton University Press.

Frankfurter, Felix, and Nathan Greene. 1930. *The Labor Injunction*. New York: Macmillan.

Friedman, Leon. 1971. *The Wise Minority*. New York: Dial Press.

Friedman, Lawrence M. 1969. "Legal Culture and Social Development." *Law & Society Review* 4:29–44.

———. 1975. *The Legal System: A Social Science Perspective*. New York: Russell Sage Foundation.

———. 1984. *American Law: An Introduction*. New York: W. W. Norton & Company.

———. 1985. *A History of American Law*. New York: Simon and Schuster.

———. 1986. "The Law and Society Movement." *Stanford Law Review* 38:763–780.

———. 1989. "Litigation and Society." *Annual Review of Sociology* 15:17–29.

———. 2004a. *Law in America: A Short History*. New York: The Modern Library.

———. 2004b. *Private Lives: Families, Individuals, and the Law*. Cambridge, MA: Harvard University Press.

———. 2005. "Coming of Age: Law and Society Enters an Exclusive Club." *Annual Review of Law and Social Science* 1:1–16.

Friedman, Lawrence M., and Jack Ladinsky. 1967. "Social Change and the Law of Industrial Accidents." *Columbia Law Review* 67:50–82.

Friedrichs, David O. 2007. *Trusted Criminals: White Collar Crime in Contemporary Society*. Belmont, CA: Wadsworth Publishing Company.

Fritsch, Jane, and David Rohde. 2001. "Lawyers Often Fail New York's Poor." *The New York Times* April 8:A1.

Frohmann, Lisa. 1991. "Discrediting Victims' Allegations of Sexual Assault: Prosecutorial Accounts of Case Rejections." *Social Problems* 38:213–226.

Fuller, Lon L. 1949. "The Case of the Speluncean Explorers." *Harvard Law Review* 62:616–645.

———. 1964. *The Morality of Law*. New Haven: Yale University Press.

Galanter, Marc. 1974. "Why the 'Haves' Come Out Ahead: Speculations on the Limits of Legal Change." *Law & Society Review* 9:95–160.

———. 1983. "Reading the Landscape of Disputes: What We Know and Don't Know (and Think We Know) About Our Allegedly Contentious and Litigious Society." *UCLA Law Review* 31:4–71.

———. 1993. "News from Nowhere: The Debased Debate on Civil Justice." *Denver University Law Review* 71:77–113.

———. 1998. "An Oil Strike in Hell: Contemporary Legends About the Civil Justice System." *Arizona Law Review* 40:717–752.

———. 2006. *Lowering the Bar: Lawyer Jokes and Legal Culture*. Madison, WI: University of Wisconsin Press.

Garcia, Michelle. 2005. "Gang Trial Tests Reach of New Laws on Terror." *The Boston Globe* February 3:A4.

Garner, David D. 2000. "Socratic Misogyny? Analyzing Feminist Criticisms of Socratic Teaching in Legal Education." *Brigham Young University Law Review* 2000: 1597–1649.

Garrow, Gail. 2006. *The Printer's Trial: The Case of John Peter Zenger and the Fight for a Free Press*. Honesdale, PA: Calkins Creek Books.

Garth, Bryant, and Joyce Sterling. 1998. "From Legal Realism to Law and Society: Reshaping Law for the Last Stages of the Activist State." *Law & Society Review* 32:409–471.

Garvey, Stephen P., Paula Hannaford-Agor, Valerie P. Hans, Nicole L. Mott, G. Thomas Munsterman, and Martin T. Wells. 2004. "Juror First Votes in Criminal Trials." *Journal of Empirical Legal Studies* 1:371–398.

Gastil, John, E. Pierre Deess, and Phil Weiser. 2002. "Civic Awakening in the Jury Room: A Test of the Connection Between Jury Deliberation and Political Participation." *The Journal of Politics* 64:585–595.

Gatrell, V. A. C. 1996. *The Hanging Tree: Execution and the English People, 1770–1868*. New York: Oxford University Press.

Gelbspan, Ross. 1991. *Break-Ins, Death Threats and the FBI: The Covert War Against the Central America Movement*. Boston: South End Press.

George, Robert P. 1999. *In Defense of Natural Law*. New York: Oxford University Press.

Getlin, Josh. 2002. "Officer Defies NYPD Over Homeless." *Los Angeles Times* November 30:A18.

Gibbs, Jack P. 1968. "Crime, Punishment, and Deterrence." *Southwestern Social Science Quarterly* 48:515–530.

Gieringer, Dale. 2007. "State's War on Drugs a 100-Year Bust." *San Francisco Chronicle* March 4:E1.

Ginger, Ann Fagan. 1972. *The Relevant Lawyers*. New York: Simon and Schuster.

Ginsburg, Tom, and Glenn Hoetker. 2006. "The Unreluctant Litigant? An Empirical Analysis of Japan's Turn to Litigation." *Journal of Legal Studies* 35:31–59.

Glaberson, William. 1999a. "The $2.9 Million Cup of Coffee: When the Verdict Is Just a Fantasy." *The New York Times* June 6:Section 4, Page 1.

———. 1999b. "Some Plaintiffs Losing Out in Texas' War on Lawsuits." *The New York Times* June 7:A1.

———. 2001. "A Study's Verdict: Jury Awards Are Not Out of Control." *The New York Times* August 6:A9.

Gleason, Sandra. 1981. "The Probability of Redress." in *Outsiders on the Inside: Women and Organizations*, edited by B. Florisha and B. Goldman. Englewood Cliffs, NJ: Prentice Hall.

Glod, Maria. 2004. "'DNA Dragnet' Makes Charlottesville Uneasy." *The Washington Post* April 14:A01.

Gluckman, Max. 1955. *The Judicial Process Among the Barotse of Northern Rhodesia*. Manchester, England: University Press for the Rhodes-Livingstone Institute.

Gold, David A., Clarence Y. H. Lo, and Erik Olin Wright. 1975. "Recent Developments in Marxist Theory of the Capitalist State, Part I." *Monthly Review* 27:29–43.

Goldberg, Carey. 2000. "Court Says a Partner Can Veto Embryo Implantation." *The New York Times* April 4:A1.

Golding, William. 1954. *Lord of the Flies*. London: Coward-McCann.

Gómez, Laura E. 2004. "A Tale of Two Genres: On the Real and Ideal Links Between Law and Society and Critical Race Theory." Pp. 453–470 in *The Blackwell Companion to Law and Society*, edited by Austin Sarat. Malden, MA: Blackwell Publishing.

Goode, Erich. 2008. *Drugs in American Society*. New York: McGraw Hill.

Goodell, Charles. 1973. *Political Prisoners in America*. New York: Random House.

Goodnough, Abby. 2005. "Schiavo Dies, Ending Bitter Case Over Feeding Tube." *The New York Times* April 1:A1.

———. 2007. "In a Break from the Past, Florida Will Let Felons Vote." *The New York Times* April 6:A14.

Gordon, Robert W. 1982. "New Developments in Legal Theory." Pp. 281–293 in *The Politics of Law*, edited by David Kairys. New York: Pantheon.

Gordon, Sarah Barringer. 2002. *The Mormon Question: Polygamy and Constitutional Conflict in Nineteenth-Century America*. Chapel Hill, NC: University of North Carolina Press.

Grana, Sheryl J. 2002. *Women and (In)justice: The Criminal and Civil Effects of the Common Law on Women's Lives*. Boston: Allyn & Bacon.

Granfield, Robert. 1992. *Making Elite Lawyers: Visions of Law at Harvard and Beyond*. New York: Routledge.

Grau, Charles W. 1982. "Whatever Happened to Politics? A Critique of Structuralist Marxist Accounts of State and Law." Pp. 196–209 in *Marxism and Law*, edited by Piers Beirne and Richard Quinney. New York: Wiley.

Gray, John Chipman. 1963 (1909). *Nature and Sources of the Law*. Boston: Beacon Press.

Green, Thomas A. 1985. *Verdict According to Conscience: Perspectives on the English Criminal Trial Jury, 1200–1800*. Chicago: University of Chicago Press.

Greenberg, David F. 1976. "On One-Dimensional Marxist Criminology." *Theory and Society* 3:610–621.

Greene, Bob. 1984. "U.S. Lawyers a New 'Poison' in Bhopal." *Chicago Tribune* December 17:C5.

Greenhouse, Carol J. 1986. *Praying for Justice: Faith, Order, and Community in an American Town*. Ithaca: Cornell University Press.

Greenhouse, Linda. 2002. "Citing 'National Consensus," Justices Bar Death Penalty for Retarded Defendants." *The New York Times* June 21:A1.

———. 2005. "Supreme Court , 5-4, Forbids Execution in Juvenile Crime." *The New York Times* March 2:A1.

Grenig, Jay E. 2005. *Alternative Dispute Resolution*. Eagan, MN: Thomson/West.

Grillo, Trina. 1991. "The Mediation Alternative: Process Dangers for Women." *Yale Law Journal* 100:1545–1581.

Grossberg, Michael. 1985. *Governing the Hearth: Law and the Family in Nineteenth-Century America*. Chapel Hill: University of North Carolina Press.

Grossman, Joel. 1966. "Social Backgrounds and Judicial Decision-Making." *Harvard Law Review* 79:1551–1564.

Gulliver, Philip H. 1979. *Disputes and Negotiations: A Cross-Cultural Perspective*. New York: Academic Press.

Gusfield, Joseph R. 1963. *Symbolic Crusade: Status Politics and the American Temperance Movement*. Urbana, IL: University of Illinois Press.

Gyorgy, Anna. 1979. *No Nukes: Everyone's Guide to Nuclear Power*. Boston: South End Press.

Hagan, John. 1974. "Extra-Legal Attributes and Criminal Sentencing: An Assessment of a Sociological Viewpoint." *Law & Society Review* 8:357–383.

Hagan, John, and Fiona Kay. 2007. "Even Lawyers Get the Blues: Gender, Depression, and Job Satisfaction in Legal Practice." *Law & Society Review* 41:51–78.

Hakim, Danny. 2006. "Judge Urges State Control of Legal Aid for the Poor." *The New York Times* June 29:B1.

Haley, John O. 1978. "The Myth of the Reluctant Litigant." *Journal of Japanese Studies* 4:359–389.

Hall, Jerome. 1952. *Theft, Law, and Society*. Indianapolis: Bobbs-Merrill.

Haltom, William, and Michael McCann. 2004. *Distorting the Law: Politics, Media, and the Litigation Crisis*. Chicago: University of Chicago Press.

Hames, Joanne Banker, and Yvonne Ekern. 2006. *Introduction to Law*. Upper Saddle River, NJ: Prentice Hall.

Hamilton, Andrea. 1995. "Gay Groups Are Spied Upon, FBI Data Show." *The Boston Globe* May 16:3.

Handler, Joel F. 1967. *The Lawyer and His Community: The Practicing Bar in a Middle-Sized City*. Madison: University of Wisconsin Press.

———. 1978. *Social Movements and the Legal System*. New York: Academic Press.

Hannaford, Paula L., Valerie P. Hans, and G. Thomas Munsterman. 2000. "Permitting Jury Discussions During Trial: Impact of the Arizona Reform." *Law and Human Behavior* 24:359–382.

Hans, Valerie P., and Neil Vidmar. 1986. *Judging the Jury*. New York: Plenum Publishing Corp.

———. 2004. "Jurors and Juries." Pp. 195–211 in *The Blackwell Companion to Law and Society*, edited by Austin Sarat. Malden, MA: Blackwell Publishing.

Harring, Sidney L. 1977. "Class Conflict and the Suppression of Tramps in Buffalo, 1892–1894." *Law & Society Review* 11:873–911.

———. 1993. "Policing a Class Society: The Expansion of the Urban Police in the Late Nineteenth and Early Twentieth Centuries." Pp. 546–567 in *Crime and Capitalism: Readings in Marxist Criminology*, edited by David F. Greenberg. Philadelphia: Temple University Press.

———. 1994. *Crow Dog's Case: American Indian Sovereignty, Tribal Law, and United States Law in the Nineteenth Century*. New York: Cambridge University Press.

Harrington, Michael. 1962. *The Other America: Poverty in the United States*. New York: Macmillan.

Harris, Beth. 1999. "Representing Homeless Families: Repeat Player Implementation Strategies." *Law & Society Review* 33:911–939.

Harris, David. 2002. *Profiles in Injustice: Why Racial Profiling Cannot Work*. New York: New Press.

Hart, H. L. A. 1958. "Positivism and the Separation of Law and Morals." *Harvard Law Review* 71:593–629.

Hay, Douglas. 1975. "Property, Authority and the Criminal Law." Pp. 17–63 in *Albion's Fatal Tree: Crime and Society in Eighteenth-Century England*, edited by Douglas Hay, Peter Linebaugh, John G. Rule, E. P. Thompson, and Cal Winslow. New York: Pantheon Books.

Heen, Sheila, and John Richardson. 2005. "'I See a Pattern Here and the Pattern Is You': Personality and Dispute Resolution." Pp. 35–51 in *The Handbook of Dispute Resolution*, edited by Michael L. Moffitt and Robert C. Bordone. San Francisco: Jossey-Bass.

Heinemann, Sue. 1996. *Timelines of American Women's History*. New York: A Roundtable Press/Perigee Book.

Heinz, John P., and Edward O. Laumann. 1982. *Chicago Lawyers: The Social Structure of the Bar*. New York: Russell Sage Foundation and American Bar Foundation.

Heinz, John P., Anthony Paik, and Ann Southworth. 2003. "Lawyers for Conservative Causes: Clients, Ideology, and Social Distance." *Law & Society Review* 37:5–50.

Heinz, John P., Robert L. Nelson, Rebecca L. Sandfeur, and Edward O. Laumann. 2005. *Urban Lawyers: The New Social Structure of the Bar*. Chicago: University of Chicago Press.

Hepburn, John R. 1977. "Social Control and the Legal Order: Legitimated Repression in a Capitalist State." *Contemporary Crises* 1:77–90.

Herbert, Bob. 2005. "On Abu Ghraib, the Big Shots Walk." *The New York Times* April 28:A25.

Herdy, Amy, and Miles Moffeit. 2004. *Betrayal in the Ranks*. Denver: The Denver Post (http://www.denverpost.com/Stories/0,0,36%257E30137%257E,00.html).

Heuer, Larry, and Steven Penrod. 1994. "Juror Notetaking and Question Asking During Trials: A National Field Experiment." *Law and Human Behavior* 18:121–150.

Heumann, Milton. 1978. *Plea Bargaining: The Experiences of Prosecutors, Judges, and Defense Attorneys*. Chicago: University of Chicago Press.

Hodos, George H. 1987. *Show Trials: Stalinist Purges in Eastern Europe*. New York: Praeger.

Hoebel, E. Adamson. 1954. *The Law of Primitive Man: A Study in Comparative Legal Dynamics*. Cambridge: Harvard University Press.

Holmes, Oliver Wendell, Jr. 1938 (1881). *The Common Law*. Boston: Little, Brown, and Company.

Holmes, Stephen T., and Ronald M. Holmes (Eds.). 2004. *Violence: A Contemporary Reader*. Upper Saddle River, NJ: Prentice Hall.

Holscher, Louis M., and Rizwana Mahmood. 2000. "Borrowing from the Shariah: The Potential Uses of Procedural Islamic Law in the West." Pp. 82–96 in *International Criminal Justice: Issues in a Global Perspective*, edited by Delbert Rounds. Boston: Allyn & Bacon.

Hoyman, Michele, and Lamont Stallworth. 1986. "Suit Filing by Women: An Empirical Analysis." *Notre Dame Law Review* 64:61–82.

Huber, Peter. 1988. *Liability: The Legal Revolution and Its Consequences*. New York: Basic Books.

Huff, C. Ronald. 2002. "Wrongful Conviction and Public Policy: The American Society of Criminology 2001 Presidential Address." *Criminology* 40:1–18.

Hull, Kathleen E. 1999. "The Paradox of the Contented Female Lawyer." *Law & Society Review* 33:687–702.

Human Rights Campaign. 2007. *Relationship Recognition in the U.S.* Washington, DC: Human Rights Campaign.

Hurst, Charles E. 2007. *Social Inequality: Forms, Causes, and Consequences*. Upper Saddle River, NJ: Prentice Hall.

Husak, Douglas. 2002. *Legalize This! The Case for Decriminalizing Drugs*. New York: Verso Books.

Hutchinson, Allan C. (Ed.). 1989. *Critical Legal Studies*. Totowa, NJ: Rowman & Littlefield.

Hyman, Herbert H., and Paul Sheatsley. 1964. "Attitudes Toward Desegregation." *Scientific American* 211:16–23.

International Court of Justice. 2005. *International Court of Justice*. New York: United Nations Office of Public Information.

Jacob, Herbert. 1978. *Justice in America: Courts, Lawyers, and the Judicial Process*. Boston: Little, Brown and Company.

———. 1988. *Silent Revolution: The Transformation of Divorce Law in the United States*. Chicago: University of Chicago Press.

Jacobs, David. 1980. "Marxism and the Critique of Empiricism: A Comment on Beirne." *Social Problems* 27:467–470.

Jacoby, Jeff. 2005. "One Nation Under Law." *The Boston Globe* April 3:D11.

James, Marlise. 1973. *The People's Lawyers*. New York: Holt, Rinehart and Winston.

Janis, Mark. 2003. *An Introduction to International Law*. New York: Aspen Publishers.

Jessop, Bob. 1977. "Recent Theories of the Capitalist State." *Cambridge Journal of Economics* 1:353–373.

Jhering, Rudolph von. 1968 (1877). *Law as a Means to an End*. New York: A.M. Kelleyy.

Johnson, Dirk. 2000. "Poor Legal Work Common for Innocents on Death Row." *The New York Times* February 5:A1.

Johnston, Lloyd D., Patrick M. O'Malley, Jerold G. Bachman, and J.E. Schulenberg. 2007. *Monitoring the Future. National Results on Adolescent Drug Use: Overview of Key Findings, 2006*. (NIH Publication No. 06-5882). Bethesda, MD: National Institute on Drug Abuse.

Jones, David A. 1986. *History of Criminology: A Philosophical Perspective*. New York: Greenwood Press.

Josephson, Matthew. 1962. *The Robber Barons: The Great American Capitalists, 1861–1901*. New York: Harcourt, Brace, & World.

Jost, Kenneth. 1999. "Plea Bargaining." *CQ Researcher* 9:115–133.

Joyner, Christopher C. 2005. *International Law in the 21st Century: Rules for Global Governance*. Lanham, MD: Romwan & Littlefield.

Kadish, Mortimer R., and Sanford H. Kadish. 1973. *Discretion to Disobey: A Study of Lawful Departures from Legal Rules*. Stanford: Stanford University Press.

Kagan, Robert A. 2001. *Adversarial Legalism: The American Way of Law*. Cambridge: Harvard University Press.

———. 2006. "The Organization of Administrative Justice Systems: The Role of Political Mistrust." Working paper #38 of the Jurisprudence and Social Policy Program, Center for the Study of Law and Society, University of California, Berkeley.

Kahn, Joseph. 2006. "A Sharp Debate Erupts in China Over Ideologies." *The New York Times* March 12:A1.

Kairys, David. 1990. "Freedom of Speech." Pp. 237–272 in *The Politics of Law*, edited by David Kairys. New York: Pantheon Books.

Kalven, Harry, and Hans Zeisel. 1966. *The American Jury*. Boston: Little, Brown and Co.

Kappeler, Victor E., and Gary W. Potter. 2005. *The Mythology of Crime and Criminal Justice*. Prospect Heights, IL: Waveland Press.

Karberg, Jennifer C., and Doris J. James. 2005. *Substance Dependence, Abuse, and Treatment of Jail Inmates, 2002*. Washington, DC: Bureau of Justice Statistics, U.S. Department of Justice.

Karmen, Andrew. 2001. *New York Murder Mystery: The True Story Behind the Crime Crash of the 1990s*. New York: New York University Press.

Kassam, Philip C. 2003. *The Discipline of Law Schools: The Making of Modern Lawyers*. Durham, NC: Carolina Academic Press.

Katz, Leo. 1987. *Bad Acts and Guilty Minds: Conundrums of the Criminal Law*. Chicago: University of Chicago Press.

Kawashima, Takeyoshi. 1969. "Dispute Resolution in Japan." Pp. 182–193 in *Sociology of Law*, edited by Vilhelm Aubert. New York: Penguin Books.

Kelley, Matt. 1999. "Governor Discusses Drug Legalization." *The Bangor Daily News* October 5:A7.

Kelsen, Hans. 1967. *Pure Theory of Law*. Berkeley: University of California Press.

Kempf, Kimberly L., and Roy L. Austin. 1986. "Older and More Recent Evidence on Racial Discrimination in Sentencing." *Journal of Quantitative Criminology* 2:29–48.

Kennedy, Duncan (Ed.). 2004. *Legal Education and the Reproduction of Hierarchy: A Polemic Against the System*. New York: New York University Press.

Kershaw, Sarah, and Monica Davey. 2004. "Plagued by Drugs, Tribes Revive Ancient Penalty." *The New York Times* January 18:Section 1, Page 1.

Kessler, Ronald C., Emil F. Coccaro, Maurizio Fava, Savina Jaeger, Robert Jin, and Ellen Walters. 2006. "The Prevalence and Correlates of DSM-IV Intermittent Explosive Disorder in the National Comorbidity Survey Replication." *Archives of General Psychiatry* 63:669–678.

Kidder, Robert L. 1983. *Connecting Law and Society*. Englewood Cliffs, NJ: Prentice Hall.

Killian, Lewis M. 1968. *The Impossible Revolution: Black Power and the American Dream*. New York: Random House.

King, Nancy. 2003. "Jury Nullification Is Unfair." Pp. 82–88 in *The Legal System: Opposing Viewpoints*, edited by Laura K. Egendorf. San Diego: Greenhaven Press.

Kivisto, Peter, and Elizabeth Hartung (Eds.). 2007. *Intersecting Inequalities: Class, Race, Sex, and Sexualities*. Upper Saddle River: Prentice Hall.

Klarman, Michael J. 2007. *Brown v. Board of Education and the Civil Rights Movement*. New York: Oxford University Press.

Kleck, Gary. 1981. "Racial Discrimination in Criminal Sentencing: A Critical Evaluation of the Evidence with Additional Evidence on the Death Penalty." *American Sociological Review* 46:783–805.

Kleivan, Inge. 1971. "Song Duels in West Greenland: Joking Relationships and Avoidance." *Folk* 13:9–36.

Kovandzic, Tomislav V., III, John J. Sloan, and Lynne M. Vieraitis. 2004. " 'Striking Out' as Crime Reduction Policy: The Impact of 'Three Strikes' Laws on Crime Rates in U.S. Cities." *Justice Quarterly* 21:207–239.

Kramer, Greg W., Melinda Wolbransky, and Kirk Heilbrun. 2007. "Plea Bargaining Recommendations by Criminal Defense Attorneys: Evidence Strength, Potential Sentence, and Defendant Preference." *Behavioral Sciences & the Law* 25:573-585.

Krenshaw, Kimberlé, Neil Gotanda, Gary Peller, and Kendall Thomas (Eds.). 1996. *Critical Race Theory: The Key Writings that Formed the Movement*. New York: New Press.

Kristof, Nicholas D. 2005. "The Rosa Parks for the 21st Century." *The New York Times* November 8:A27.

Kritzer, Herbert M. 1991. "Propensity to Sue in England and the United States of America: Blaming and Claiming in Tort Cases." *Journal of Law and Society* 18:452–479.

———. 2004. "American Adversarialism." *Law & Society Review* 38:349–383.

Kuersten, Ashlyn K. 2003. *Women and the Law: Leaders, Cases, and Documents*. Santa Barbara: ABC-CLIO.

Labaton, Stephen. 2005. "Senate Approves Measure to Curb Big Class Actions." *The New York Times* February 11:A1.

Lacey, Nicola. 2004. "The Constitution of Identity: Gender, Feminist Legal Theory, and the Law and Society Movement." Pp. 471–486 in *The Blackwell Companion to Law and Society*, edited by Austin Sarat. Malden, MA: Blackwell Publishing.

Lane, Charles. 2007. "Court Backs School on Speech Curbs." *The Washington Post* June 26:A6.

Langbein, John H. 1992. "On the Myth of Written Constitutions: The Disappearance of the Jury Trial." *Harvard Journal of Law and Public Policy* 15:119–128.

Lanoue, David J. 1988. *From Camelot to the Teflon President: Economics and Presidential Popularity Since 1960*. New York: Greenwood Press.

Large, Donald W. 1972. "Is Anybody Listening? The Problem of Access in Environmental Litigation." *Wisconsin Law Review* 1972:62–113.

Leaf, Clifton. 2002. "Enough Is Enough." *Fortune* March 18:60–68.

Lee, Erika. 2003. *At America's Gates: Chinese Immigration During the Exclusion Era, 1882–1943*. Chapel Hill: University of North Carolina Press.

Lee, Harper. 1960. *To Kill a Mockingbird*. New York: Harper & Row.

Lefcourt, Robert (Ed.). 1971. *Law Against the People*. New York: Vintage Books.

Leipold, Andrew D. 1996. "Rethinking Jury Nullification." *Virginia Law Review* 82:253–324.

Lemmings, David. 2000. *Professors of the Law: Barristers and English Legal Culture in the Eighteenth Century*. New York: Oxford University Press.

Lemon, Nancy. 1996. *Domestic Violence Law: A Comprehensive Overview of Cases and Sources*. San Francisco: Austin and Winfield.

Lens, Sidney. 1973. *The Labor Wars: From the Molly Maguires to the Sitdowns*. Garden City, NY: Doubleday.

Lerner, Michael A. 2007. *Dry Manhattan: Prohibition in New York City*. Cambridge, MA: Harvard University Press.

Levine, F., and E. Preston. 1970. "Community Resource Orientation Among Low Income Groups." *Wisconsin Law Review* 1970:80–113.

Levine, James P. 1983. "Jury Toughness: The Impact of Conservatism on Criminal Court Verdicts." *Crime and Delinquency* 29:71–87.

———. 1992. *Juries and Politics*. Pacific Grove, CA: Brooks/Cole Publishing Co.

Levitt, Steven D., and Thomas J. Miles. 2006. "Economic Contributions to the Understanding of Crime." *Annual Review of Law and Social Science* 2:147–164.

Lewis, Anthony. 2004. "A President Beyond the Law." *The New York Times* May 7:A25.

Lewis, Anthony, and writers for the New York Times. 1966. *The Second American Revolution: A First-Hand Account of the Struggle for Civil Rights*. London: Faber and Faber.

Lewis, Neil A. 2002. "U.S. Rejects All Support for New Court on Atrocities." *The New York Times* May 7:A11.

Lewis, Peter H. 1996. "160 Nations Meet to Weigh Revision of Copyright Law." *The New York Times* December 2:A1.

Lieberman, Jethro K. 1981. *The Litigious Society*. New York: Basic Books.

Lipschultz, Sybil (Ed.). 2003. *Women, the Law, and the Workplace*. New York: Routledge.

Lipset, Seymour Martin. 1996. *American Exceptionalism: A Double-Edged Sword*. New York: W.W. Norton.

Liptak, Adam. 2002. "Pain-and-Suffering Awards Let Juries Avoid New Limits." *The New York Times* October 28:A14.

———. 2005. "Go Ahead, Test a Lawyer's Ingenuity. Try to Limit Damages." *The New York Times* March 6:Section 4, Page 5.

———. 2007. "Court Ruling Expected to Spur Convictions in Capital Cases." *The New York Times* June 9:A1.

Llewellyn, Karl. 1930. *The Bramble Bush*. New York: Oceana.

Llewellyn, Karl, and E. Adamson Hoebel. 1941. *The Cheyenne Way: Conflict and Case Law in Primitive Jurisprudence*. Norman, OK: University of Oklahoma Press.

Loftin, Colin, and David McDowall. 1981. "'One with a Gun Gets You Two': Mandatory Sentencing and Firearms Violence in Detroit." *Annals of the American Academy of Political and Social Science* 455:150–167.

Lomax, Louis. 1962. *The Negro Revolt*. New York: Harper and Row.

Loomis, Carol J. 2002. "The Odds Against Doing Time." *Fortune* March 18: http://www.fortune.com/indext.jhtml?channel=print_article.jhtml&doc_id=206660.

Lowy, Michael, and Craig Haney. 1980. "The Creation of Legal Dependency: Law School in a Nutshell." Pp. 36–41 in *The People's Law Review*, edited by Ralph Warner. Reading, MA: Addison-Wesley.

Lundman, Richard J. 2001. *Prevention and Control of Juvenile Delinquency*. New York: Oxford University Press.

Mabee, Carleton. 1969. *Black Freedom: The Nonviolent Abolitionists from 1830 Through the Civil War*. New York: Macmillan.

Macaulay, Steward. 1963. "Non-Contractual Relations in Business: A Preliminary Study." *American Sociological Review* 28:55–66.

MacCoun, Robert J. 2005. "Voice, Control, and Belonging: The Double-Edged Sword of Procedural Fairness." *Annual Review of Law and Social Science* 1:171–201.

MacKinnon, Catharine A. 2005. *Women's Lives, Men's Laws*. Cambridge, MA: Belknap Press of Harvard University Press.

Maguire, Kathleen, and Ann L. Pastore (Eds.). 2007. *Sourcebook of Criminal Justice Statistics* [Online]. Available at http://www.albany.edu/sourcebook.

Maine, Sir Henry Sumner. 1864. *Ancient Law: Its Connection with the Early History of Society, and Its Relation to Modern Ideas*. New York: C. Scribner.

Malinowski, Bronislaw. 1926. *Crime and Custom in Savage Society*. New York: Harcourt, Brace & Company.

Mann, Coramae Richey. 1993. *Unequal Justice: A Question of Color*. Bloomington, IN: Indiana University Press.

Mannheim, Karl. 1936. *Ideology and Utopia: An Introduction to the Sociology of Knowledge*. New York: Harcourt, Brace, and Company.

Mansnerus, Laura. 1994. "Law; Mediation as a Route to Lower Divorce Costs." *The New York Times* May 28:Section 1, Page 34.

Manza, Jeff, and Christopher Uggen. 2006. *Locked Out: Felon Disenfranchisement and American Democracy*. New York: Oxford University Press.

Marshall, Anna-Maria. 2005a. "Directions in Research on Law and Social Movements." *Amici: Newsletter of the Sociology of Law Section of the American Sociological Association* 13:1–6.

———. 2005b. "Idle Rights: Employees' Rights Consciousness and the Construction of Sexual Harassment Policies." *Law & Society Review* 39:83–123.

Marshall, Brent K., J. Steven Picou, and Jan R. Schlichtmann. 2004. "Technological Disasters, Litigation Stress, and the Use of Alternative Dispute Resolution Mechanisms." *Law & Policy* 26:289–307.

Marvell, Thomas B., and Carlisle E. Moody. Jr. 1994. "Prison Population Growth and Crime Reduction." *Journal of Quantitative Criminology* 10.

———. "The Impact of Enhanced Prison Terms for Felonies Committed with Guns." *Criminology* 33:247–281.

———. 1996. "Specification Problems, Police Levels, and Crime Rates." *Criminology* 34:609–646.

Marx, Gary T. 1988. *Undercover: Police Surveillance in America*. Berkeley: University of California Press.

Marx, Karl, and Friedrich Engels. 1962 (1848). "The Communist Manifesto." Pp. 21–65 in *Marx and Engels: Selected Works*, vol. 2. Moscow: Foreign Language Publishing House.

Mather, Lynn M. 1973. "Some Determinants of the Method of Case Disposition: Decisionmaking by Public Defenders in Los Angeles." *Law & Society Review* 8:187–215.

Mather, Lynn M., Craig A. McEwen, and Richard J. Maiman. 2001. *Divorce Lawyers at Work: Varieties of Professionalism in Practice*. New York: Oxford University Press.

Maxwell, Carol J. C. 2002. *Pro-Life Activists in America: Meaning, Motivation, and Direct Action*. New York: Cambridge University Press.

Mayhew, Leon, and Albert J. Reiss. 1969. "The Social Organization of Legal Contacts." *American Sociological Review* 34:309–318.

Mazie, Steven V. 2006. "You Say You Want a Constitution." *The New York Times* March 30:A25.

McCann, Michael. 1994. *Rights at Work: Pay Equity Reform and the Politics of Legal Mobilization*. Chicago: University of Chicago Press.

———. 2006. "Law and Social Movements: Contemporary Perspectives." *Annual Review of Law and Social Science* 2:17–38.

McGill, Christa. 2006. "Educational Debt and Law Student Failure to Enter Public Service Careers: Bringing Empirical Data." *Law & Social Inquiry* 31:677–708.

McIntosh, Wayne V. 1990. *The Appeal of Civil Law: A Political Economic Analysis of Litigation*. Urbana, IL: University of Illinois Press.

Megivern, James J. 1997. *The Death Penalty: An Historical and Theological Survey*. New York: Paulist Press.

Meier, Robert F., and Gilbert Geis. 2007. *Criminal Justice and Moral Issues*. New York: Oxford University Press.

Merry, Sally Engle. 1979. "Going to Court: Strategies of Dispute Management in an American Urban Neighborhood." *Law & Society Review* 13:891–925.

———. 1982. "The Social Organization of Mediation in Non-Industrial Societies: Implications for Informal Community Justice in America." Pp. 17–45 in *The Politics of Informal Justice*, vol. 2, edited by Richard L. Abel. New York: Academic Press.

———. 1990. *Getting Justice and Getting Even: Legal Consciousness Among Working-Class Americans*. Chicago: University of Chicago Press.

Merry, Sally Engle, and Susan S. Silbey. 1984. "What Do Plaintiffs Want? Reexamining the Concept of Dispute." *Justice System Journal* 9:151–178.

Merryman, John Henry, and Rogelio Pérez-Perdomo. 2007. *The Civil Law Tradition: An Introduction to the Legal Systems of Western Europe and Latin America*. Stanford: Stanford University Press.

Meyer, David S. 2007. *The Politics of Protest: Social Movements in America*. New York: Oxford University Press.

Michalowski, Raymond L., and Edward W. Bohlander. 1975. "Repression and Criminal Justice in Capitalist America." *Sociological Inquiry* 46:95–106.

Miliband, Ralph. 1969. *The State in Capitalist Society*. New York: Basic Books.
Mill, John Stuart. 1999 (1859). *On Liberty*. Peterborough, Canada: Broadview Press.
Miller, Jerome G. 1996. *Search and Destroy: African American Males in the Criminal Justice System*. New York: Cambridge University Press.
Miller, J. Mitchel, Christopher J. Schreck, and Richard Tewksbury. 2006. *Criminological Theory: A Brief Introduction*. Boston: Allyn & Bacon.
Miller, Richard E., and Austin Sarat. 1980–1981. "Grievances, Claims, and Disputes: Assessing the Adversary Culture." *Law & Society Review* 15:525–566.
Mills, Steve, Ken Armstrong, and Douglas Holt. 2000. "Flawed Trials Lead to Death Chamber." *Chicago Tribune* A1.
Milsom, S. F. C. 1981. *Historical Foundations of the Common Law*. Toronto: Butterworths.
Mintz, John. 2005. "Infighting Cited at Homeland Security." *The Washington Post* February 2:A1.
Mitford, Jessica. 1969. *The Trial of Dr. Spock*. New York: Knopf.
Mokdad, Ali H., James S. Marks, Donna F. Stroup, and Julie L. Gerberding. 2004. "Actual Causes of Death in the United States, 2000." *Journal of the American Medical Association* 291:1238–1245.
Moore, Lloyd E. 1973. *The Jury: Tool of Kings, Palladium of Liberty*. Cincinnati: W.H. Andrews.
Moore, Samuel K. 1999. "Vitamins Makers Settle U.S. Civil Suit for $1.17 Billion." *Chemical Week* November 10:15.
Moorehead, Caroline. 1999. *Dunant's Dream: War, Switzerland, and the History of the Red Cross*. St. Paul, MN: West Group.
Morgan, Phoebe A. 1999. "Risking Relationships: Understanding the Litigation Choices of Sexually Harassed Women." *Law & Society Review* 33:67–92.
Morris, Aldon. 1984. *The Origins of the Civil Rights Movement: Black Communities Organizing for Change*. New York: Free Press.
Moulton, Beatrice A. 1969. "The Persecution and Intimidation of the Low Income Litigant as Performed by the Small Claims Court in California." *Stanford Law Review* 21:1657–1684.
Muir, William K., Jr. 1967. *Prayer in the Public Schools: Law and Attitude Change*. Chicago: University of Chicago Press.
Mumola, Christopher J. 1999. *Substance Abuse and Treatment, State and Federal Prisoners*, 1997. Washington, DC: Bureau of Justice Statistics, U.S. Department of Justice.
Munger, Frank. 2004. "Rights in the Shadow of Class: Poverty, Welfare, and the Law." Pp. 330–353 in *The Blackwell Companion to Law and Society*, edited by Austin Sarat. Malden, MA: Blackwell Publishing.
Muraskin, Roslyn (Ed.). 2007. *It's a Crime: Women and Justice*. Upper Saddle River, NJ: Prentice Hall.
Murphy, Walter F. 1966. "Courts as Small Groups." *Harvard Law Review* 79:1565–1572.
Musto, David F. 1999. *The American Disease: Origins of Narcotic Control*. New York: Oxford University Press.
Myers, Martha A. 1979. "Rule Departures and Making Law: Juries and Their Verdicts." *Law & Society Review* 13:781-798.
———. 1981. "Judges, Juries, and the Decision to Convict." *Journal of Criminal Justice* 9:289–303.
———. 2000. "The Social World of America's Courts." Pp. 447–471 in *Criminology: A Contemporary Handbook*, edited by Joseph F. Sheley. Belmont, CA: Wadsworth.

Myers, Steven Lee. 2004. "Judging Abu Ghraib; Why Military Justice Can Seem Unjust." *The New York Times* June 6:Section 4, Page 3.

Myre, Grey. 2006. "Israel Raids West Bank Towns, Killing 6." *The New York Times* May 15:A3.

Nadelmann, Ethan A. 1992. "Drug Prohibition in the United States: Costs, Consequences, and Alternatives." Pp. 299–322 in *Drugs, Crime, and Social Policy: Research, Issues, and Concerns*, edited by Thomas Mieczkowski. Boston: Allyn & Bacon.

Nader, Laura (Ed.). 1969. *Law in Culture and Society*. Berkeley: University of California Press.

Nader, Laura. 2002. *The Life of the Law: Anthropological Projects*. Berkeley: University of California Press.

Nader, Laura, and Harry F. Todd Jr. (Eds.). 1978a. *The Disputing Process—Law in Ten Societies*. New York: Columbia University Press.

———. 1978b. "Introduction." Pp. 1–40 in *The Disputing Process—Law in Ten Societies*, edited by Larua Nader and Jr. Harry F. Todd. New York: Columbia University Press.

Nagin, Daniel S. 1978. "General Deterrence: A Review of the Empirical Evidence." in *Deterrence and Incapacitation: Estimating the Effects of Criminal Sanctions on Crime Rates*, edited by Alfred Blumstein, Jacqueline Cohen, and Daniel S. Nagin. Washington, DC: National Academy of Sciences.

———. 1998a. "Criminal Deterrence Research at the Outset of the Twenty-first Century." *Crime and Justice: A Review of Research* 23:1–42.

———. 1998b. "Deterrence and Incapacitation." Pp. 345–368 in *The Handbook of Crime and Punishment*, edited by Michael Tonry. New York: Oxford University Press.

Neilands, J. B. 1972. *Harvest of Death: Chemical Warfare in Vietnam and Cambodia*. New York: Free Press.

Nelken, David. 2004. "Comparing Legal Cultures." Pp. 113–127 in *The Blackwell Companion to Law and Society*, edited by Austin Sarat. Malden, MA: Blackwell Publishing.

Nielsen, Laura Beth. 2004a. *License to Harass: Law, Hierarchy, and Offensive Public Speech*. Princeton: Princeton University Press.

———. "The Work of Rights and the Work Rights Do: A Critical Empirical Approach." Pp. 63–79 in *The Blackwell Companion to Law and Society*, edited by Austin Sarat. Malden, MA: Blackwell Publishing.

Nielsen, Marianne O., and James W. Zion (Eds.). 2005. *Navajo Nation Peacemaking: Living Traditional Justice*. Tucson: University of Arizona Press.

Nolan-Haley, Jacqueline M. 2001. *Alternative Dispute Resolution in a Nutshell*. St. Paul, MN: West Group.

Nonet, Philippe. 1969. *Administrative Justice: Advocacy and Change in a Government Agency*. New York: Russell Sage Foundation.

O'Driscoll, Patrick. 2005. "Denver Votes to Legalize Marijuana Possession." *USA Today* November 3:A1.

Olson, Walter K. 2002. *The Rule of Lawyers: How the New Litigation Elite Threatens America's Rule of Law*. New York: Truman Talley Books.

O'Meara, Daniel P. 2002. *Arbitration of Employment Disputes*. Philadelphia: Center for Human Resources, Warton School, University of Pennsylvania.

Ostrowski, James. 2001. "The Rise and Fall of Jury Nullification." *Journal of Libertarian Studies* 15:89–115.

Packer, Herbert L. 1964. "Two Models of the Criminal Process." *University of Pennsylvania Law Review* 113:1–68.

———. 1968. *The Limits of the Criminal Sanction*. Stanford: Stanford University Press.

Paternoster, Raymond. 1991. *Capital Punishment in America*. New York: Lexington Books.

Paulson, Amanda. 2004. "Modern Life Stretching Family Law: US Courts Grapple with Nontraditional Custody Issues." *The Christian Science Monitor* August 10:1.

Paulson, Amanda, and Daniel B. Wood. 2005. "California Court Affirms Gay Parenting." *The Christian Science Monitor* August 25:1.

Pear, Robert. 2005. "Bush Begins Drive to Limit Malpractice Suit Awards." *The New York Times* January 6:A1.

Pease, Jane H., and William H. Pease. 1974. *They Who Would Be Free: Blacks' Search for Freedom, 1830–1861*. New York: Antheneum.

Penrod, Steven, and Larry Heuer. 1998. "Improving Group Performance: The Case of the Jury." Pp. 127–152 in *Theory and Research on Small Groups*, edited by R. Scott Tindale et al. New York: Plenum.

Perez-Pena, Richard. 2000. "The Death Penalty: When There's No Room for Error." *The New York Times* February 13:WK3.

Petersilia, Joan. 1985. "Racial Disparities in the Criminal Justice System: A Summary." *Crime and Delinquency* 31:15–34.

Piehl, Anne Morrison, and Shawn D. Bushway. 2007. "Measuring and Explaining Charge Bargaining." *Journal of Quantitative Criminology* 23:105–125.

Pierce, Glenn L., and William J. Bowers. 1981. "The Bartley-Fox Gun Law's Short-Term Impact on Crime." *Annals of the American Academy of Political and Social Science* 455:120–137.

Polletta, Francesca. 2000. "The Structural Context of Novel Rights Claims: Southern Civil Rights Organizing, 1961–1966." *Law & Society Review* 34:367–407.

Pontell, Henry N. 1984. *A Capacity to Punish: The Ecology of Crime and Punishment*. Bloomington, IN: Indiana University Press.

Poulantzas, Nicos. 1973. *Political Power and Social Classes*. London: New Left Books.

Pound, Roscoe. 1930. *Criminal Justice in America*. New York: Henry Holt.

———. 1997 (1930). *Criminal Justice in America*. New Brunswick, NJ: Transaction Publishers.

Powers, Stephen P., and Stanley Rothman. 2002. *The Least Dangerous Branch? Consequences of Judicial Activism*. Westport, CT: Praeger Publishers.

Pring, George W., and Penelope Canan. 1996. *SLAPPs: Getting Sued for Speaking Out*. Philadelphia: Temple University Press.

Pritchett, C. Herman. 1948. *The Roosevelt Court*. New York: Macmillan.

Pusey, Allen. 2000. "Judges Rule in Favor of Juries." *Dallas Morning News* May 7:15.

Quinney, Richard. 1977. *Class, State, and Crime*. New York: David McKay.

Radcliffe-Brown, A. R. 1952. *Structure and Function in Primitive Society*. London: Cohen & West.

Radelet, Michael L., Hugo Adam Bedau, and Constance E. Putnam. 1992. *In Spite of Innocence: Erroneous Convictions in Capital Cases*. Boston: Northeastern University Press.

Radsch, Courtney C. 2005. "Driver-Cellphone Laws Exist, But Their Value Is Disputed." *The New York Times* January 18:A18.

Reichel, Philip L. 2005. *Comparative Criminal Justice Systems: A Topical Approach*. Upper Saddle River, NJ: Prentice Hall.

Reiman, Jeffrey. 2007. *The Rich Get Richer and the Poor Get Prison: Ideology, Class, and Criminal Justice*, 8th edition. Boston: Allyn & Bacon.

Reinarman, Craig, Peter D. A. Cohen, and Kaal L. Hendrien. 2004. "The Limited Relevance of Drug Policy: Cannabis in Amsterdam and in San Francisco." *American Journal of Public Health* 94:836–842.

Reuters. 2006. "Reese Witherspoon Sues Magazine Over Baby Story." *The Washington Post* June 21: http://www.washingtonpost.com/wp-dyn/content/article/2006/06/21/AR2006062101764.html.

Rheinstein, Max (Ed.). 1954. *Max Weber on Law in Economy and Society*. Cambridge: Harvard University Press.

Rhode, Deborah L. 2001. *The Unfinished Agenda: A Report on the Status of Women in the Legal Profession*. Chicago: American Bar Association.

Richards, Leonard L. 2002. *Shays's Rebellion*: The American Revolution's Final Battle. Philadelphia: University of Pennsylvania Press.

Richland, Justin B., and Sarah Deer. 2004. *Introduction to Tribal Legal Studies*. Walnut Creek, CA: AltaMira Press.

Riksheim, Eric, and Steven M. Chermak. 1993. "Causes of Police Behavior Revisited." *Journal of Criminal Justice* 21:353–382.

Rimelspach, Rene L. 2001. "Mediating Family Disputes in a World with Domestic Violence: How to Devise a Safe and Effective Court-Connected Mediation Program." *Ohio State Journal on Dispute Resolution* 17:95–111.

Rinzler, Carol E. 1979. "How To Be the Oldest Kid in Your Law School Class." *Ms.* February:102.

Risen, James, and Eric Lichtblau. 2005. "Bush Lets U.S. Spy on Callers Without Courts." *The New York Times* December 16:A1.

Ritter, Alison, and Jacqui Cameron. 2006. "A Review of the Efficacy and Effectiveness of Harm Reduction Strategies for Alcohol, Tobacco, and Illicit Drugs." *Drug & Alcohol Review* 25:611–624.

Rockwell, David N. 1971. "The Education of the Capitalist Lawyer: The Law School." Pp. 90–104 in *Law Against the People: Essays to Demystify Law, Order and the Courts*, edited by Robert Lefcourt. New York: Vintage Books.

Roper, Robert, and Victor Flango. 1983. "Trials Before Judges and Juries." *The Justice System Journal* 8:186–198.

Rosenberg, Gerald N. 1991. *The Hollow Hope: Can Courts Bring About Social Change?* Chicago: University of Chicago Press.

Rosenberg, Tina. 1995. *The Haunted Land: Facing Europe's Ghosts After Communism*. New York: Random House.

Rosenfeld, Richard. 2000. "Patterns in Adult Homicide: 1980–1995." Pp. 130–163 in *The Crime Drop in America*, edited by Alfred Blumstein and Joel Wallman. Cambridge: Cambridge University Press.

Rosenfeld, Richard, Joel Wallman, and R. Fornango. 2005. "The Contribution of Ex-prisoners to Crime Rates." Pp. 80–104 in *Prisoner Reentry and Crime in America*, edited by Jeremy Travis and Christy Visher. New York: Cambridge University Press.

Rosett, Arthur, and Donald R. Cressey. 1976. *Justice by Consent: Plea Bargains in the American Courthouse*. Philadelphia: Lippincott.

Rosoff, Stephen M., Henry N. Pontell, and Robert Tillman. 2007. *Profit Without Honor: White Collar Crime and the Looting of America*. Upper Saddle River, NJ: Prentice Hall.

Ross, H. Laurence. 1992. *Confronting Drunk Driving: Social Policy for Saving Lives*. New Haven: Yale University Press.

Rostain, Tanina. 2004. "Professional Power: Lawyers and the Constitution of Professional Authority." Pp. 146–169 in *The Blackwell Companion to Law and Society*, edited by Austin Sarat. Malden, MA: Blackwell Publishing.

Rothe, Dawn, and Christopher W. Mullins. 2006. *Symbolic Gestures and the Generation of Social Control: The International Criminal Court*. Lanham, MD: Lexington Books.

Rothenberger, John E. 1978. "The Social Dynamics of Dispute Settlement in a Sunni Muslim Village in Lebanon." Pp. 152–180 in *The Disputing Process—Law in Ten Societies*, edited by Laura Nader and Harry F. Todd Jr. New York: Columbia University Press.

Rubenstein, Arie M. 2006. "Verdicts of Conscience: Nullification and the Modern Jury Trial." *Columbia Law Review* 106:959–993.

Rueschemeyer, Dietrich. 1973. *Lawyers and Their Society*. Cambridge, MA: Harvard University Press.

Sabol, William J., Todd D. Minton, and Paige M. Harrison. 2007. *Prison and Jail Inmates at Midyear 2006*. Washington, DC: Bureau of Justice Statistics, U.S. Department of Justice.

Salama, M. M. 1982. "General Principles of Criminal Evidence in Islamic Societies." Pp. 109–123 in *The Islamic Criminal Justice System*, edited by M. Cherif Bassiouni. London: Oceana Publications.

Salas, Luis. 1983. "Emergence and Decline of Cuban Popular Tribunals." *Law & Society Review* 17:588–612.

Samuels, Suzanne. 2006. *Law, Politics, and Society*. Boston: Houghton Mifflin Company.

San Francisco Chronicle. 1978. "Judge Hears Case of the Broken Date." *San Francisco Chronicle* July 25:5.

Saporito, Bill. 2004. "Kobe Rebounds." *Time* September 13:72–73.

Sarat, Austin. 2004. "Legal Scholarship in the Liberal Arts." *The Chronicle of Higher Education* September 3:B20.

Sarat, Austin, and Joel B. Grossman. 1975. "Courts and Conflict Resolution: Problems in the Mobilization of Adjudication." *American Political Science Review* 69:1200–1217.

Sarat, Austin, and Stuart A. Scheingold (Eds.). 2006. *Cause Lawyers and Social Movements*. Stanford: Stanford University Press.

Sarat, Austin, and William L. F. Felstiner. 1995. *Divorce Lawyers and Their Clients: Power and Meaning in the Legal Process*. New York: Oxford University Press.

Savelsberg, Joachim J. 2000. "Contradictions, Law, and State Socialism." *Law and Social Inquiry* 25:1021–1048.

Sax, Joseph L. 1968. "Conscience and Anarchy: The Prosecution of War Resisters." *Yale Review* 57:481–494.

Schapera, Isaac. 1955. *A Handbook of Tswana Law and Custom*. London: Oxford University Press.

Scheflin, Alan W. 1972. "Jury Nullification: The Right to Say No." *Southern California Law Review* 45:168–226.

Scheingold, Stuart A. 1974. *The Politics of Rights: Lawyers, Public Policy, and Political Change*. New Haven: Yale University Press.

Scheingold, Stuart A., and Austin Sarat. 2004. *Something to Believe In: Politics, Professionalism, and Cause Lawyering*. Stanford: Stanford University Press.

Schmitt, Erica Leah, Patrick A. Langan, and Matthew R. Durose. 2002. *Characteristics of Drivers Stopped by Police, 1999*. Washington, DC: Bureau of Justice Statistics, U.S. Department of Justice.

Schneider, Elizabeth M. 2000. *Battered Women & Feminist Lawmaking*. New Haven: Yale University Press.

Schubert, Glendon. 1963. *Judicial Decision-Making*. New York: Free Press.

Schur, Edwin M. 1968. *Law and Society: A Sociological View*. New York: Random House.

Schur, Edwin M., and Hugo Adam Bedau. 1974. *Victimless Crimes: Two Sides of a Controversy*. Englewood Cliffs, NJ: Prentice-Hall.

Schwartz, John. 2007. "Court Clears Way for Suit on New Orleans Flooding." *The New York Times* February 3:A10.

Schwartz, Richard D., and Jerome C. Miller. 1964. "Legal Evolution and Societal Complexity." *American Journal of Sociology* 70:159–169.

Seabury, Seth, Nicholas M. Pace, and Robert T. Reville. 2004. "Forty Years of Civil Jury Verdicts." *Journal of Empirical Legal Studies* 1:1–24.

Selznick, Philip. 1961. "Sociology and Natural Law." *Natural Law Forum* 6:84–108.

Seron, Carroll. 1996. *The Business of Practicing Law: The Work Lives of Solo and Small-Firm Attorneys*. Philadelphia: Temple University Press.

Seron, Carroll, and Frank Munger. 1996. "Law and Inequality: Race, Gender ... and, of Course, Class." *Annual Review of Sociology* 22:187–212.

Shaffer, Thomas L., and Mary M. Shaffer. 1991. *American Lawyers and Their Communities: Ethics in the Legal Profession*. Notre Dame, IN: University of Notre Dame Press.

Shanor, Charles A., and L. Lynn Hogue. 1996. *Military Law in a Nutshell*. St. Paul, MN: West Publishing Company.

Shapira, Amos, and Keren C. DeWitt-Arar (Eds.). 1995. *Introduction to the Law of Israel*. The Hague: Kluwer Law International.

Shelden, Randall G. 1982. *Criminal Justice in America: A Sociological Approach*. Boston: Little, Brown and Company.

———. 2001. *Controlling the Dangerous Classes: A Critical Introduction to the History of Criminal Justice*. Boston: Allyn & Bacon.

Sherrill, Robert. 1970. *Military Justice Is to Justice as Military Music Is to Music*. New York: Harper and Row.

Shinkle, Peter. 2005. "Get Used to Law Jokes, Author Tells Students." *St. Louis Post-Dispatch* September 15:C6.

Siegel, Reva B. 1996. "'The Rule of Love': Wife Beating as Prerogative and Policy." *Yale Law Journal* 105:2117–2207.

Silberman, Matthew. 1985. *The Civil Justice Process*. New York: Academic Press.

Silbey, Susan S. 2005. "After Legal Consciousness." *Annual Review of Law and Social Science* 1:323–368.

Silverstein, Helena. 1996. *Unleashing Rights: Law, Meaning, and the Animal Rights Movement*. Ann Arbor: University of Michigan Press.

Singer, Rena. 2001. "The Double-Edged Sword of Nigeria's Sharia." *The Christian Science Monitor* 1.

Smigel, Erwin O. 1969. *The Wall Street Lawyer: Professional Organization Man?* Bloomington, IN: Indiana University Press.

Smith, Bruce P. 2005. "Plea Bargaining and the Eclipse of the Jury." *Annual Review of Law and Social Science* 1:131–149.

Smith, Douglas A. 1986. "The Neighborhood Context of Police Behavior." Pp. 313–341 in *Communities and Crime*, edited by Albert J. Reiss Jr., and Michael Tonry. Chicago: University of Chicago Press.

Smith, Douglas A., Christy Visher, and Laura A. Davidson. 1984. "Equity and Discretionary Justice: The Influence of Race on Police Arrest Decisions." *Journal of Criminal Law and Criminology* 75:234–249.

Smith, Erica L., and Matthew R. Durose. 2006. *Characteristics of Drivers Stopped by Police, 2002*. Washington, DC: Bureau of Justice Statistics, U.S. Department of Justice.

Smith, Mark S. 1985. "Attorneys Fight Age-Old Image Battle." *Bangor Daily News* July 20:22.

Smith, Watson, and John M. Roberts. 1954. *Zuni Law: A Field of Values*. Cambridge: Harvard University Press.

Snyder, Eloise C. 1958. "The Supreme Court as a Small Group." *Social Forces* 36:232–238.

Sokoloff, Natalie J., Barbara Raffel Price, and Jeanne Flavin. 2004. "The Criminal Law and Women." Pp. 11–29 in *The Criminal Justice System and Women: Offenders, Prisoners, Victims, & Workers*, edited by Barbara Raffel Price and Natalie J. Sokoloff. New York: McGraw-Hill.

Songer, Donald R., Reginald S. Sheehan, and Susan Brodie Haire. 1999. "Do the 'Haves" Come Out Ahead Over Time? Applying Galanter's Framework to Decisions of the U.S. Courts of Appeals, 1925–1988." *Law & Society Review* 33:811–832.

Songer, Michael J., and Isaac Unah. 2006. "The Effect of Race, Gender, and Location on Prosecutorial Decision to Seek the Death Penalty in South Carolina." *South Carolina Law Review* 58:161–205.

Soss, Joe, Laura Langbein, and Alan R. Metelko. 2003. "Why Do White Americans Support the Death Penalty?" *The Journal of Politics* 65:397–421.

Spelman, William. 2000. "The Limited Importance of Prison Expansion." Pp. 97–129 in *The Crime Drop in America*, edited by Alfred Blumstein and Joel Wallman. Cambridge: Cambridge University Press.

Spohn, Cassia, and David Holleran. 2000. "The Imprisonment Penalty Paid by Young, Unemployed Black and Hispanic Male Offenders." *Criminology* 38: 281–306.

Spohn, Cassia, and Jerry Cederblom. 1991. "Race and Disparities in Sentencing: A Test of the Liberation Hypothesis." *Justice Quarterly* 8:305–327.

Starr, June, and Barbara Yngvesson. 1975. "Scarcity and Disputing: Zeroing-in on Compromise Solutions." *American Ethnologist* 2:553–566.

Steffensmeier, Darrell, and Stephen Demuth. 2001. "Ethnicity and Judges' Sentencing Decisions: Hispanic–Black–White Comparisons." *Criminology* 39:145–178.

Stewart, Richard B. 2003. "Administrative Law in the Twenty-First Century." *New York University Law Review* 78:437–460.

Stolberg, Sheryl Gay. 2006. "Senate Rejects Award Limits in Malpractice." *The New York Times* May 9:A25.

Stonehill, Bill. 2007. "Law in Japan." http://www.eyesonjapan.com/jp5.htm.

Strick, Anne. 1978. *Injustice for All*. New York: Penguin Books.

Sudnow, David. 1965. "Normal Crimes: Sociological Features of the Penal Code in a Public Defender's Office." *Social Problems* 12:255–276.

Sykes, Charles J. 1992. *A Nation of Victims: The Decay of the American Character*. New York: St. Martin's Press.

Sykes, Gresham M. 1958. *The Society of Captives: A Study of a Maximum Security Prison*. Princeton: Princeton University Press.

Taft, Philip, and Philip Ross. 1990. "American Labor Violence: Its Causes, Character, and Outcome." Pp. 174–186 in *Violence: Patterns, Causes, Public Policy*, edited by Neil Alan Weiner, Margaret A. Zahn, and Rita J. Sagi. San Diego: Harcourt Brace Jovanovich.

Tagliabue, John. 2007. "Law Firms from U.S. Invade Paris." *The New York Times* July 25:C1.

Taub, Nadine, and Elizabeth M. Schneider. 1998. "Women's Subordination and the Role of Law." Pp. 328–355 in *The Politics of Law: A Progressive Critique*, edited by David Kairys. New York: Basic Books.

Tay, Alice Erh-Soon. 1990. "Communist Visions, Communist Realities, and the Role of Law." *Journal of Law & Society* 17:155–169.

The New York Times. 2004. "Police Officer Is Put on Year's Probation." *The New York Times* October 29:B6.

The Sentencing Project. 2007. *Felony Disenfranchisement Laws in the United States*. Washington, DC: The Sentencing Project.

Thompson, E. P. 1975. *Whigs and Hunters: The Origin of the Black Act*. London: Allen Lane.

Tierney, Kevin. 1979. *Darrow: A Biography*. New York: Crowell.

Tillman, Robert, and Henry N. Pontell. 1992. "Is Justice 'Collar-Blind'?: Punishing Medicaid Provider Fraud." *Criminology* 30:547–573.

Tittle, Charles R. 1969. "Crime Rates and Legal Sanctions." *Social Problems* 16:409–423.

Tocqueville, Alexis de. 1994(1835). *Democracy in America*. New York: Knopf.

Tonry, Michael. 1994. *Malign Neglect: Race, Crime, and Punishment in America*. New York: Oxford University Press.

Toth, Stephen A. 2003. "The Desire to Deport: The Recidivist of Fin-de-Siècle France." *Nineteenth-Century Contexts* 25:147-160.

Travis, Jeremy, and Christy Visher (Eds.). 2005. *Prisoner Reentry and Crime in America*. New York: Cambridge University Press.

Trubek, David M. 1989. "Where the Action Is: Critical Legal Studies and Empiricism." *Stanford Law Review* 36:575–622.

Trubek, David M., Austin Sarat, William L. F. Felstiner, Herbert M. Kritzer, and Joel B. Grossman. 1983. "The Costs of Ordinary Litigation." *UCLA Law Review* 31:72–127.

Tunnell, Kenneth D. 2006. *Living Off Crime*. Lanham, MD: Rowman & Littlefield.

Turow, Scott. 1977. *One L*. New York: Putnam.

Tushnet, Mark. 1986. "Critical Legal Studies: An Introduction to Its Origins and Underpinnings." *Journal of Legal Education* 36:505–517.

———. 2005. "Survey Article: Critical Legal Theory (Without Modifiers) in the United States." *The Journal of Political Philosophy* 13:99–112.

Tyler, Tom R. 2004. "Procedural Justice." Pp. 435–452 in *The Blackwell Companion to Law and Society*, edited by Austin Sarat. Malden MA: Blackwell Publishing.

———. 2006. *Why People Obey the Law*. Princeton: Princeton University Press.

Tyler, Tom R., and Y. J. Huo. 2002. *Trust and the Rule of Law*. New York: Russell Sage Foundation.

Unger, Roberto Mangabeira. 1976. *Law in Modern Society: Toward a Criticism of Social Theory*. New York: Free Press.

———. 1986. *The Critical Legal Studies Movement*. Cambridge, MA: Harvard University Press.

Unnever, James D., and Francis T. Cullen. 2005. "Executing the Innocent and Support for Capital Punishment: Implications for Public Policy." *Criminology & Public Policy* 4:3–37.

———. 2007. "Reassessing the Racial Divide in Support for Capital Punishment." *Journal of Research in Crime and Delinquency* 44:124–158.

Upham, Frank K. 1976. "Litigation and Moral Consciousness in Japan: An Interpretive Analysis of Four Japanese Pollution Suits." *Law & Society Review* 10:579–619.

Vago, Steven. 2006. *Law and Society*. Upper Saddle River, NJ: Prentice Hall.

Van Dyke, Jon M. 1970. "The Jury as a Political Institution." *The Center Magazine* 3:17–26.

Van Hoy, Jerry. 1997. *Franchise Law Firms and the Transformation of Personal Legal Services*. Westport, CT: Quorum Books.

Venables, Robert W. 2004. *American Indian History: Five Centuries of Conflict & Coexistence*. Santa Fe, NM: Clear Light Publishers.

Vidmar, Neil, and Valerie P. Hans. 2007. *American Juries: The Verdict*. Amherst, NY: Prometheus Books.

Visher, Christy A. 1983. "Gender, Police Arrest Decisions, and Notions of Chivalry." *Criminology* 21:5–28.

Vold, George, Thomas Bernard, and Jeffrey B. Snipes. 2002. *Theoretical Criminology*. New York: Oxford University Press.

Wald, Patricia M. 1965. *Law and Poverty*. Washington, DC: U.S. Government Printing Office.

Walker, Samuel. 1998. *Popular Justice: A History of American Criminal Justice*, 2nd edition. New York: Oxford University press.

———. 2006. *Sense and Nonsense About Crime and Drugs: A Policy Guide*, 6th edition. Belmont, CA: Wadsworth Publishing Company.

Walker, Samuel, Cassia Spohn, and Miriam DeLone. 2007. *The Color of Justice: Race, Ethnicity, and Crime in America*, 4th edition. Belmont, CA: Wadsworth Publishing Company.

Wallace, Ruth A., and Alison Wolf. 2006. *Contemporary Sociological Theory: Expanding the Classical Tradition*. Upper Saddle River, NJ: Prentice Hall.

Wanner, Craig. 1975. "The Public Ordering of Private Relations: Part I: Initiating Civil Cases in Urban Trial Courts." *Law & Society Review* 8:421–440.

Ware, Stephen J. 2001. *Alternative Dispute Resolution*. St. Paul, MN: West Group.

Warr, Mark. 2002. *Companions in Crime: The Social Aspects of Criminal Conduct*. New York: Cambridge University Press.

Washington Post. 2007. "House Passes Ban on Job Bias Gainst Gays." November 8:A07.

Weber, Max. 1978(1921). *Economy and Society: An Outline of Interpretive Sociology*. Berkeley: University of California Press.

Weinstein, Henry. 2007. "VA Failing Mideast Vets, Lawsuit Contends." *Los Angeles Times* July 24:A1.

Weinstein, Jack B. 1993. "Considering Jury 'Nullification': When May and Should a Jury Reject the Law to Do Justice?" *American Criminal Law Review* 30:239–254.

Wilbanks, William. 1987. *The Myth of a Racist Criminal Justice System*. Monterey: Brooks/Cole Publishing Company.

Wildenthal, Bryan H. 2002. "Fighting the Lone Wolf Mentality: Twenty-first Century Reflections on the Paradoxical State of American Indian Law." *Tulsa Law Review* 38:113–145.

Wilgoren, Jodi. 2003. "Governor Assails System's Errors as He Empties Illinois Death Row." *The New York Times* January 12:A1.

Wilson, James Q., and Allan Abrahamse. 1992. "Does Crime Pay?" *Justice Quarterly* 9:359–377.

Wilson, Scott. 2006. "Israelis Kill Islamic Jihad Leader in Arrest Attempt." *The Washington Post* May 15:A10.

Wissler, Roselle L. 1995. "Mediation and Adjudication in Small Claims Court." *Law & Society Review* 29:323–358.

Wolfe, M., D. Lichtenstein, and Singh Gurkipal. 1999. "Gastrointestinal Toxicity of Nonsteroidal Anti-inflammatory Drugs." *The New England Journal of Medicine* 340:1888–1889.

Wright, Bradley R. E., Avshalom Caspi, Terrie E. Moffitt, and Ray Paternoster. 2004. "Does the Perceived Risk of Punishment Deter Criminally Prone Individuals? Rational Choice, Self-Control, and Crime." *Journal of Research in Crime and Delinquency* 41:180–213.

www.nolo.com. 2007. "Defendants' Incentives for Accepting Plea Bargains." http://www.nolo.com/article.cfm/ObjectID/4E8D6815-1797-46FC-8F8AB242FFE6391A/catID/D4C65461-8D33-482C-92FCEA7F2ADED29A/104/143/272/ART/.

Yale Law Journal. 1964. "Note: The Changing Role of the Jury in the Nineteenth Century." *Yale Law Journal* 74:170–192.

Yardley, Jim. 2005. "In Worker's Death, View of China's Harsh Justice." *The New York Times* December 31:A1.

Zahn, Margaret A., Henry H. Brownstein, and Shelly L. Jackson (Eds.). 2004. *Violence: from Theory to Research*. Cincinnati: Anderson Publishing.

Zane, J. Peder. 2005. "Tell It to the Judge ... But Only If You Feel You Really Must." *The New York Times* July 16:Section 3, Page 8.

Zatz, Marjorie Sue. 1994. *Producing Legality: Law and Socialism in Cuba*. New York: Routledge.

Zeller, Tom, Jr. 2006. "Despite Laws, Stalkers Still Roam on the Internet." *The New York Times* April 17:A1.

Zemans, Frances Kahn. 1983. "Legal Mobilization: The Neglected Role of the Law in the Political System." *American Political Science Review* 77:690–703.

Zitrin, Richard, and Carol M. Langford. 1999. *The Moral Compass of the American Lawyer: Truth, Justice, Power, and Greed*. New York: Ballantine Books.

GLOSSARY

Absolute deterrence the effect of having some law (in terms of arrest, punishment, and other legal sanctions) or of having a legal system versus the effect of no law or of no legal system

Adjudication a method of dispute processing in which the third party (judge) has the authority to make a binding decision in a dispute regardless of whether the disputants wish the third party's involvement

Administrative law the rules and regulations of administrative agencies

Adversarial model the view that prosecutors and defense attorneys vigorously contest the evidence of a defendant's guilt

Alternative dispute resolution measures of dispute processing that avoid adjudication

Arbitration a method of dispute processing involving an impartial third party who imposes a decision with the prior agreement of the disputants

Avoidance a method of dispute processing in which an aggrieved party ends the relationship that produced the problem or by physically removes oneself from the situation or location in which the problem is located

Bourgeoisie the ruling class that owns the means of production

Brutalization effect an increase in the number of homicides after executions occur

Case law law made by appellate court judges through their rulings

Case study method teaching in law school that relies on the reading of casebooks, or collections of actual cases and court opinions designed to illustrate various legal principles

Cause lawyering the practice of law on behalf of a principle or to achieve a social change goal

Cause lawyers lawyers who practice cause lawyering

Certainty in the study of legal punishment, the likelihood in a location of being arrested; typically measured as the number of arrests for a given type of crime divided by the number of UCR-measured offenses for that type of crime

Charismatic authority Max Weber's term for power that stems from an individual's extraordinary personal qualities and from that individual's hold over followers because of these qualities

Civil law (a) the family of law consisting of detailed codes or collections of laws and found in continental European nations and some other countries (also called code law); (b) law governing private transactions among individuals and organizations

Code law the family of law consisting of detailed codes or collections of laws and found in continental European nations and some other countries (also called civil law)

Coercion a method of dispute processing that involves the use of threats or pressure to compel a change in someone's behavior or thinking

Collective incapacitation the mass imprisonment of offenders without regard to their history and likely future of offending (also called gross incapacitation)

Common law the family of law emphasizing case law and found in Great Britain and its former colonies

Conflict theory a theoretical perspective that says law helps perpetuate inequality by maintaining the control of the powerful

Consensual crime crime that does not involve unwilling victims because people engage in it voluntarily; common examples include illegal drug use, prostitution, and illegal gambling

Consensual model the view that the criminal courts characterized by a good deal of cooperation between prosecution and defense

Conviction rate the percentage of all prosecuted criminal cases that end in a conviction

Corporate attorneys attorneys who work directly for corporations or in large law firms with corporate clients

Courtroom workgroup the prosecutor, defense attorney, and judge who are said to cooperate to process criminal cases efficiently based on shared understandings of typical crimes and punishments

Coverture the common law concept that wives are the property of their husbands

Crime control model the view that because most criminal suspects are guilty, their cases should be expedited through the courts

Criminal law the body of law that prohibits acts seen as so harmful to the public welfare that they deserve to be punished by the state, and that governs how these acts are handed by official state procedures

Critical legal studies a perspective stemming from the late 1970s that law is inherently political and is used to legitimize existing inequality

Critical race theory a perspective holding that the United States is a racist society and that the law both manifests this racism and contributes to it

Dangerous classes a popular label applied to the poor during the nineteenth century

Dependent variable something that is affected or influenced by an independent variable

Deterrence a reduction in crime because individuals do not want to risk the many disadvantages incurred by arrest, punishment, and other legal sanctions

Deterrence theory the belief that law can be used to deter criminal behavior by potential offenders in the general population or by actual offenders after they have been convicted and punished

Deviance behavior that violates social norms and arouses negative social reactions

Dialectical view the Marxist belief that the state and its legal system are relatively autonomous from the ruling class and that legal and political victories by the have-nots can be real and consequential and not just sham

Discretionary model a view that the sequence of any case in the criminal justice system can be understood as a series of many decisions by legal personnel

Dispute settlement a function of law involving its use to help resolve disputes among two or more parties

Due process model a view of the criminal courts that both the innocent and the guilty in a democracy deserve to have their rights observed in order to preserve individual liberty

Executive orders laws passed and actions undertaken by the executive branch of the government

Families of law the major legal systems characterizing nations around the world, usually delineated as: (1) common law; (2) civil law (also called code law); (3) theocratic law; (4) socialist law; and (5) traditional law

Fellow-servant rule a nineteenth-century legal doctrine that an employer was not responsible for injuries to an employee caused by the actions of another employee

Feminist legal theory the view that law both reflects and contributes to fundamental sexism in the larger society

Formal law rules and regulations that are written and enforced according to written guidelines

Functionalist theory the view law is an essential social institution that makes important contributions to the proper working of society

General deterrence the ability of law to deter offending in the general population by sending a message that potential offenders will be apprehended and punished if they break the law

Gross incapacitation the mass imprisonment of offender without regard to their history and likely future of offending (also called collective incapacitation)

Historical school the name given to the work of several nineteenth and early twentieth century scholars who discussed how law and legal systems changed over the centuries as societies developed from ancient times to the time the scholars were writing

Incapacitation the prevention of crime resulting from the incarceration of convicted offenders

Independent variable something that affects or influences a dependent variable

Individual deterrence the ability of law to deter criminal behavior by offenders who have already been arrested and punished because they do not want to incur the risk of additional punishment (also called specific deterrence)

Informal law the use of discretion regarding whether to enforce rules and sanctions even if an violation of the law has occurred Informal

Instrumentalist view the Marxist view that law and the state are tools that can fairly easily be used by the ruling class to oppress the working class and in other respects to maintain its own superior position

Intermediate scrutiny standard a legal doctrine that differential legal treatment is permitted as long as it is "substantially related" to the achievement of important governmental goals

International law the body of law that governs the relationships among the nations of the world and among organizations with international dealings

Judicial review the principle in U.S. law that appellate courts may overturn legislative statutes and invalidate executive actions if they are deemed to violate Constitutional standards

Jury nullification jury verdicts that clearly contradict the evidence, especially jury acquittals despite clear evidence of defendants' legal guilt

Legal consciousness everyday understandings of and experiences with the law

Legal culture the beliefs and values that people hold about law and the legal system

Legal formalism the nineteenth-century view, still popular today, that law is a self-contained system of logic that is independent of social and moral considerations

Legal mobilization the use of law by social movements

Legal order in Roberto Mangabeira Unger's work, law that is autonomous from the state and general (or applicable to everyone) regardless of any individual's power, wealth, or personal connections

Legal procedures how and why laws get enacted, disputes get settled, and cases get handled

Legal realism a popular school of thought in the United States in the early to mid-twentieth century that studied how law really works as opposed to how it is supposed to work

Legal repression the use of law to control social movements and, more generally, dissent

Legality in Philip Selznick's work, the rule of law

Lumping it the ignoring of a dispute by doing nothing and thus taking "one's lumps"

Marginal deterrence the effect on criminal behavior of increasing criminal sanctions versus not increasing such sanctions

Means of production tools, factories, and other sources of economic production whose ownership, according to Karl Marx and Friedrich Engels, divided capitalist society into two classes, the bourgeoisie and the proletariat

Mechanical solidarity Emile Durkheim's term for the type of social order that exists in small, traditional societies because of their homogeneity

Mediation a method of dispute processing involving a third party who helps the disputants reach a resolution, usually through a compromise; either party to the dispute is free to decline any solution or change in thinking or behaving a mediator might suggest

Natural law legal principles derived from nature and thought to be binding on human society

Negotiation a method of dispute processing in which the disputants try to persuade each other of their way of thinking about the grievance underlying the dispute

Normal crime a conception that the courtroom workgroup has of the characteristics of a typical crime that allows the workgroup to decide on the appropriate punishment for the defendant

Official law governmental rules and regulations

One-shotters parties who use the law and courts very rarely or at best only occasionally

Organic solidarity Emile Durkheim's term for the type of social order that exists in large, modern societies because of their interdependence of roles

Organization of the legal system the structure and functioning of the components of the legal system

Personal services attorneys attorneys at the bottom of the legal professional hierarchy who work by themselves or in small firms and represent clients in criminal cases, divorces, real estate transactions, personal injury cases, and wills and other estate planning

Philosophical question in the study of consensual crimes, the issue of the extent to which the state should prohibit behaviors in which people want to engage because the behaviors are perceived to be immoral and/or socially harmful

Plea bargaining negotiations between the prosecution and defense in which the defendant agrees to plead guilty, usually in return for reduced charges and/or reduced punishment

Private law the body of law that governs relationships among individuals and organizations (also called *civil law*)

Procedural justice the result achieved if a case during its various stages follows the many procedural rules that govern the judicial process

Procedural law the body of law governing how substantive law is implemented and enforced

Proletariat the working class that does not own the means of production

Public law the body of law that outlines the structure of government, the powers and responsibilities of public officials, and the government's relationship to private citizens

Rational choice theory the view that people act after weighing the potential consequences of their actions and calculating whether a given act is more likely to be beneficial or costly

Rationality logical and impersonal decision-making in which the means have a reasonable connection to the ends and vice versa

Rational-legal authority Max Weber's term for power that stems from an law and is based on a belief in the legitimacy of a society's laws and rules and in the right of leaders acting under these rules to make decisions and set policy

Repeat players parties who use the law and courts on a recurring basis

Repressive law Emile Durkheim's term for the type of punishment characterizing small, traditional societies

Restitutive law Emile Durkheim's term for the type of punishment characterizing large, modern societies

Rule of law the idea that no one in a democracy, no matter how wealthy or powerful the individual may be, is above the law

Selective incapacitation a policy involving the targeting of chronic offenders for incarceration

Separate sphere the belief that that women belong at home and men belong in the working world outside the home

Severity in the study of legal punishment, the harshness of punishment after conviction of a crime

Social change in the study of law and society, one of the functions of law involving the bringing about of some change in some aspect of society

Social control in the study of law and society, one of the functions of law involving the maintaining of social order

Social institutions established patterns of behavior and relationships for addressing and supplying a society's fundamental needs

Social science question in the study of consensual crimes, the issue of whether laws against consensual crime do more good than harm or more harm than good

Social science view of law the belief that law is a social institution that both affects and is affected by society

Social stratification social inequality

Socialist law the family of law characterizing Communist nations

Sociological jurisprudence a twentieth-century school of thought that argued that law has social underpinnings and impact

Socratic method in law schools, a vigorous question-and-answer process about legal cases that students have read in casebooks

Specific deterrence the ability of law to deter criminal behavior by offenders who have already been arrested and punished because they do not want to incur the risk of additional punishment (also called individual deterrence)

Statutory law the body of law passed and enacted by Congress at the federal level and legislatures at the state level

Stratification in reference to the legal profession, the ranking of the legal profession according to dimensions of wealth, power, and prestige

Structuralist view the Marxist view that the state and legal system are relatively autonomous from the ruling class and that short-term victories by the have-nots ultimately preserve the existing order by contributing to its legitimacy

Substantive justice as one of the dimensions of law highlighted in the study of law and society, the various kinds of rules that help guide behavior; more formally, the body of law governing which behaviors are permitted and which behaviors are prohibited

Substantive law as one of the dimensions of the law highlighted in the study of law and society, the various kinds of rules that help guide behavior; more formally, the body of law governing which behaviors are permitted and which behaviors are prohibited

System capacity the ability of the legal system to handle criminal cases

Theocratic law the family of law that rests heavily on religious belief

Tort a violation of a civil law; a civil wrong

Traditional authority Max Weber's term for the type of power that is rooted in traditional, or long-standing, beliefs and practices of a society

Traditional law the family of law found in small, traditional societies

Traditional view of law the view that law is, and should be, a self-contained system of logic

Tramp acts a series of laws in the post–Civil War period that were used to suppress unemployed workers

Unofficial law nongovernmental rules and regulations

Utilitarianism a perspective developed during the eighteenth and nineteenth centuries that argued that because people act to maximize their pleasure and reduce their pain, legal punishment needed only to be sufficiently harsh to deter potential offenders from committing crime.

Wrongful executions executions of innocent people

INDEX